PURKOVIĆ ALEKSANDAR

PRINCIPLES OF DATA COMMUNICATION

PRINCIPLES OF DATA COMMUNICATION

R. W. LUCKY

J. SALZ

E. J. WELDON, JR.

Bell Telephone Laboratories, Inc.

McGRAW-HILL BOOK COMPANY

New York San Francisco Toronto London Sydney

PRINCIPLES OF DATA COMMUNICATION

 Library of Congress Catalog Card Number 67-28302

ISBN 07-038960-8

567890 KPKP 798765

Preface

In any rapidly developing branch of applied science there is an inevitable and unfortunate time lag between the conception of an idea and its implementation. This lag results in part from the normal delays involved in the publication of technical papers and in the design, construction, and testing of hardware. A second and frequently more serious delay is a result of the fact that it is often inconvenient or impractical for engineers to extract the pertinent information from the mass of published material. Data communications, developing as rapidly as it is, is especially prone to this latter malady.

This book is intended to hasten the adoption of recently discovered, economically attractive data-communications techniques by presenting them in a single, coherent framework. Thus the book is addressed primarily to the data-communications system designer-analyst. As such the authors have assumed a working knowledge of elementary probability theory, Fourier analysis, and matrix theory, as well as a certain level of general technical maturity.

The basic approach of the book is theoretical; however, because of its engineering direction and space limitations, proofs are included only when it is felt that they contribute sufficient insight. In all cases the theoretical approach is tempered by considerations of practical realizability and instrumentation difficulties. Both the approach and the choice of subject matter reflect the current interest in achieving more efficient utilization of real transmission channels, i.e., higher speeds and greater accuracies. Thus a considerable amount of material is included on optimum systems, automatic and adaptive equalization, and error control.

In the last decade the telegraph art has matured into the data communications of today. During this time the authors and their coworkers at the Bell Telephone Laboratories have been concerned with the theoretical and applied aspects of data transmission over facilities such as exist in the telephone plant. This background, reflected in all sections of the book, is particularly evident in the concrete examples employed.

The authors were motivated and encouraged by their colleagues at the Bell Telephone Laboratories during the course of bringing this book into being and were especially inspired by the pioneering book "Data Transmission" by their distinguished colleagues, W. R. Bennett and J. R. Davey.

R. W. LUCKY
J. SALZ
E. J. WELDON, JR.

Contents

PRINCIPLES OF DATA COMMUNICATION

1

Introduction

Man has devised numerous machines to perform tasks that are either too difficult or too tedious for him to do unaided. These devices have grown more complex over the years so that today it is necessary for machines to communicate not only with men but also with other machines. Communications between computers, between business machines and computers, and between spacecraft and ground stations are examples of machine-to-machine communication. Although this type of communication is important today, it seems likely that its most significant effects on society will be felt in the not-too-distant future. Like earlier major improvements in communications, widespread machine-to-machine communication will undoubtedly lead to dramatic changes in man's way of life.

Almost without exception, communication between machines employs electrical signals, and the information to be conveyed is digital, i.e., numerical. The discipline of data communications deals with the analysis and design of systems for transmitting this digital information. Although digital communication has been used in various rudimentary forms for centuries, the science of data communications is relatively new and developing rapidly.

This development has been spurred by two related demands. The amount of data to be transmitted is increasing almost exponentially with time and, because of the complex ways in which the data are used, higher transmission accuracies are necessary. Neither of these two demands can be met solely by building more and better communication facilities. Rather it is necessary to employ existing facilities with as high a data rate and an accuracy as possible. Thus the emphasis in the discipline of data communications is on the efficient use of transmission facilities.

This book is about efficient digital communication. In dealing with this subject the fields of information theory, statistical communication theory, and coding theory are particularly appropriate; thus the book necessarily contains much mathematical material. However, it has been written primarily from an engineering standpoint. Consequently questions of economic viability and physical realizability have dictated the choice and emphasis of material.

Also, mathematical theorems and results are frequently given without proof or in an heuristic manner. On the other hand, we do not treat specific circuits or detailed design problems. Throughout, the emphasis is on the analytical techniques which can be applied to the study of data-communications systems.

Because of the rapid expansion of the data-communications field, it has become nearly impossible for the engineer to keep abreast of current research in areas even closely allied with his own. Our purpose here is to make the most important of these results available in a mathematically palatable form.

From a technical point of view we can regard the data-communications system as consisting of three basic blocks: the transmitter, the channel, and the receiver (see Fig. 1.1). The transmitter has the task of assigning an electrical waveform to each possible sequence of digits received as input from the data source. The electrical waveform is then passed through the channel, which may typically be a telephone connection, a satellite link, a microwave system, or a radio link. In passage through the channel the transmitted waveform is invariably corrupted by unwanted, random signals known as *noise*. Because of these random signals, the received waveform does not correspond exactly to any of the possible transmitted waveforms. Nevertheless, the receiver must make a decision as to which of the data sequences is most likely to have given rise to the particular received waveform.

Fig. 1.1. Block diagram of a data communications system.

To minimize the probability of making a wrong decision at the receiver, the waveforms selected for transmission must be ones which are not likely to be confused by the receiver. One obvious way to accomplish this is to choose signals which have a great deal more power than the average noise waveform. However, the amount of signal power which can be employed is always limited (sometimes by the power available at the transmitter, sometimes by restrictions imposed by the channel). Thus it is necessary to seek more subtle means of avoiding transmission errors within the constraint of allowed signal power.

Conceptually, the solution to the problem posed by the channel noise lies in the law of large numbers. In effect this law says that while the outcome of a single chance event may fluctuate wildly, the overall result of many repetititions of this chance event is subject to quite-accurate prediction. In data communications the law of large numbers can be used to advantage when the duration of the input signals is made very long. Certain statistical properties of the noise, such as its power, then become much more predictable than if a short signal had been used to convey the information.

Unfortunately this gain in noise immunity comes at a great cost in complexity of the transmitter and receiver since for a given rate of data transmission, the number of possible messages grows exponentially with the length of the signal. For example, assume that binary data, the familiar 0s and 1s, are to be transmitted at a rate of 1 bit per second (bps). If the transmitter assigns waveforms once per second, only two waveforms are required, one to represent the "1" and another to represent the "0." To employ longer basic signals the transmitter might assign waveforms only once every three sec. But then eight waveforms are required to represent the combinations 000, 001, 010, 011, 100, 101, 110, and 111. Similarly if the transmitter waits N seconds, 2^N sequences must all be allocated particular waveforms. To take advantage of the law of large numbers with this approach, N would have to be quite large and the resulting transmitter and receiver hopelessly complex.

1.1 Modulation and Coding

Because of the exponential growth of complexity with signal duration, the straightforward assignment of long message waveforms in the direct manner mentioned above has not been attempted in practice. Instead, the waveforms are selected at frequent intervals so that only a small number of basic signals are needed (often two). The process of mapping the input digital sequence into a train of short analog waveforms is known as *modulation*. At the receiver the basic waveforms are interpreted individually so that the output of the detector is a sequence of digits representing best guesses of the transmitted data. This classic method of transmitting data foregoes any possibility of obtaining virtually "error-free" transmission through the use of long, basic signaling intervals.

Employing the above general type of modulation enables us to view the communication system of Fig. 1.1 in its entirety as a strictly digital channel, which we term the *coding channel*. In the typical binary case this channel accepts successive 0s and 1s (bits) at its inputs and usually reproduces them at its output. Occasionally, however, because of noise and other channel impairments, the output digits do not agree with the input digits. The *digital error rate* is the average number of disagreements between output and input digits.

Whereas on the modulation channel each possible message was associated with a signal waveform, on the coding channel each message (or block of digits from a message sequence) must be associated with a sequence of digits. Just as our goal in choosing signals was to minimize the probability that waveforms would be confused, so now do we desire to associate with messages digit sequences which are as different as possible from one another. (This notion of "differentness" of sequences is a very fundamental idea in coding theory and will be explored thoroughly in the text.) For example, suppose that one of the 2^k possible equally likely messages is to be transmitted over a binary channel. The binary-digit (bit) sequence associated with each message must be at least k digits long. If it is exactly this long, every possible received k-bit sequence corresponds to a sequence associated with some message. Therefore any change (error) introduced into a sequence by the channel will cause the received message to correspond to a different message from the transmitted sequence. That is, every channel error will cause a message to be misinterpreted. This vulnerability to error can be reduced substantially by alloting a larger number of bit intervals, say n, to the transmission of each of the 2^k messages. Since usually only a small fraction $(2^k/2^n)$ of the possible transmitted sequences correspond to messages, it is quite possible to choose these sequences to be quite unlike one another. This makes the communications system considerably less vulnerable to channel errors.

The process of associating an n-bit with a k-bit message is termed *encoding*. The inverse operation, *decoding*, is performed at the receiver and consists of associating a message with each received block. The general data-communications system of Fig. 1.1 now has the form of Fig. 1.2

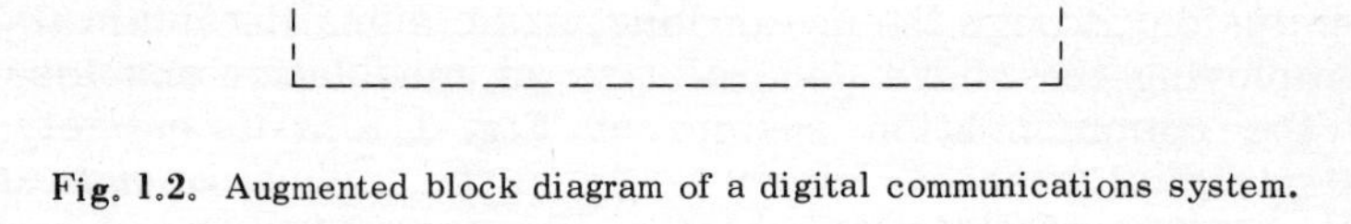

Fig. 1.2. Augmented block diagram of a digital communications system.

Because n is greater than k, employing coding results in a net reduction in data rate. This trading of data rate for error rate is a property common to all coding systems. However, we shall see that by using very long sequences, say up to thousands or tens of thousands of bits in length, it is often possible to attain a very low error rate at the expense of only a slight reduction in data rate.

At this point the reader may ask why we have introduced the idea of coding since the problem of associating one of a large set of signals with the message at the transmitter and the inverse operation at the receiver remains. Have we not just avoided the construction of a hopelessly complex analog machine by having to build an equally costly digital machine? Happily, the answer is no. First of all, the encoding process now typically consists of associating with a k-bit message sequence a somewhat longer n-bit sequence. Because of the digital nature of this association, it can be performed according to a very simple set of rules. Consequently instead of having to search through the entire set of signals to find the one associated with a particular message, the encoder can calculate the sequence to be transmitted from the message sequence. Similarly, decoding is also simpler in the digital domain, although it is more complex than encoding. A second advantage of digital coding is that it is simpler and more economical to process and store than analog signals.

In summary, the reason for separating the general communication problem into two parts, modulation and coding, is the following. In order to use communication channels efficiently, long signals must be employed. Since the number of signals grows exponentially with their length, the transmitter and receiver must be capable of handling an enormous number of signals. The only practical way these signals can be processed and stored is by quantizing the allowed signal values and using digital circuitry. Fortunately this quantization does not necessarily result in much loss in capability of the system. Furthermore, simple digital coding rules are known which circumvent the problem of exponential growth in complexity with message length.

1.2 The Structure of the Book

Even though the functions of coding and modulation are closely related in concept, entirely different mathematical disciplines have come into being in each of these fields. The digital coding theorist primarily employs the tools of modern and linear algebra, while the person concerned with modulation is more likely to use, say, Fourier integrals. Thus in this book, as in a communications system, the coding and modulation sections appear quite dissimilar even though physically adjacent. In the basic structure of the book, Chaps. 4 through 9 are concerned with various aspects of modulation and demodulation, while in Chaps. 10 through 12 methods of coding and decoding are treated. These two fields have common ground in Chaps. 2 and 3, which deal with channel characterization and fundamental bounds on communication.

In the chapters common to coding and modulation we first consider the characteristics of the channels typically used for data transmission. To facilitate the analytical design and evaluation of data communications systems, a few simple mathematical models are mentioned which describe the most important physical characteristics commonly found in real channels. Then in Chap. 3 the fundamental bounds which these physical characteristics impose on our ability to transmit data accurately are discussed. These bounds are important to the design of data systems in that they allow the engineer to decide when a further increase in equipment complexity can be expected to improve performance substantially or when such additional cost would be virtually wasted.

The material on modulation and demodulation in Chaps. 4 though 9 is divided into the treatment of two allied topics, baseband transmission in Chaps. 4 though 6 and passband transmission in Chaps. 7 through 9. The design and evaluation of a data system almost always involve both baseband and passband aspects. Generally the modulator transmits its symbols in the form of pulses of energy. The design of these pulses is accomplished at baseband as described in Chap. 4. The effect of the channel on these pulses and the performance of conventionally designed systems are also described in this chapter. In Chap. 5 design improvements based on exact knowledge of channel characteristics are considered. Chapter 6 then completes the treatment of baseband transmission by discussing methods of automatic equalization whereby the data system "learns" the channel characteristics and attempts to bring the performance of the system closer to the ideals evaluated in the previous chapter.

Since few realistic channels transmit energy near zero frequence (dc), it is usually necessary to shift the baseband channel to a more appropriate location in the frequency spectrum. Techniques which accomplish this shift while preserving the linearity of the baseband system, e.g., vestigal sideband transmission, are treated in Chap. 7. While these linear techniques are generally the most efficient, practical considerations frequently dictate the use of nonlinear modulation. The two most important techniques of this type, frequency modulation and phase modulation, are treated in Chaps. 8 and 9, respectively.

In Chap. 10 we turn our attention to the coding problem and discuss the basic mathematical properties of error-correcting and detecting codes. Almost immediately it becomes apparent that there are two fundamentally different, practical families of codes: block codes in which messages are associated with n-bit sequences, or blocks, and convolution codes in which information is not segmented into blocks but rather is processed continuously. Although the two types of codes have much in common, it is mathematically convenient to examine them separately. This is done in Chaps. 11 and 12. Classes of codes of both types are presented: those which are capable of correcting errors that occur at random and/or in bursts and those which can be implemented simply. It is seen that although block codes are better understood and are more highly developed than convolutional codes, the two types of codes have essentially the same capabilities and are subject to the same inherent limitations. Extensive lists of the most important types of block and convolutional codes are given in appendices at the end of the relevant chapter.

This book, like nearly every other reference book, necessarily touches on several areas which cannot be treated as thoroughly as they have been elsewhere. While the properties of random signals and noise are used extensively, they are not explained at an elementary level. Similarly, although the disciplines usually known as statistical communication theory and information theory permeate the present work, no attempt has been made to present these subjects tutorially. Also, an effort has been made to present coding theory with a minimum of emphasis on modern algebra in order to make the material more accessible to the classically educated communications engineer. This has necessarily limited the depth to which certain subjects, e.g., random-error-correcting cyclic codes, can be investigated.

To aid the uninitiated and encourage the ambitious reader we have included a list of pertinent references.

2

Channel Characterization

In Chap. 1 we saw that a data communications system consists of three major parts: the transmitter, the channel, and the receiver. In the remaining chapters of this book we consider the problem of designing efficient transmitters and receivers. Before this problem can be discussed intelligently, it is necessary to say something about the third major component in the communications system, the channel. In this chapter we shall discuss various simple mathematical models which accurately represent the basic properties of many real communication channels.

We have pointed out that in a practical data-communications system it is necessary to split the transmitter and receiver into an encoder and modulator and the receiver into a demodulator and decoder. These dichotomies are reflected in our characterization of communications channels. Thus even though both the coding system and the modulation system are a part of the same data-transmission facility, a different model of the communication channel is appropriate to each of these topics. From the point of view of modulation and demodulation, the channel generally includes the transmission medium and whatever transducers are needed to couple the signal to this medium. On the other hand, for coding and decoding the channel is treated as a digital link comprising the modulator, the modulation channel, and the demodulator.

In the two major sections of this chapter the modulation and coding channels are discussed separately. In each case the channel is modeled in a simple form which is more or less appropriate depending on the specific environment and type of communication system of interest to the designer. Many of the common types of channels, such as are employed in voiceband and wideband telephone systems, high-frequency radio systems, troposcatter systems, and deep-space probes, exhibit the same basic characteristics, differing chiefly in the degree to which each of the various factors is present. The models presented account for the main impairments common to all of these systems, while effects of lesser importance (and generally lesser mathematical tractability) are discarded.

2.1 Characterization for Modulation

The modulation channel is that portion of the communications system which includes the transmission medium and its associated transducers. This transmission medium typically includes cable, microwave, or high-frequency radio links. In any event the channel is analog, and its function is to reproduce at its output the message presented to it at its input. A real channel accomplishes this only approximately. Its output typically differs from its input in two ways. First, the channel may modify the message in a deterministic, although not necessarily known, fashion. Examples of this type of modification are dispersion, frequency offset, and nonlinearities. The channel can also corrupt the transmitted signal statistically. In this category are various types of additive and multiplicative noise such as thermal and impulse noise, and fades.

Throughout the sections of this book which deal with modulation and detection at baseband and passband, the channel will be modeled by a linear transfer function followed by an additive gaussian noise source. The model therefore takes account of the most important deterministic impairment, *dispersion*, as well as the most prevalent and mathematically tractable statistical impairment, *gaussian noise*. The reasons for excluding other types of impairment are varied and will be discussed in more detail in the following sections; basically, however, the exclusions involve relative importance and insight weighted by ease of analysis. Although the effects of time variation of the channel are not specifically considered, it is clear that the analysis of a time-invariant model is pertinent as long as the rate of time variation is small compared with the rate of data transmission.

The following sections discuss briefly the typical impairments encountered in the modulation-demodulation channel.

2.1.1 Noise

Generally two kinds of noise can be distinguished in a communication channel. The first, gaussian noise, will be our chief concern in designing and evaluating modulators-demodulators (modems) for data transmission. The sources of gaussian noise are many and unavoidable. They include thermal and shot noise from transmitting and receiving equipment, thermal noise introduced in the transmission medium, and radiation picked up by the receiving antenna. According to the central-limit theorem, any noise which arises from the combined effects of many independent sources tends to be gaussian distributed. Because of this omnipresence, and because of the relative ease with which it can be handled mathematically, gaussian noise has been employed almost exclusively in modulation-channel models throughout the growth of modern communications theory.

A second type of noise, known as *impulse noise*, is characterized by long, quiet intervals followed by bursts of much higher amplitudes than would be predicted by a normal- or gaussian-distribution law. This type of noise has many natural and man-made sources including lightning, switching transients, microwave fades, and accidental hits during maintenance work. In the telephone network, for example, impulse noise causes most errors in slow- and medium-speed data transmission.

The difficulty in measuring and mathematically characterizing this type of noise has prevented any significant analytical work in this area, except with practically atypical isolated, identical impulses and with impulses occurring with high density. (In the latter case the central-limit theorem begins to apply so that the noise approaches gaussian with small correction terms.) However, the limited analytical work which has been done has always shown that if data-transmission systems are ordered by relative performance in the presence of gaussian noise, this ordering will also be valid for impulse noise. This same ordering phenomenon has been observed in the laboratory and used as a justification for the practice of testing with Gaussian noise.

The reasons for not including impulse noise in the model of the channel for modulation go beyond the obvious considerations of proper characterization and mathematical tractability. For the most part the sections of this book which deal with modulation and detection stress efficient use of the communication channel. Because of the long quiet intervals inherent to impulse noise, the speed limitation of the channel is determined by the gaussian background noise rather than by the more spectacular bursts of impulse noise. Since the impulses cause errors with high probability whether the system operates at low or high speeds, the concern in efficient design is to protect against the ultimate limitation, i.e., gaussian noise.

It is possible to conceive of various modulation and detection methods, such as the so-called *smear-desmear technique*, which in theory protect against impulse noise. These systems have often not performed well in practice for a reason which points up a basic misunderstanding of the characteristics of impulsive noise. This misunderstanding arises naturally from the thought that an impulse is necessarily a short-duration burst of high energy. Many such short-duration bursts are in fact observed, but so are many very long bursts. It is these long bursts (often of lengths of upwards of several hundreds or even thousands of symbols) which cause most errors attributed to impulse noise. Long bursts result from such phenomena as dropouts, accidental hits, and deep microwave fades. Modulation techniques are impotent in dealing with these phenomena; only digital coding holds any hope of handling long bursts. Therefore from the point of view of modulation theory, impulses are regarded as catastrophic events which are passed on as digital errors to be dealt with by the coding portion of the communications system.

2.1.2 Linear Distortion

All real channels exhibit some form of time dispersion. In a high-frequency radio channel this dispersion may be due to multipath transmission, while in a telephone channel the dispersion can be attributed to the imperfect transfer characteristics of the transmission system. In the telephone channel the transfer function is measured in terms of attenuation and delay as a function of frequency as shown in Fig. 2.1.[1] The *envelope delay*, which is the derivative of the phase characteristic, represents the relative time of arrival of various frequency components of the input signal. Since neither the envelope delay nor the attenuation is generally constant with frequency, the input signal suffers a distortion in passage through the channel. In data transmission this distortion takes the form of an overlap in time between successive symbols known as intersymbol interference.

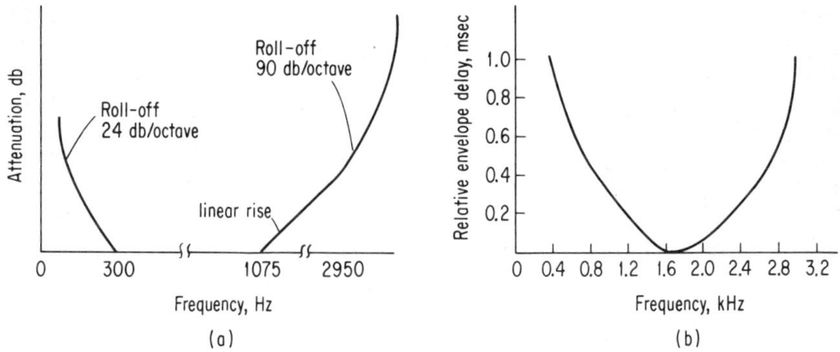

Fig. 2.1. (a) Amplitude and (b) delay characteristics for a telephone channel. *(Source: From Alexander, Gryb, and Nast.[1])*

Intersymbol interference represents a kind of deterministic channel impairment. If the channel characteristics are known, it is usually possible in theory to remove intersymbol interference. However, the transmission channel is never known exactly; thus it is impossible to eliminate completely the effects of intersymbol interference in a real data system. For example, in calls over the switched telephone network, a channel of the type characterized in Fig. 2.1 is chosen randomly at the beginning of each transmission from an ensemble which is so large as to be functionally infinite. Naturally large variation in the transmission characteristics results from this random selection. The effects of intersymbol interference may be so large on some channels as to preclude reliable operation of the data system, while on other channels near perfect transmission is experienced.

Unfortunately the effects of intersymbol interference are difficult to include in a mathematical analysis of a communications system. Nevertheless, these effects represent an important and unavoidable facet of the data-communications problem. Thus an arbitrary linear network is included in the channel model in the chapters dealing with modulation and detection.

2.1.3 Other Impairments

A number of causes other than noise and linear distortion can result in the output of a channel being different from the input. While some of these impairments are important design considerations (and, as such, will be mentioned from time to time), none will be consistently used as part of the channel model. Among the miscellaneous impairments are nonlinearities, frequency offset, and phase jitter [incidental frequency modulation (FM)].

Nonlinearities are always present in a communications system to some small extent because of the impossibility of achieving truly linear amplification or filtering. These types of nonlinearities are largely negligible, but occasionally significant effects result when amplifiers are overloaded into operation in a highly nonlinear region. Significant nonlinearities also occur on the switched telephone network owing to the action of voice compandors (circuits designed to compress and later expand the dynamic range of speech signals).

Frequency offset and phase jitter are other phenomena associated with telephone transmission. Both effects result from the use of a carrier system within the telephone channel. The voice frequency band, nominally 0 to 3kHz, is heterodyned or shifted in frequency to higher frequencies and then multiplexed with other voiceband signals to form a portion of a wideband signal. At a distant point this signal is demultiplexed and the original voice channels are separated. In heterodyning the voiceband back to baseband, the reference carrier may differ in frequency and phase from the modulating carrier. Thus at the receiver the voiceband lies between ϵ to $3 + \epsilon$ Hz, where ϵ is a frequency shift of typically a few cycles. This frequency offset makes the telephone channel technically a time-varying system since the response to an applied impulse is a function of the time at which the impulse was applied. However, the offset is unimportant from a theoretical point of view since it represents a simple and constant transformation of the transmitted wave. In practice it can be simply removed at the receiver.

In addition to the frequency offset the instability of the modulating- and demodulating-carrier generators causes a random jitter in the phase of the received signal. This jitter is equivalent to a low-index, random-frequency modulation of the transmitted signal and is consequently termed *incidental FM*. The severity of the incidental FM depends in large part upon the kind of carrier system used on a particular connection. Technical improvements are tending to eliminate this problem in the telephone plant.

2.2 Characterization for Coding

In the preceding section we saw that the modulation channel consists of the transmission medium and the necessary transducers. In this section we consider the coding channel. This channel includes the modulation channel as well as the modulator and demodulator (see Fig. 1.3).

The most important difference between these two channels is that while the modulation channel is analog, the coding channel is digital. Its input is a sequence of digits chosen from some typically small alphabet; its output is another sequence which is usually, but not always, identical to the input sequence. The difference between the two sequences can be thought of as an *error sequence*. It is this sequence which we attempt to characterize in this section.

Since it contains the modulation channel, the coding channel also suffers from both probabilistic and deterministic degradations. However, as we shall see the deterministic effects can usually be taken into account by the modem. Thus if the modem is designed reasonably well, errors due to deterministic causes are usually unimportant as compared to those caused by noise. For this reason, only errors due to probabilistic causes are considered in this section.

In Chap. 3 it is pointed out that binary codes have several important practical advantages over more general codes. For this reason they are considered almost exclusively in Chaps. 10 through 12. Also, for this reason in this section we characterize coding channels in a manner most appropriate for binary coding. This does not restrict the modem to employ binary signals, however. If $L = 2^r$, an L-ary modem signal can be represented as a binary r-tuple, and binary coding can be employed.

There are numerous ways in which the 2^r modem signals can be associated with binary r-tuples. In general we wish to associate signals with r-tuples so as to make the binary error rate as small as possible. In most situations (such as when pulse amplitude modulation is used) it is possible to order the 2^r modem signals so that the channel is more likely to cause adjacent-signal errors than other types. In this situation the best mapping is a Gray code, one of which is illustrated below for $r = 3$.

Table 2-1. Gray Code for $r = 3$.

Channel Signal	Associated Binary 3-Tuples
1	0 0 0
2	0 0 1
3	0 1 1
4	0 1 0
5	1 1 0
6	1 1 1
7	1 0 1
8	1 0 0

With this code an adjacent-signal error always causes only one binary digit to be incorrect. The reader should note the construction of this code. The eight-signal Gray code is constructed from the four-signal Gray code (in the dashed rectangle in Table 2-1) by reflecting the set of 2-tuples about a horizontal line and then prefixing the original set by zeros and its reflection by ones, as shown. Gray codes for larger values of r can be constructed iteratively in this manner from the table. Throughout this book it is always assumed that a Gray code is used to associate modem signals with sequences of binary digits.

In Chaps. 11 and 12 it will be seen that it is possible to design sets of binary sequences which are relatively immune to demodulation errors caused by gaussian noise, impulse noise, or both. That is, binary codes can be constructed which are capable of correcting errors occurring independently at random and/or errors which occur in "bursts." In order to evaluate the performance of these codes it is necessary either to have a fairly detailed set of observed channel error statistics or to have a mathematical model of the channel whose pertinent statistics are known to be those of the real channel.

In the following paragraphs several digital communication channel models are presented. These models are chosen to have two basic characteristics:

1. With a suitable choice of parameters, the most important statistics of the error sequence emitted by the model can be made to match closely the corresponding statistics of a wide variety of real channels.
2. The models are inherently simple enough so that meaningful performance estimates can often be calculated in a straightforward manner. Also, where information about a channel is desired which is too detailed to be calculated directly, a model can be used as a basis for a computer simulation of the channel.

Generally speaking, it is convenient to classify channels as being either *memoryless*, in which case demodulation errors are independent, random events, or as having *memory*, in which case demodulation errors exhibit some statistical dependence.

2.2.1 Memoryless Channels

When only white gaussian noise is added to the transmitted signal, the channel is usually, for all practical purposes, memoryless. A deep-space-probe channel, for example, is approximately memoryless.

If the channel signals are chosen from an alphabet of size L, the effect of the channel on the transmitted symbol can be represented by an L-by-L *transition matrix*. The entry (i, j) in this matrix is $P(j/i)$, the conditional probability that the symbol corresponding to the jth signal is received when the ith signal is transmitted. When only two symbols are employed and if the probability of correct transmission is identical for both, the channel is said to be a *binary symmetric channel* (BSC).[2] Such a channel can be completely characterized by a single parameter p, the probability of a bit error.

On such a channel all patterns of length n which contain exactly m errors occur with equal probability, $p^m q^{n-m}$, where $q = 1 - p$. Since p never exceeds 0.5 on the average (since otherwise the decoder could interpret a 1 as a 0 and vice versa), this probability decreases with increasing m. Thus a well-designed random-error-correcting code should correct as many patterns as possible which contain relatively few errors. Since there are

$$\binom{n}{m} = \frac{n!}{m!\,(n-m)!} \tag{2.1}$$

patterns which contain exactly m errors, the probability of exactly m errors being added to n transmitted digits, the $P(m,n)$, is given by the well-known binomial distribution

$$P(m,n) = \binom{n}{m} p^m q^{n-m} \tag{2.2}$$

In Chaps. 10 and 11 we shall see that it is possible to design codes which correct any pattern, containing up to a given number of errors. In Sec. 11.5 this $P(m,n)$ distribution will prove useful in evaluating the probability of erroneous decoding for random-error-correcting codes.

If the modem employs an L-symbol alphabet where $L = 2^r$, each L-ary signal can be represented as a binary r-tuple. In this case a memoryless L-ary channel can be viewed as a binary channel with an r-bit memory. It is not difficult to compute the exact $P(m,n)$ distribution for this channel, given the transition matrix and the form of the Gray code employed. However, it is often more convenient to treat the channel as a BSC for which p is the average bit error rate on the L-ary channel. The $P(m,n)$ distribution obtained in this way typically has noticeably different fine structure than the true distribution; however, its overall shape is similar.

2.2.2 Channels with Memory

We have pointed out that on many real channels, not just gaussian noise but also impulse noise is added to the signal. On such channels demodulation errors are not independent random events, and the channel is said to have memory. Telephone channels, afflicted as they are by switching transients, dropouts, lightning strokes, etc., and high-frequency radio channels, which are subject to fading, are examples of channels of this type.

Perhaps because of the myriad of underlying physical phenomena involved, the characterization of channels with memory is more difficult than that of memoryless channels. Ideally such a characterization should be mathematically tractable and should represent the pertinent aspects of the error process accurately. Unfortunately no completely satisfactory characterization of channels with memory has been found.

In the preceding section it was pointed out that the $P(m,n)$ distribution is useful in evaluating the performance of codes on memoryless channels. In Sec. 11.5 we shall see that this distribution is also useful for some channels with memory. Another useful quantity is the $P(l,n)$ distribution. The probability $P(l,n)$ is defined as the probability of the occurrence of a burst of errors of length l in a block of length n. A third distribution which has been used to evaluate the performance of codes is the distribution of *error-free intervals*. All three of these distributions can be obtained from the various channel models discussed in the following paragraphs. In some cases, however, analytical difficulties make computer solutions necessary.

The simplest characterization of an error-causing channel is a BSC. However, the BSC is a very poor channel model if the impulse noise dominates the gaussian noise and most errors occur in infrequent long bursts. For example, the $P(m,n)$ curves for the real channel and the BSC will differ widely even though the average error rates are the same. This is illustrated in Fig. 2.2 in which the average $P(m, 31)$ distribution of an ensemble of switched voiceband telephone[3] channels is compared with the corresponding distribution for a BSC with the same average bit error rate. It can be seen that the two curves differ substantially. The error-free interval and $P(l,n)$ distributions of the BSC are also significantly different from those of most bursty channels. It seems clear that this simple model is inadequate to characterize channels with memory.

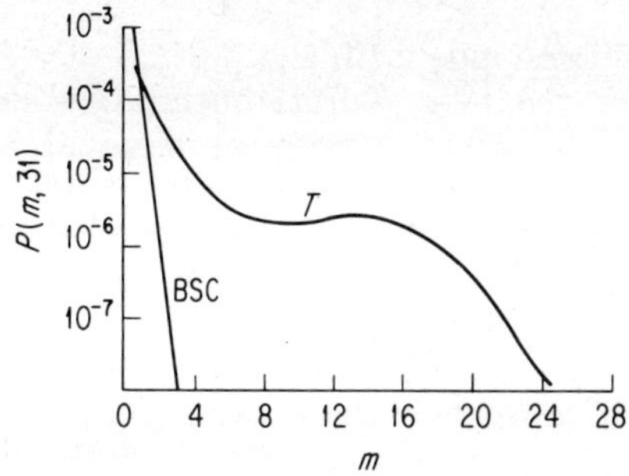

Fig. 2.2. $P(m, 31)$ distribution for telephone channel (T) and binary symmetric channel model with same average bit error rate (5×10^{-5}).

A more complex model in which the error-generating mechanism occupies one of a given number of states with a fixed probability has been successful in characterizing channels on which errors occur in bursts. This *Gilbert model*[4] considers the channel to be composed of a small number of BSCs. The state of the channel, i.e., the BSC through which a given bit is actually transmitted, is controlled by a Markov process. (Besides being tractable mathematically, this model is intuitively satisfying in that error bursts on real channels often are the result of a temporary change in the macroscopic physical state of the channel.) Typically the probability of remaining in a given state is considerably greater than the probability of leaving. Since one of the BSCs is usually taken to have a bit error rate of 0.5, errors tend to occur in bursts. Ordinarily two or three states are sufficient to define a model whose $P(m,n)$ or intererror interval distribution is similar to that of a given real channel. Figure 2.3 depicts a three-state Gilbert model for a typical telephone channel.

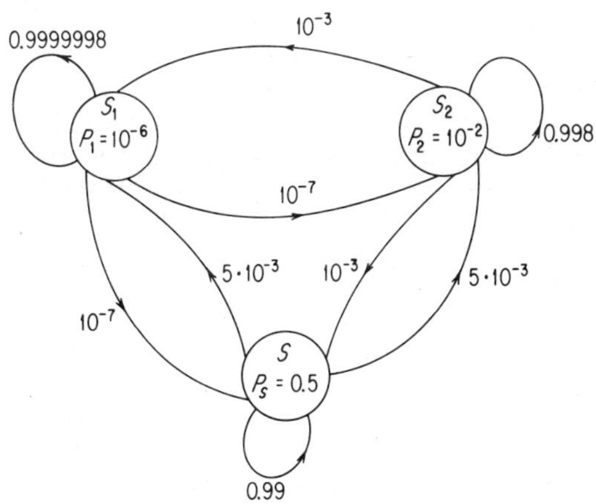

Fig. 2.3. A three-state Gilbert model for a typical telephone channel.

If the average time spent in each state of a Gilbert channel is longer than n bits, the $P(m,n)$ distribution can be approximated by the sum of the $P(m,n)$ distributions of several states weighted by the fraction of the time spent in each state. The $P(l,n)$ and error-free interval distributions can be approximated in a similar manner. Exact expressions for the $P(m,n)$ and error-free interval distributions have also been obtained.[5]

A modification of the Gilbert model which yields an excellent match to recorded telephone data has been proposed. This model retains the basic idea of decomposing a real channel into a small number of different mathematically tractable channels. However, these channels are taken to be not BSCs but rather renewal-process channels. The lengths of the intererror intervals on such channels are taken to be independent random variables distributed according to some empirically determined distribution. A three-state model of this type whose intererror interval distributions are defined by approximately a dozen parameters yields a very good fit to observed data, in terms of both the error-free interval and $P(m,n)$ distributions. For this channel the latter distribution can be determined from the error-free interval distribution alghough the process is to tedious to be carried out by hand.[6]

The intererror interval distribution of several telephone channels has been observed to be closely matched by a Pareto distribution,[7,8] that is,

$$p(x) = cx^{-\alpha-1} \tag{2.3}$$

where $p(x)$ denotes the probability that an intererror interval is exactly x bits long. The parameter α typically lies between 0 and 1. This single-parameter model, of which the preceding three-state model is a generalization, is based on the assumption that intererror intervals are independent random events.

Some of these models seem to fit the intererror interval distribution of telephone facilities quite well. However, the $P(m,n)$ distribution of the model usually differs significantly from that of the real channel. For this reason the multiple-state models discussed earlier seem preferable in most situations.

2.2.3 Other Channels

In the preceding sections we considered channels which alter the transmitted digit sequence by adding errors. Besides causing additive errors, some channels corrupt transmitted data by occasionally causing the demodulator to misframe a symbol, i.e., to interpret parts of two adjacent transmitted symbols as one received symbol. More drastic alterations of this type may cause the demodulator to skip several bits or insert bits where none belongs.

unless protected against such occurrences ordinarily end meaningful communication. Fortunately, methods of combatting this sort of channel disturbance with coding have been devised.

From the point of view of this kind of coding, the most pertinent mathematical model is simply a channel which occasionally deletes and/or inserts bits into the transmitted sequence. Since both additive and timing errors occur on many real channels, it is often necessary to use this model in conjunction with the more complex additive-error models of the preceding sections to represent real channels accurately.

2.3 Summary

In this chapter we discussed the characteristics of communication channels which are relevant to the topics in the forthcoming chapters. In accordance with the basic philosophy espoused in Chap. 1, we considered the modulation and coding aspects of communication channels separately. With respect to the former, we saw that although other types of impairments are often present, two of the most important limitations on the performance of the modulator-demodulator are additive gaussian noise and linear distortion. Because of their importance and mathematical tractability, these two are included in the channel model which will be employed in the analyses of modulation in Chaps. 4 through 9. It was also pointed out that although nongaussian, i.e., impulse, noise is often present on real channels, the modulation equipment is usually incapable of handling it. For this reason no attempt was made to include impulse noise in the modulation-channel model. Rather it was taken into account by the coding channel, since long codes can be designed to correct the resulting error bursts as well as the random errors due to gaussian noise. A number of models, typically employing several channel states, have been proposed to characterize channels on which both random and burst errors occur. Finally, it was pointed out that in addition to causing additive errors, real channels tend to cause demodulator timing errors. From the point of view of coding this can result in bits being inserted into or deleted from the transmitted sequence.

Notes

Data pertaining to the characteristics of telephone media are given in Alexander, Gryb, and Nast.[1] Error statistics in data communication over the switched telephone network are presented in Townsend and Watts.[3] Part III of Schwartz, Bennett, and Stein[9] contains a detailed discussion of high-frequency fading communications media and also includes a considerable amount of experimental data on high-frequency channel characteristics and attainable error rates.

3

Performance Bounds for Digital Communication

In the previous chapter some simple mathematical models were used to describe the most important characteristics attributed to real channels. It is quite natural to consider now the ultimate limitations which these characteristics impose on the rates and accuracies of data transmission. A great deal of effort has been expended by communications theorists in deriving such performance bounds for digital communication systems. The purpose of this chapter is to discuss in an heuristic fashion the chief content and meaning of these bounds as they relate to the problems of modulation and coding.

We are concerned here with the two channels discussed in the preceding chapter: (1) the channel for modulation which we have modeled by a linear transfer function followed by an additive gaussian noise source and (2) the channel for coding which will be assumed here to be memoryless and thus describable by means of a matrix of the probabilities of transitions between possible input and output signals. Although coding channels which have memory can often be characterized satisfactorily (see Chap. 2), it is only for the memoryless channels that meaningful performance bounds have been derived.

As will be seen, the bounds for the modulation and coding channels are quite similar in form. Both relate the total signal duration to the transmission rate and probability of error. However, the fundamental difference between modulation and coding—the former employs very much shorter signals than the latter—dictates that the bounds be employed in the design of modulation and coding systems in different fashions.

The coding bounds relate the code rate, or efficiency, k/n, to the code length n in symbols and to the probability of decoding erroneously. (Recall that the encoder maps sequences of k information symbols onto n-symbol sequences for transmission.) For present purposes the cost of the equipment required for encoding and decoding can be thought of as being roughly proportional to the code length. The system designer uses these bounds to relate parameters dictated by external constraints to parameters which he is free to choose. For example, in a particular situation the data rate and probability of error may be given; the bounds show that this level of performance can be obtained with a code of a particular length. This information, coupled with a knowledge of specific types of codes and methods of implementation, can serve as a guide in the initial stages of system design.

In the design of the modulation system bounds enter in a different fashion. The methods and types of modulation and demodulation are much more constrained than is the selection of coding techniques. Thus the theoretical bounds which can be derived are seldom useful as a tool in design but rather find their value as a means of evaluation for design proposals. First the designer is concerned with what constraints in performance are imposed in theory by the analog modulation channel. Second, he is concerned with what constraints are imposed on the coding system by a particular modulation technique. The difference between these two sets of constraints is due to inefficient modulation. The engineer naturally looks to find a modulation technique which minimizes this loss.

The remainder of this chapter is divided into two sections, bounds for modulation and bounds for coding. In the section on modulation we shall discuss what level of performance can be obtained from the modulation channel by ideal means. In the coding section the idealized relationships among the code rate, the code length, and the probability of error are examined. In each case a great deal of theoretical analysis and justification will be, of necessity, omitted in favor of a simple, heuristic presentation of central ideas and results.

3.1 Bounds for Modulation

We assume that the modulation channel consists of a linear filter with transfer function $C(\omega)$ followed by an additive source of gaussian noise which, for simplicity, is taken to be white with double-sided spectral density N_0. The problem we must approach is the following. Given a transmission rate of R_c bps and messages of length T sec, what is the minimum average probability of error which can be obtained over this channel using an average transmitted power of P? For concreteness we shall also assume that the transmitted waveform begins at time zero and ends at time T, while the receiver is allowed to observe and process the received waveform over an infinite interval. (The observation time at the receiver would ideally have to be greater than the transmission time because of dispersion in the channel tending to "spread out" the transmitted waveform in time.)

First of all we observe that there are $M = 2^{R_c T}$ possible messages which we must be prepared to transmit. The ideal modulator assigns a waveform $s_i(t)$, where $i = 1, 2, \ldots, M$, to each of the possible messages. Just how these waveforms are optimally selected is a problem we pass by for the moment. We expect to observe one of these M waveforms at the receiver after the waveform has been passed through the channel and masked by the addition of a noise signal $\xi(t)$. The recieved signal $z(t)$ can then be written

$$z(t) = y_i(t) + \xi(t) \tag{3.1}$$

where

$$y_i(t) = \int_{-\infty}^{\infty} s_i(t - \tau)\, c(\tau)\, d\tau \tag{3.2}$$

and $c(t)$ is the impulse response of the channel characteristic $C(\omega)$.*

$$c(t) = \frac{1}{2\pi} \int_{-\infty}^{\infty} C(\omega)\, e^{i\omega t}\, d\omega \tag{3.3}$$

This notation is shown in a diagram of the modulation-channel model in Fig. 3.1.

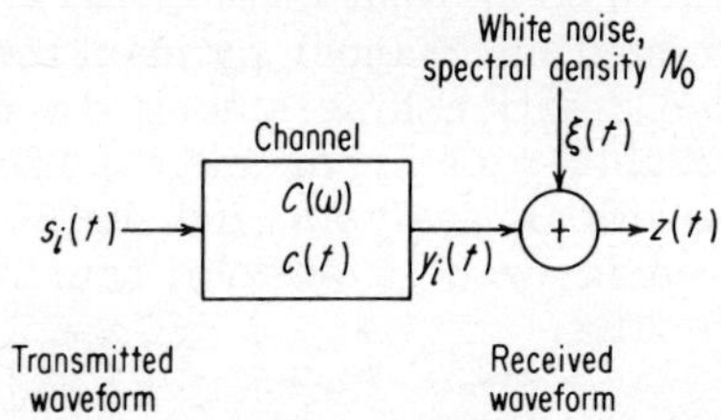

Fig. 3.1. Modulation channel model.

3.1.1 Signal Representation and Detection

In communications-systems analysis it is nearly always convenient to represent waveforms in terms of sets of orthonormal basis functions. Thus we might choose to write the signals $s_i(t)$ in the form

$$s_i(t) = \sum_{j=1}^{N} s_{ij} \phi_j(t) \qquad 0 \le t \le T \tag{3.4}$$

*Unless otherwise noted, Fourier transform pairs will be represented by lowercase letters in the time domain and capital letters in the frequency domain.

where $i = 1, 2, \ldots, M$ and the $\phi_j(t)$ are orthonormal on the interval $[0, T]$. The number of coefficients N needed to provide representation for the set of M input signals might ordinarily be infinite, although this depends on the particular choice for the basis set $\{\phi_j\}$.

For the channel output signals $y_i(t)$ and noise $\xi(t)$ we choose a different set of basis functions $\{\theta_j(t)\}$—this time orthonormal on the infinite interval.

$$y_i(t) = \sum_{j=1}^{\infty} y_{ij}\theta_j(t) \tag{3.5}$$

$$\xi(t) = \sum_{j=1}^{\infty} \xi_j\theta_j(t) \tag{3.6}$$

The signals $s_i(t)$, $y_i(t)$, and $z(t)$ can now be thought of as vectors $\mathbf{s}_i$, $\mathbf{y}_i$, and $\mathbf{z}$ whose components are the coefficients in their respective expansions. It can be observed that the action of the channel filter on the input vectors, Eq. (3.2), can be expressed in a matrix form

$$\mathbf{y}_i = \mathbf{C}\mathbf{s}_i \tag{3.7}$$

Since the input signals have equal energy PT, the M vectors $\mathbf{s}_i$ all lie on the surface of an N-dimensional sphere. The effect of the channel as expressed in Eq. (3.7) is to rotate and stretch or shrink each signal vector $\mathbf{s}_i$ in some fashion as to produce $\mathbf{y}_i$. Thus any particular properties of arrangement and distance inherent in the original signal set are changed by the channel. Finally, a noise vector $\boldsymbol{\xi}$, whose components are independent, gaussian, random variables, is added to $\mathbf{y}_i$ to give the received vector $\mathbf{z}$.

Ideally, the receiver waits the entire duration of the received signal and then makes a single decision as to which of the M messages was most likely to have been transmitted, given the particular received vector $\mathbf{z}$. With equally likely signals, this is equivalent to selecting the signal $\mathbf{s}_i$ with maximum likelihood $p(\mathbf{z} \mid \mathbf{s}_i)$. In the case under discussion this merely amounts to selecting the signal $\mathbf{s}_i$ for which $\mathbf{y}_i$ is closest (in euclidean norm) to the received vector $\mathbf{z}$. In Chap. 5 the topic of optimum reception is dealt with in somewhat greater detail, and it is shown that the optimum receiver can, at least in theory, be implemented using a bank of M filters as shown in Fig. 3.2. Each filter is matched to one of the expected received signals; thus the kth matched filter has the response $y_k(T_0 - t)$, where T_0 is some arbitrary delay chosen to ensure realizability. Notice particularly that the filters are matched to the *received* signals $y_i(t)$, rather than the transmitted signal $s_i(t)$. Thus each of the M matched filters contains a channel-matching filter $C^*(\omega)$ in tandem with a transmitted signal-matching filter $S^*(\omega)$, the complex conjugates of $C(\omega)$ and $S(\omega)$, respectively.

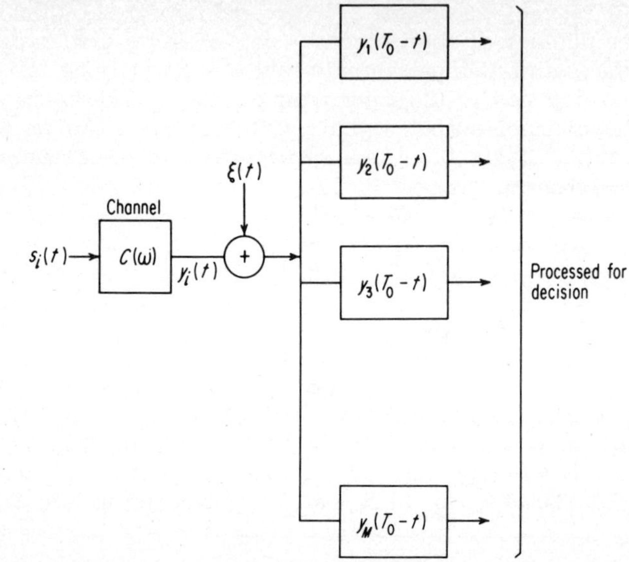

Fig. 3.2. Optimum receiver.

The probability of error for the system is a function of the positions of the M vectors y_i at the output of the channel. Generally speaking, we wish to make these vectors as large and as far apart as possible subject to the input-signal power limitation. The relationship between the set of input vectors $\{s_i\}$ and the set of output vectors $\{y_i\}$ is important to understand so that the input vectors can be chosen for minimum probability of error. This relationship can be considerably simplified if the basis functions $\{\phi_i\}$ and $\{\theta_i\}$ are chosen so as to diagonalize the channel matrix C in Eq. (3.7). Holsinger[1] has shown that this in fact is possible if the functions $\phi_i(t)$ are chosen as eigenfunctions of the integral equation

$$\lambda_j \phi_j(t) = \int_0^T \phi_j(\tau) K(t,\tau)\, d\tau \tag{3.8}$$

whose kernel $K(t,\tau)$ is

$$K(t,\tau) = \int_0^\infty c(s-t)\, c(s-\tau)\, ds \tag{3.9}$$

The functions $\theta_j(t)$, used for expansion of the received signal, are then determined by passing $\phi_j(t)$ through the channel filter $C(\omega)$. Using these functions for representation of input and output vectors results in the simple expression

$$z_j = \sqrt{\lambda_j}\, s_{ij} + \xi_j \tag{3.10}$$

for the jth component of the output vector when signal s_i has been transmitted. In other words the channel characteristic $C(\omega)$ can be represented by a set of N parallel, nondispersive channels, the jth of which has gain $\sqrt{\lambda_j}$ and additive white noise of density N_0.

The interesting interpretation of Eqs. (3.9) and (3.10) is that the kernel $K(t,\tau)$ is the impulse response of the portion of the optimum system consisting of the channel characteristic $C(\omega)$ and the common, channel-matching component $C^*(\omega)$ of the matched filters. The functions $\phi_j(t)$ are therefore those functions that are self-reproducing to within the gain factor $\sqrt{\lambda_j}$ when passed through this part of the optimum system. If the eigenvalues λ_j are ordered such that $\lambda_1 \geq \lambda_2 > \lambda_3 \geq \cdots$, then $\phi_1(t)$ is the function with maximum energy transfer through the optimum system, $\phi_2(t)$ has next-largest energy transfer, etc. In this representation the dilemma involved in choosing a good set of input vectors is clearly evident. To obtain the most energy through the system, the low-order eigenfunctions should be heavily emphasized. On the other hand, we would like to make these signals as unlike as possible in order to minimize the probability of error. The latter consideration would seem to call for the use of as many different eigenfunctions as possible. In general the problem of signal selection is too complicated to solve except in very special cases. However, we are able to obtain bounds by the introduction of a random-coding argument.

3.1.2 The Random-coding Bound

In theory the probability of error can be evaluated for any particular set of input vectors. This probability of error could then be minimized over all the choices of sets of input vectors to obtain a lower bound on attainable probability of error for the channel $C(\omega)$ and noise N_0. Unfortunately it is impossible in any case of practical interest even to evaluate the probability of error for a specific set of input vectors—not to speak of minimizing with respect to their placement. The random-coding bound involves an ingenious circumvention of this difficulty. In obtaining this bound, the set of input vectors is picked at random, and the average probability of error for the ensemble of such randomly chosen sets is evaluated. Since there must exist a set of input vectors with probability of error at least as low as the average of all such sets, the random-coding argument provides an upper bound for the best attainable probability of error for the channel.

Inasmuch as it is practically impossible to compute the probability of error for any specific set of input vectors, it may seem unreasonable to expect that this quantity can be averaged over an ensemble of choices. However, Shannon[2] discovered that by interchanging the order of obtaining probability of error and averaging over the ensemble, the problem could be made mathematically tractable. Gallager[3] later simplified the mathematical process with the introduction of a particular inequality which led to a concise formulation of the problem. The analysis of Gallager was applied by Holsinger to the case of the dispersive channel which is considered here. In this formulation the components s_{ij} of the input vectors are chosen independently from gaussian distributions of zero mean and variances σ_j^2. Thus the probability density which defines the input vectors is

$$p(\mathbf{s}_i) = \prod_{j=1}^{N} \frac{1}{\sqrt{2\pi}\,\sigma_j} e^{-s_{ij}^2/2\sigma_j^2} \tag{3.11}$$

where $i = 1, 2, \ldots, M$. [We are assuming that the components s_{ij} are with respect to the particular basis functions of Eq. (3.8).]

The signal energy constraint becomes

$$\sum_{j=1}^{N} \sigma_j^2 = PT \tag{3.12}$$

Now an upper bound for the average probability of error for the set of Eq. (3.11) can be derived using the inequality of Gallager. This inequality utilizes a parameter ρ, $0 \leq \rho \leq 1$, which can be adjusted in any given situation to obtain the tightest (lowest-possible) bound. When this inequality is used, an expression for the average probability of error can be written in the form

$$\mathrm{P}_e \leq 2^{-TE(R_c,\,\rho,\,\boldsymbol{\sigma})} \tag{3.13}$$

in which $\boldsymbol{\sigma}$ denotes the N-dimensional vector of the standard deviations $(\sigma_1, \sigma_2, \ldots, \sigma_N)$. This expression can then be minimized over $\boldsymbol{\sigma}$ and ρ to result in the bound

$$\mathrm{P}_e \leq 2^{-TE(R_c)} \tag{3.14}$$

The function $E(R_c)$, called the *error exponent*, is of the general form depicted in Fig. 3.3. Starting from $E(0)$, the zero-rate exponent, the curve is a straight line of slope -1 until the critical rate R_{crit} is encountered. From R_{crit} the slope of the exponent gradually increases until it becomes zero at the point where the exponent itself becomes zero. This point is called the *capacity* C of the channel. The capacity is the highest rate where the exponent $E(R_c)$ is positive. From Eq. (3.14) we see that the probability of error can be made arbitrarily small for any rate where $E(R_c)$ is positive by increasing the signal duration T. Although only an upper bound has been set for P_e, it can also be shown that arbitrarily small error rates cannot be obtained above capacity C. The origin of the concept of channel capacity is in the classic paper of Shannon.[4]

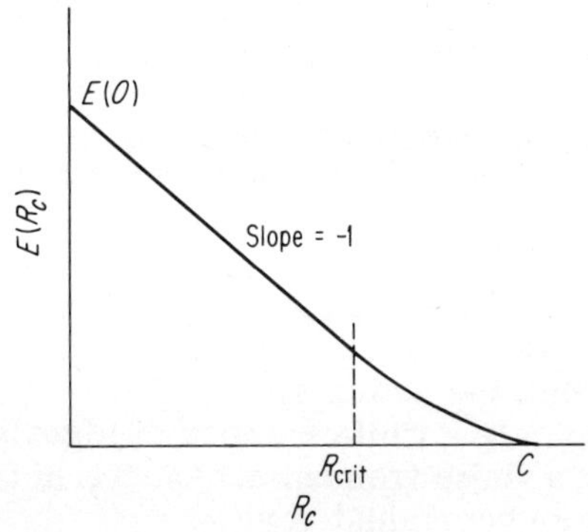

Fig. 3.3. Random-coding bound. (See Eq. 3.14.)

The reason vanishingly small error rates can be obtained should be understood as a manifestation of the law of large numbers mentioned in Chap. 1. Let us envision the $M = 2^{R_c T}$ channel output vectors y_i and the noise vector ξ as T is increased. The lengths of these vectors are proportional to the square roots of their energies, which are increasing linearly with T. Thus to keep all the vectors from growing as T is increased, we normalize by $1/\sqrt{T}$. With this normalization the lengths of the signal vectors are constant. It is also clear that the distances between the signal vectors y_i do not grow without bound as T increases. Why then does P_e decrease so dramatically? The answer to this question lies in the phenomenon of *sphere hardening* for the noise vector ξ.

Although the expected length of the normalized noise vector is constant with increasing T, the *variance* of its length decreases. Thus for large T the received vector with high probability lies close to the surface of a (many dimensional) sphere whose center is the transmitted vector. As $T \to \infty$ this sphere-hardening interpretation of the law of large numbers dictates that the received vector will always lie on one of the spheres surrounding the possible signal vectors. If the transmission rate is less than capacity, the spheres do not intersect and there is no difficulty in determining the correct transmitted signal. In other words the signals can be detected with zero probability of error using the minimum distance (maximum-likelihood) decoding criterion.

In deriving the random-coding bound, Eq. (3.14), the minimization over σ shows how much energy must be assigned (statistically) to each eigenfunction. Interestingly, the components of σ all vanish for eigenfunctions of higher-order than, say, the kth. This means that beyond this point it becomes preferable to crowd more vectors into the available dimensionality (k), rather than to increase the dimensionality by the addition of more eigenfunctions to the basis. This statement is only true with respect to the particular basis functions described by Eq. (3.8). In any other basis for representation an infinite set of functions would generally be required.

The exact expression for exponent $E(R_c)$ depends in a rather complicated fashion on the eigenvalues of integral equation (3.8). In any practical situation the number of bits to be transmitted R_cT would be a relatively large number; thus the exponent would be closely approximated by its value for the limiting case $T \to \infty$. For this limiting case a much simpler expression can be obtained for the exponent $E(R_c)$. Here the eigenfunctions tend to sine waves, the variances σ_j^2 become the power spectral density of the signal, and the notion of using only a finite number of eigenfunctions becomes that of using only a finite frequency band W. In terms of this band W the capacity C may be evaluated as

$$C = \int_W \log_2 \left[\frac{|C(f)|^2 P}{2WN_0} + \frac{|C(f)|^2}{W} \int_W \frac{d\nu}{|C(\nu)|^2} \right] df \qquad (3.15)$$

The band W and the optimum signal-energy distribution at capacity are determined by the "water-pouring" technique shown in Fig. 3.4. The positive half of the signal power P is "poured" into the curve $N_0/|C(\omega)|^2$. The final level reached gives the bandwidth W, while the "depth" at each point denotes the spectral density at that frequency. Notice that the signal power density, apart from a constant, is proportional to $|C(\omega)|^2$. Also observe that the phase characteristic of the channel $C(\omega)$ is immaterial.

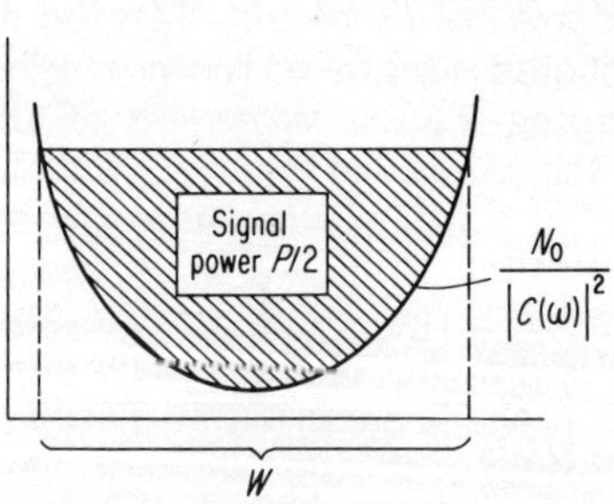

Fig. 3.4. "Water-pouring" analogy for optimum signal-power distribution.

When the channel has unity gain over a bandwidth W and zero transmission elsewhere, the capacity C reduces to the well-known formula of Shannon[4]

$$C = W \log_2 \left[\frac{P}{2WN_0} + 1 \right] \tag{3.16}$$

Continuing the description of the exponent $E(R_c)$ shown in Fig. 3.3, it can be shown that

$$E(0) = \int_W \log_2 \left[\frac{|C(f)|^2 P}{4WN_0} + \frac{|C(f)|^2}{W} \int_W \frac{d\nu}{|C(\nu)|^2} \right] df \tag{3.17}$$

and

$$R_{\text{crit}} = E(0) - W \left[2 + \frac{8N_0}{P} \int_W \frac{1}{|C(f)|^2} \, df \right]^{-1} \log_2 e \tag{3.18}$$

In these cases the band W is given by the water-pouring analogy of Fig. 3.4 with the exception that only one-half the signal power is used (one-fourth in positive frequencies). Now Eqs. (3.15) to (3.18) define $E(R_c)$ except in the region $R_{\text{crit}} \leq R_c \leq C$. Although expressions parametrically defining the exponent in this region are available, for most practical purposes the exponent can be drawn quite accurately using a French curve and the knowledge of the values and slopes at R_{crit} and C.

There naturally exists a question as to the tightness of the random-coding bound of Eq. (3.14). This is a hard question to answer in the case of the dispersive channel, but results from analysis of the gaussian channel (nondispersive and with white noise) give us excellent guideposts. Here it has been shown that a lower bound for P_e, derived using a sphere-packing argument, coincides (in exponent) with the random-coding bound in the range $R_{\text{crit}} \leq R_c \leq C$. Thus it is often stated that the exponent is known exactly within this range of rates. For rates lower than R_{crit} the random-coding bound is not especially tight. A much better upper bound for P_e (lower bound for the exponent) can be obtained by removing "bad" choices for the input signal set from the random ensemble. This bound is called the *expurgated bound*.[2] A number of other bounds also exist in different portions of the range $0 \leq R \leq R_{\text{crit}}$.

Ideally the bounds we have been discussing tell the system designer how long the input signals must be in order to reduce the probability of error to some specified value. Since the alphabet size is $2^{R_c T}$, even very modest values of T require the generation and detection of a great many different signals. Thus at present it is impractical to consider the direct attainment of low error rates by the specific methods discussed in this section. The bound of Eq. (3.14) has its greatest value as a standard against which the more realistic data systems considered in the following chapters can be measured.

3.2 Bounds for Coding

We have seen that the probability of error in a data-transmission system can be made to decrease exponentially with the signal duration, provided the signals are constrained only in average power and bandwidth. Unfortunately the types of signals which allow us to obtain these low error rates [Eq. (3.4)] are not very easily implemented. Thus it becomes necessary to recognize an additional constraint on the problem of signal design—that of ease of implementation. This leads us to employ *digital signals*, i.e., signals composed of the superposition of time translates of a few basic signals.

In this section we show that digital signals can be employed to obtain an exponential decrease in P_e with increasing signal duration. Not surprisingly the error exponent in this case is smaller than in the less-constrained case discussed in the preceding section. Thus on a given channel a higher error rate is obtained with signals of a given length if the signals are constrained to be digital. This point is illustrated in Sec. 3.3, where a detailed example of the use of the bounds presented in this section and the preceding one is examined. We now turn our attention to the derivation of the random-coding bound for digital signals.

3.2.1 The Discrete Memoryless Channel

Let M represent the size of the set of equally likely messages from which one is chosen at random, and let T denote the time allotted to the transmission of each message. Now, however, the size L of the alphabet employed by the modem is constrained to be relatively small, and messages consist of sequences of n alphabet symbols. This set of messages will be referred to as an *M-message code of length n*.

We now assume that the coding channel, which includes the modulator and demodulator, is memoryless. (The reason for this latter restriction is that meaningful results can be obtained for the memoryless channel which cannot, because of their mathematical intractability, be obtained when the channel has memory.) The effect of the channel on a transmitted symbol can then be represented by an L-by-L transition matrix. The entry (i, j) in this matrix is $p(j|i)$, the probability that the jth symbol is received when the ith is transmitted. Such a channel is usually referred to as a *discrete memoryless channel* (DMC).† We are concerned now with the lowest error rate attainable by coding on a given DMC, that is,

†In general a DMC need not have the same number of inputs and outputs. However, throughout this book only L-by-L channels are considered.

with the performance of the best code in the class of M-message codes of length n.

Earlier in this chapter the point was made that although it may be practically impossible to determine the performance of a given set of signals, it is nonetheless possible to evaluate the average performance of a large class of signal sets. This random-coding argument applies to digital signals as well. The error rate attainable with the best code in a given class cannot exceed the average error rate for the class, a quantity which can be evaluated.

We shall state the results of this evaluation, the well-known random-coding bound, without proof. The simplest known derivation of this bound is given in Gallager.[3] To begin with, the following assumptions and definitions are required:

1. A code is constructed as follows: To each of the M equally likely messages is assigned a particular n-place vector (henceforth called a *code word*), whose elements are chosen randomly from an L-element alphabet $(a_1, a_2, \ldots, a_L)$ according to the distribution $p(a_k)$, where $k = 1, 2, \ldots, L$.
2. *Maximum likelihood decoding* is employed. That is, a received n-tuple $y = (y_1, y_2, \ldots, y_n)$ is interpreted as (decoded into) the code word $x = (x_1, x_2, \ldots, x_n)$ for which

$$Pr(y \mid x) = \prod_{i=1}^{n} Pr(y_i \mid x_i) \tag{3.19}$$

is largest. The case in which more than one such code word exists is practically unimportant and is not considered further.

3. For a code constructed as in step 1 and decoded as in step 2 we shall be concerned with the average probability of erroneous decoding where the average is across the M equally likely code words. In particular, the lowest probability of errorneous decoding attainable with any M-message code of length n constructed with any alphabet distribution $p(a_k)$ is of interest.

It can be shown[3] that for any distribution $p(a_k)$ the average (over all possible codes in the class of interest) probability of erroneous decoding P_e has an upper bound

$$P_e \leq 2^{-n\left\{-\rho R - \log \Sigma_{j=1}^{L}\left[\Sigma_{k=1}^{L} p(a_k)(P(j\mid k))^{1/(1+\rho)}\right]^{1+\rho}\right\}} \tag{3.20}$$

where, as earlier, ρ is an arbitrary number, $0 \leq \rho \leq 1$, and

$$R = \frac{1}{n}\log M \tag{3.21}$$

This latter quantity, the code rate, is measured here in bits per channel symbol. Since the best code of a given length and rate clearly cannot have a higher probability than the average of all such codes, Eq. (3.20) also provides an upper bound on the performance of the best code.

For a given code length and rate one is naturally interested in obtaining the lowest-possible upper bound. The exponent in Eq. (3.20) which minimizes the bound is

$$E(R) = \max_{\rho,\, p(a_k)} \left\{ -\rho R - \log \sum_{j=1}^{L} \left[\sum_{k=1}^{L} p(a_k)(P(j \mid k))^{1/(1+\rho)} \right]^{1+\rho} \right\} \tag{3.22}$$

The exponent is a convex function of ρ and the probability distribution $p(a_k)$; therefore the maximization on the two parameters can be performed independently. Although a digital computer must be employed for channels of any complexity, the maximization is not especially difficult conceptually. Then with $E(R)$ known, Eqs. (3.20) and (3.22) yield

$$P_e \leq 2^{-nE(R)} \tag{3.23}$$

As before, $E(R)$ is a monotone decreasing function of R for which $E(C) = 0$. It has, roughly speaking, the same shape as that shown in Fig. 3.3.

Equation (3.23) shows that the exponential decrease of P_e with increasing signal duration is retained despite the restriction to digital signals. This fact and the implementation advantages of digital coding strongly suggest the use of L-ary codes. This point is discussed in the following section.

3.2.2 The Use of Nonbinary Codes

The random-coding bound of the preceding section shows that it is possible to obtain an exponential decrease in error rate with signal duration while employing L-ary codes. The price paid for this significant step toward implementability is a slight overall lowering of the $E(R)$ curve. This, of course, results in a slight reduction in channel capacity and an increase in the bound on error probability attainable with a given signal set of a given rate and duration.

It would be desirable at this point to compare the probability of error for the gaussian channel and the various discrete memoryless channels which can be imbedded in it by using, say, pulse-amplitude modulation (PAM). Unfortunately, the expressions for the probability of error are too complex to permit simple analytical comparisons to be made. However, it should be clear that the degradation caused by quantizing the original gaussian channel signal must decrease as the number of levels increases. The largest degradation is suffered if only two PAM levels are employed; doubling the number of levels decreases this degradation substantially. Further increases in the number of levels improve performance, although by ever-decreasing amounts.

Since performance improves with the number of PAM levels employed, one may ask why we should not use an arbitrarily large number of levels since then we have digital signals and yet are quite close to the performance attainable with analog signals. The answer is this. Nothing is magical about multilevel digital signals. They have the advantage of being easily stored and processed, and especially in the binary case, efficient ways of encoding and decoding are known. However, as the alphabet size increases, digital signals become increasingly difficult to implement. Thus one seeks an engineering compromise—enough levels to ensure that performance is not seriously degraded, not too many levels so that the encoder and decoder can be built.

3.2.3 Binary Codes on Nonbinary Channels

With the exception of the binary case, it has proved difficult to find codes which perform well on the discrete memoryless channel and which can be implemented simply. This is particularly unfortunate since the random-coding argument implies that nearly all codes in a given class perform quite well. Although a few practical nonbinary codes are known, for the most part binary coding is considerably simpler to implement. This suggests using binary codes on L-ary channels.

In real data-communication systems it is convenient to choose the number of channel symbols to be a power of 2. Then each L-ary symbol can be represented as a binary r-tuple, and the DMC can then be viewed as a binary channel with a r-bit memory. If the probability of adjacent-level errors is considerably greater than that of other errors, most L-ary symbol errors result in only a single-bit error. (See the discussion of Gray coding in Chap. 2.) In this case the channel looks approximately like a binary symmetric channel for which p is slightly greater than $1/r$ times the L-ary error rate. For practical values of r, the performance of binary codes on the Gray-coded DMC will not differ appreciably from their performance on this BSC with the same average bit error rate. This point is illustrated in Sec. 3.3.

The $E(R)$ curve for this BSC can be obtained by specializing the results of the previous section to the binary case. This curve then provides an indication of the probability of error attainable by coding for the DMC as if it were a BSC. In this case Eqs. (3.22) and (3.23) reduce to

$$P_e \leq 2^{-nE(R)} \tag{3.24}$$

where

$$E(R) = H(p) + (\lambda - p)H'(p) - H(\lambda) \qquad C \geq R \geq R_{\text{crit}}$$
$$E(R) = 1 - 2\log(\sqrt{p} + \sqrt{q}) - R \qquad R_{\text{crit}} \geq R \geq 0 \tag{3.25}$$

and where

$$H(x) = -x\log x - (1 - x)\log(1 - x) \tag{3.26}$$

$$H'(x) = \frac{dH(x)}{dx} \tag{3.27}$$

$$R_{\text{crit}} = 1 - H\,\frac{\sqrt{p}}{\sqrt{p} + \sqrt{q}} \tag{3.28}$$

$$H(\lambda) = 1 - R \tag{3.29}$$

The $E(R)$ curve given by Eq. (3.25) is plotted in Fig. 3.5. Again it is difficult to compare this result with that for the DMC in the general case. Certain qualitative statements can be made, however. For a given signal-to-noise ratio the $E(R)$ curve for the DMC approaches that of the gaussian model as L increases. For small L the DMC can be approximated quite well by a BSC, and the two $E(R)$ curves are rather close together. As L increases, however, the BSC approximation breaks down and as a result, the separation between the two curves increases. A point is reached at which increasing L causes the BSC curve to be lowered even though the DMC curve is raised. If binary coding is to be employed, it is approximately at this point that the best compromise between cost and performance is attained. To utilize the channel more efficiently non-binary coding techniques, e.g., probabilistic decoding (Chap. 12), must be used.

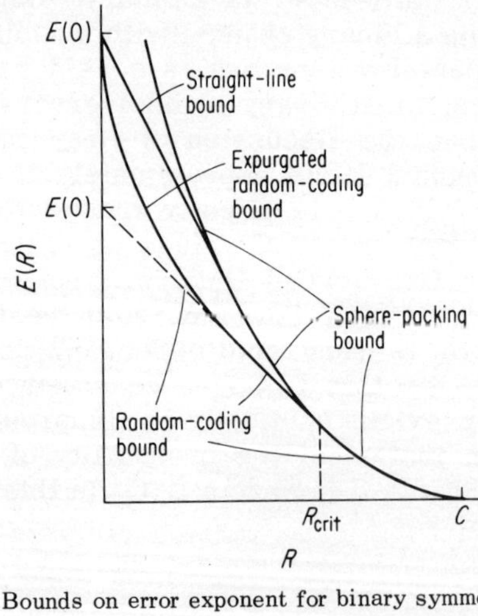

Fig. 3.5. Bounds on error exponent for binary symmetric channel.

In addition to the random-coding bound, several other bounds on P_e have also been derived. As before, by removing the restriction that ρ [in Eq. (3.22)] be not greater than 1, it is possible to improve on the random-coding bound for low rates. This improvement results from expurgating from the randomly selected ensemble of codes those code words which are much more likely to be decoded erroneously than the average.[2,3] The expurgated random-coding bound is also plotted in Fig. 3.5 for the binary channel. Upper bounds on the error exponent, i.e., lower bounds on the error rate attainable with the best code of a given length and efficiency, can also be derived. For rates above R_{crit}, one of these bounds (the sphere-packing bound) is exponentially identical to the random-coding bound (see Fig. 3.5). Thus for this range of rates the error exponent is known exactly. In the binary case another lower bound on P_e which improves on the sphere-packing bound for low rates has been derived. This "straight-line bound"[5] is also plotted in Fig. 3.5. For $R = 0$ it coincides with the expurgated random-coding bound and extends on a straight line to a point where it is tangent to the sphere-packing bound. Figure 3.5 shows that while the error exponent is not known exactly for rates below R_{crit}, the difference between the upper and lower bounds is relatively small.

Although the exponential bounds on P_e discussed above have an attractively simple form, they are often not as accurate as one would like them to be. This can also be said of the bounds for the gaussian and discrete memoryless channels, but because of the very general use for which the latter are intended, a slight looseness is unimportant. In the binary case, however, this may no longer be true. Since extensive classes of practical binary codes are known, it may be desirable to have accurate knowledge of just how low P_e can be made for given values of R and n. Thus the more accurate and, of course, more complex versions of the random-coding and sphere-packing bounds given in Shannon[2] and Gallager,[3] are of interest.

3.3 Example of the Use of Bounds on a Telephone Channel

We have pointed out that the restriction to digital coding and then to binary coding degrades performance. Unfortunately it is difficult to make a broad, meaningful statement about the magnitude of these degradations. An example is presented to give the reader some insight into this matter.

Let us consider a good-quality voiceband telephone channel whose attenuation and delay curves are the typical ones shown in Fig. 3.6. The signal-to-noise ratio on such telephone facilities is typically on the order of 30 db, the figure assumed here.

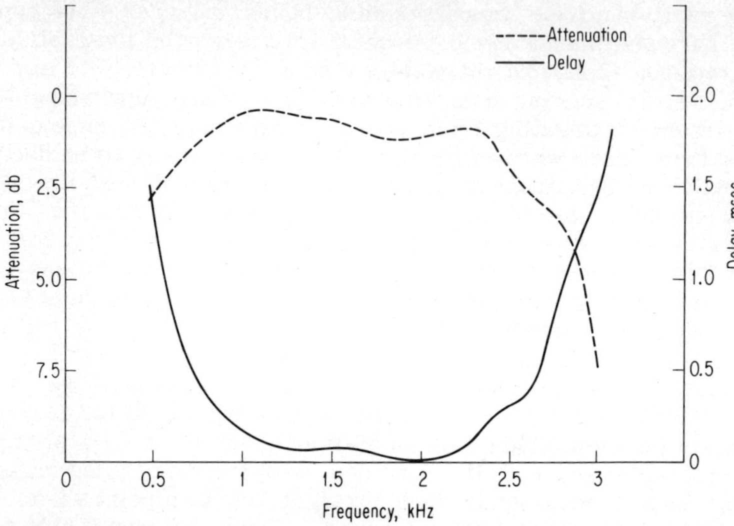

Fig. 3.6. Delay and attenuation characteristics for the example telephone channel.

Application of Eqs. (3.15), (3.17), and (3.18) enables us to compute $E(0)$, R_{crit}, and C, respectively. For this gaussian channel the resulting error exponent is plotted as curve G of Fig. 3.7. As explained earlier, we cannot construct a modem to take advantage of the arbitrarily small error rates promised by this curve because of the exponentially large number of signals required.

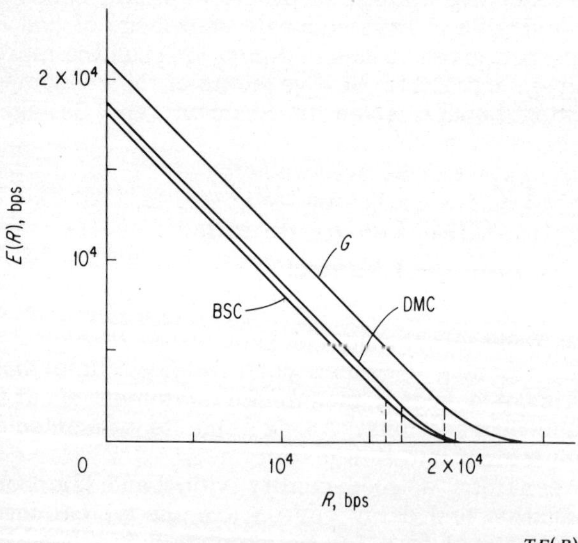

Fig. 3.7. $E(R)$ curves for example telephone channel, where $P_e \leq 2^{-TE(R)}$. G = gaussian channel, DMC = discrete memoryless channel, and BSC = binary symmetric channel.

Therefore, we must restrict the modem to the transmission of relatively short signals, each of which carries only a few bits of information. One type of modem which is suitable here employs vestigal-sideband pulse-amplitude modulation. (We shall study this type of modulation in later chapters.) A judicious choice of parameters for this modem results in a symbol rate of 5,000 per sec with each symbol containing four bits of information. Thus only 16 signals must be generated by the modem.

If it is assumed that the channel noise affects each symbol independently, the coding channel can be regarded as memoryless. In this situation for an optimized modem the channel is describable by the 16-by-16 transition matrix

Input level	Output level 1	2	3	4	5	...	13	14	15	16
1	0.9976	0.0024	$\approx 10^{-10}$							
2	0.0024	0.9952	0.0024	$\approx 10^{-10}$						
3	$\approx 10^{-10}$	0.0024	0.9952	0.0024	$\approx 10^{-10}$					
4		$\approx 10^{-10}$	0.0024	0.9952						
5			$\approx 10^{-10}$							
.			. . .							
.										
.										
13									$\approx 10^{-10}$	
14								0.9952	0.0024	$\approx 10^{-10}$
15							$\approx 10^{-10}$	0.0024	0.9952	0.0024
16								$\approx 10^{-10}$	0.0024	0.9976

where blank entries represent quantities considerably smaller than 10^{-10}. With this matrix and Eq. (3.22) we can compute the $E(R)$ curve for the coding channel. This is curve DMC of Fig. 3.7.

The 16-ary channel can be viewed as a binary channel with a 4-bit memory. If Gray coding is employed, the dominant adjacent-level transitions result in single-bit errors, and the much less probable two-level changes result in 2-bit errors. The average bit error rate is then approximately

$$\begin{aligned} p &\cong \frac{1}{16 \cdot 4} \quad 2[0.002 + (2 \cdot 10^{-10})] + 14\{2[0.0024 + 10^{-10}]\} \\ &= 1.12 \cdot 10^{-3} \end{aligned} \tag{3.30}$$

The $E(R)$ curve for the BSC with this average bit error rate is plotted as curve BSC of Fig. 3.7. Table 3-1 lists the values of channel capacity, R_{crit}, and $E(0)$ for the three channels.

Table 3-1 and the curves of Fig. 3.7 illustrate several important points. First, the larger of the two degradations from the ideal performance of the gaussian channel results from employing digital signaling; the degradation caused by employing memoryless binary coding is less important. Second, in the 15,000-to-20,000 bps range the degradation caused by digital signaling is very large—

Table 3-1 Comparison of Channels.

Characteristic	Channel		
	Gaussian	Discrete memoryless	Binary symmetric
Capacity	23,500	19,800	19,700
R_{crit}	19,200	16,700	15,900
$E(0)$	21,000	18,800	18,200

so large, in fact, that transmission at these rates seems impractical with today's technology. Third, for rates below 15,000 bps little is to be gained by employing ideal channel-symbol coding. Thus in this case the present lack of practical, nonbinary coding techniques does not seriously reduce our ability to transmit data accurately.

This concludes the introductory portion of the book in which we have stated the problems encountered in data communication, described the types of channels over which data is transmitted, and established limits on the performance that can be attained on these channels. We now turn our attention to modulation techniques.

3.4 Summary

In this chapter we have been concerned with the best performance attainable over various types of memoryless data-communication channels. First we examined the dispersive gaussian channel described in Chap. 2. It was seen that although it is difficult to evaluate the performance of a given set of signals by averaging over the ensemble of such signal sets, it is possible to show that the error rate attainable with the best signal set can be made to decrease exponentially with increasing signal duration. We also found that the amount of equipment necessary to implement this type of signaling is impractically large. Consequently, we turned our attention to a more practical signal format, i.e., digital signals. It was shown that if the modem employs a relatively small alphabet and if signal sets (codes) are composed of sequences of modem signals, the exponential decrease in error rate with signal duration is retained, although at a slower rate. Finally in order to simplify further the still-formidable problem of implementation, we examined the performance of binary codes on discrete memoryless channels. A detailed example served to illustrate that this additional restriction on the signaling format has very little effect on performance in most practical situations.

Notes

The material on the gaussian channel is taken from Holsinger[1] who had extended the work of Gallager[3] on the nondispersive gaussian channel to the dispersive case. Gallager's work in turn is a simplification and generalization of the original work of Shannon.[2] A more recent and tighter random-coding bound for the nonwhite gaussian channel is available in the report of Ebert.[6]

Slepian[7] has computed Shannon's upper and lower bounds for the gaussian channel over a wide range of rates, probabilities of error, and block lengths. The discussion of the discrete memoryless channel and the binary symmetric channel is also taken from Gallager.[3] Finally, for a study of information theory the reader is referred to the original paper by Shannon[4] and the books of Ash,[8] Fano,[9] and Wozencraft and Jacobs.[10]

4

Baseband Pulse Transmission

The previous chapter discussed theoretical limitations on the ability to transmit digital signals over dispersive, noisy channels. The mathematical bounds are most useful in an engineering sense in providing a standard of comparison for actual data systems. Unfortunately the theoretical bounds and their derivations do not indicate how a realistic digital communications system can be built so as to approximate optimum performance.

The optimum system requires signal sets and receiver complexities which grow exponentially with time. In fact, it is intuitively evident that signal waveforms must be chosen from a high-dimensionality space in order to reduce the probability of error due to gaussian noise. As the dimensionality of the signal space becomes larger, the perturbations due to noise from each transmitted signal tend to cluster with high probability about a sphere whose radius is the standard deviation of the noise. This effect, known as *sphere hardening*, in the limit enables reception at vanishingly small error rates, provided, of course, that the information rate is less than capacity.

A practical system realizes the large dimensionality signals through the digital encoding of a small set of possible waveshapes generated by the modulator. Corresponding to the many possible selections of these basic waveshapes are many known varieties of modulators. The very existence of all these varieties attests to the fact that no one modulation method has been found which combines the desired attributes of efficiency, simplicity, and flexibility.

The basic techniques of modulation involve varying the phase, frequency, position, or amplitude of transmitted pulses. Several of these methods may even be used at one time. In this book we shall not consider hybrid systems using a combination of these methods nor, because of their relative unimportance in data communication, shall we discuss time or position modulation. The greater portion of our material will be devoted to amplitude modulation.

We have chosen to emphasize amplitude modulation techniques because (1) the most efficient modulation methods are within this class, (2) these techniques are simpler to analyze than phase- or frequency-modulation methods, (3) they are better understood than other methods, and (4) the analytical approaches used in studying amplitude modulation can often be approximately applied to phase- and frequency-modulation systems.

Phase- and frequency-modulation methods, although generally less efficient than comparable amplitude-modulation methods, are quite widespread in practice. They have the advantages of relative simplicity of instrumentation and of an inherent rigidity against certain types of prevalent channel impairments. We shall devote a chapter to each of these methods in a later portion of this book. Because of the inherent nonlinearity of phase and frequency modulation, their analysis is more difficult than amplitude modulation and cannot be accomplished to the extent with which we shall deal with the latter.

In pulse-amplitude modulation the information is encoded into the amplitude values for a single pulse shape $g(t)$, transmitted at T-sec intervals. Thus the 2^{RNT} message vectors for transmission at rate R bps for NT sec may be chosen in the form

$$s_k(t) = \sum_{n=1}^{N} a_{kn} g(t - nT) \qquad a_{kn} \subset I_L \tag{4.1}$$

where $k = 1, \ldots, 2^{RNT}$. Here each of the signals is constructed from time translates of a single pulse $g(t)$ using coefficients a_{kn} chosen from a discrete, generally small set of L values I_L.

If we envision the signals described by Eq. (4.1) in terms of the signal vectors of Chap. 3, we see that a possibly severe constraint has been placed on their selection. The basis functions $\{\phi_n\}$ must be the T-sec time translates of $g(t)$. Generally the set $\{g(t - nT)\}$ will not be complete in the input-signal space; thus in effect we sacrifice dimensionality in return for relative ease in signal generation. Also, the coefficients a_{kn} can only take on a small set of discrete values—sometimes only two values. Thus the input space has been quantized. It is this quantization which enables us to use digital coding and decoding techniques and build practical, high-performance data-communications systems.

The block diagram of the complete PAM system is shown in Fig. 4-1. The encoder accepts input data in blocks of RNT bits.

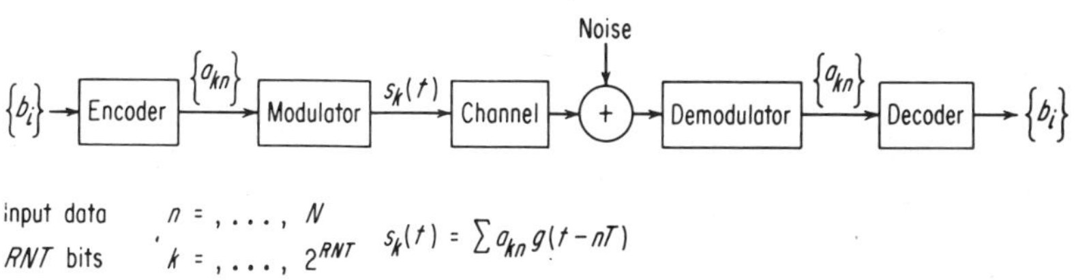

Fig. 4.1. Pulse-amplitude modulation.

This information is then used to select the signal vector $s_k(t)$. The encoder determines the N coefficients a_{kn}, where $n = 1, \ldots, N$, while the modulator produces $s_k(t)$ from the amplitude-modulated superposition of the N successive pulses. At the receiver the demodulator attempts the symbol-by-symbol classification of the N amplitudes $a_{k1}, \ldots, a_{kN}$. Finally, the decoder determines which of the 2^{RNT} messages was most likely to have been transmitted and passes on to the data sink the corresponding RNT bits of information.

Since the encoding and decoding operations will be studied in later chapters, it will be convenient to drop the double-subscript notation and deal only with the modulator-demodulator (modem) portion of the system. The transmitter has the conceptually simple task of transmitting amplitudes a_n, chosen from an L-letter alphabet, at times nT. Thus the transmitted signal is written

$$s(t) = \sum_n a_n g(t - nT) \qquad a_n \subset I_L \tag{4.2}$$

The demodulator then has the task of classifying the amplitudes a_n from the received signal.

In this chapter, as well as in Chaps. 5 and 6, we assume that the channel is at baseband, i.e., transmits energies to zero frequency. In a real communication system this is seldom the case; some auxiliary processing is necessary to transport the signal energy to the position in the frequency spectrum best suited for transmission and thence, at the receiver, back to baseband again. The most efficient means of accomplishing this translation are linear (described in Chap. 7). The effect of linear translation can therefore be considered as a portion of the baseband channel. In later chapters we see how "equivalent" baseband channels can be derived for practical systems. It is to our present advantage, however, to keep the model as uncluttered as possible while we review the analytical methods useful in baseband pulse transmission.

Chapter 4 begins with a description of the classical design of PAM systems for ideal channels. This classical design is based on a determination of transmitter and receiver which eliminates any influence of the pulse $a_n g(t - nT)$ on any other pulse $a_m g(t - mT)$. When this design is used on a real channel, the inevitable perturbation from the ideal causes intersymbol interference. The resulting degradation in performance of the PAM system is examined in the second major section of this chapter. Finally, some signal designs are described which take advantage of certain predetermined intersymbol influences.

In the succeeding baseband chapters (5 and 6) we shall examine the optimum structure of the PAM system when the channel characteristics are known exactly. Here the prior knowledge of the channel enables optimum design from the standpoint of minimizing probability of error. In the present chapter, we assume that the channel is not known and is therefore considered nominally ideal for design purposes.

4.1 Design and Performance of PAM for Ideal Channels

The simplest type of PAM system employs fixed transmitting and receiving filters, $G_T(\omega)$ and $G_R(\omega)$, respectively, and a sampler and threshold detector as shown in Fig. 4.2. For representational simplicity the input symbols a_n (the notation a_n is used for both the symbol and the amplitude at time nT) are envisioned as a sequence of narrow impulses whose shape $g(t - nT)$ at the channel input is determined by the transmitting filter $G_T(\omega)$.

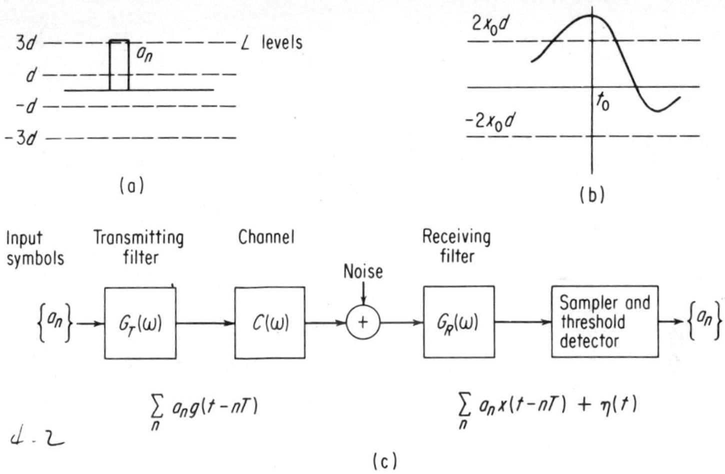

Fig. 4.2. (a) Allowed transmitter levels. (b) Decision thresholds (indicated by dashed lines). (c) PAM system with simple threshold-detection.

Usually the number of levels L is a power of 2, and the spacing between levels is uniform. For a given average power the spacing is maximized by using equal positive and negative ranges of values for a_n. Thus the allowed transmitter levels are $\pm d$, $\pm 3d$, . . . , $\pm(L-1)d$ where $2d$ is the distance between adjacent levels.

In passage through the channel the pulses $g(t)$ suffer time dispersion and added noise. At the receiver a filter $G_R(\omega)$ effects signal and noise shaping. This filter is followed by a sampler and a slicer (threshold detector) which look at the received waveform at times $kT + t_0$ and determined in each case which of the L allowed symbols is most likely. (The reference time t_0 is inserted to account for the delay in transmission through the channel.)

At the output of the receiving filter, $G_R(\omega)$, a series of amplitude-modulated, overlapping pulses appears. The shape of these pulses, called $x(t)$, is determined by the channel and the transmitting and receiving filters, that is,

$$x(t) = \frac{1}{2\pi}\int_{-\infty}^{\infty} G_T(\omega)C(\omega)G_R(\omega)e^{i\omega t}\,d\omega \tag{4.3}$$

With the addition of noise, the output waveform $y(t)$ becomes

$$y(t) = \sum_n a_n x(t - nT) + \eta(t) \tag{4.4}$$

where $\eta(t)$ is the additive noise process. At time $kT + t_0$, the desired output voltage is a_k; however, the actual value is

$$y(kT + t_0) = \sum_n a_n x(kT - nT + t_0) + \eta(kT + t_0) \tag{4.5}$$

Throughout the text expressions such as $y(kT + t_0)$ will be abbreviated by y_k. On received waveforms the reference delay t_0 is understood to be added to the number of pulse intervals denoted by the subscript. Thus Eqs. (4.5) is written

$$y_k = \sum_n a_n x_{k-n} + \eta_k \tag{4.6}$$

To isolate the desired amplitude a_k, Eq. (4.6) is rewritten in the form

$$y_k = x_0\left(a_k + \frac{1}{x_0}\sum_{\substack{n \\ n\neq k}} a_n x_{k-n} + \frac{\eta_k}{x_0}\right) \tag{4.7}$$

The factor x_0 represents the gain (or attenuation) of the signal in passing through the system. This simple change of scale is usually compensated by a variable gain control at the receiver. It is convenient to think of the threshold detector slicing y_k/x_0 at levels $0, \pm 2d, \pm 4d, \ldots, \pm(L-2)d$ in order to determine which L possible value of a_k is closest to the received, normalized sample. Alternatively, the receiver can set decision thresholds at $0, \pm 2x_0 d, \pm 4x_0 d, \ldots, \pm(L-2)x_0 d$ and deal directly with y_k. An error occurs whenever

$$\left|\sum_{\substack{n \\ n\neq k}} a_n x_{k-n} + \eta_k\right| > x_0 d \tag{4.8}$$

The second and third terms of Eq. (4.7) represent intersymbol interference and noise, respectively. The intersymbol interference arises from the overlapping tails of other pulses adding to the particular pulse $a_k g(t - kT)$ which is examined at the kth sampling time. The challenge in the design of a PAM system is to manipulate the transmitting and receiving filter characteristics to minimize the combined effects of intersymbol interference and noise in order to achieve minimum probability of error.

The problem is complicated by the possibility that, as in telephone transmission, the designer may not know the channel's transmission characteristics. Usually the design then proceeds on the basis that the channel has nominally perfect transmission over the band of interest. This, the classical Nyquist design problem, will be discussed next.

4.1.1 Baseband Shaping and the Nyquist Criterion

We have seen that the performance of the PAM system depends on the samples x_n of the system response $x(t)$. When the channel is assumed ideal, the Fourier transform $X(\omega)$ of this response is given by the product of the transmitting- and receiving-filter characteristics $G_T(\omega)G_R(\omega)$. In design and analysis of baseband-pulse-transmission systems it is convenient, and often necessary, to be able to relate the samples x_n of the time response to the properties of the frequency characteristic $X(\omega)$. The fundamental analytical tool which we shall employ for this purpose is the sampling theorem. The sampling theorem for band-limited functions is probably familiar to the majority of readers. (Those unfamiliar with its substance are referred to Middleton[1] for details.) Briefly the theorem enables us to express the time response $x(t)$ and the frequency response $X(\omega)$ for a function band-limited to $[-W, W]$ Hz in terms of samples taken at $1/2W$-sec intervals. That is, $X(\omega)$ has the Fourier series representation

$$X(\omega) = \begin{cases} \dfrac{1}{2W} \displaystyle\sum_n x\left(\frac{n}{2W}\right) e^{-in\omega/2W} & |\omega| \leq 2\pi W \\ 0 & \text{elsewhere} \end{cases} \tag{4.9}$$

which implies that

$$x(t) = \sum_n x\left(\frac{n}{2W}\right) \frac{\sin \pi(2Wt - n)}{\pi(2Wt - n)} \tag{4.9'}$$

The interval $1/2W$ sec is known as the *Nyquist interval* or, alternatively, the frequency $1/2T$ Hz for a sampling interval of T sec is known as the *Nyquist frequency*. The point of the theorem is that a function band-limited to the frequency range $[-W, W]$ Hz has exactly $2W$ degrees of freedom per second. When these are specified, the function is known exactly through Eq. (4.9).

In baseband pulse transmission we are concerned with samples of $x(t)$ taken at T-sec intervals. If $X(\omega)$ is band-limited to the Nyquist frequency $1/2T$ Hz, these samples uniquely determine the function $x(t)$. Invariably, however, $X(\omega)$ is band-limited to some frequency W greater than $1/2T$. In this case redundancy is involved in the specification of $X(\omega)$, that is, a great many frequency characteristics lead to the same sample sequence $\{x_n\}$. As far as the performance of the PAM system is concerned, these characteristics are *equivalent*. Only one of this class of equivalent channels is band-limited to the Nyquist band. This one we call the *equivalent Nyquist channel*. For any given characteristic $X(\omega)$ we shall see that a simple "folding" operation can be used to derive the equivalent Nyquist channel $X_{eq}(\omega)$. This channel then has the same sample values in its time response as does $X(\omega)$; yet, since $X_{eq}(\omega)$ is band-limited to the Nyquist band $[-\pi/T, \pi/T]$ rad per sec, the sampling theorem can be used directly. As the first instance of the usefulness of this concept we shall derive the conditions on $X(\omega)$ which result in elimination of intersymbol interference. These conditions were first derived by Nyquist[2] in 1928, and the resulting specification is known as the *Nyquist criterion*.

Notice from Eq. (4.7) that intersymbol interference can only be eliminated by making $x_n = 0$ for all $n \neq 0$. In other words the requirement is that the precursors and tails of $x(t)$ pass through 0 at regular T-sec intervals. An example of such a pulse is shown in Fig. 4.3. It should be clear that these pulses can be amplitude-modulated and transmitted at T-sec intervals without overlap at the sampling instants.

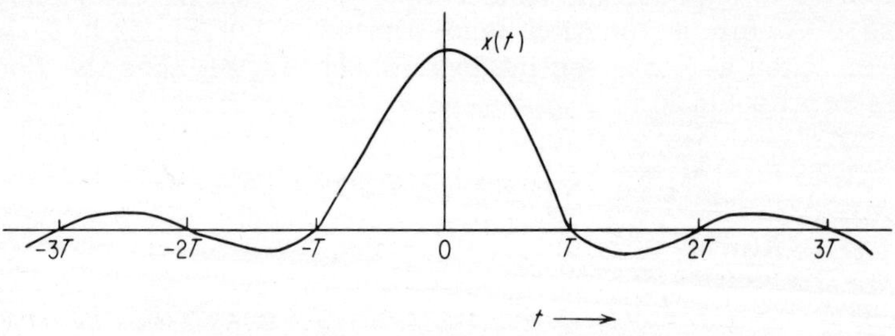

Fig. 4.3. Nyquist pulse.

The reference time delay t_0 and the pulse amplitude x_0 are arbitrary. In order to be specific we shall set $x_0 = 0$ and $t_0 = 1$. (To change these values it is only necessary to add ωt_0 to the phase of of $X(\omega)$ and to scale $X(\omega)$ by the desired gain x_0.) These requirements that $X(\omega)$ must meet can now be written in the form

$$x_n = \frac{1}{2\pi} \int_{-\infty}^{\infty} X(\omega)\, e^{i\omega nT}\, d\omega = \delta_{n0} \tag{4.10}$$

where

$$\delta_{n0} = \begin{cases} 0 & n \neq 0 \\ 1 & n = 0 \end{cases}$$

The integrals in Eq. (4.10) can be divided into segments corresponding to Nyquist bandwidths of the frequency range

$$x_n = \frac{1}{2\pi} \sum_k \int_{\pi/T(2k-1)}^{\pi/T(2k+1)} X(\omega)\, e^{i\omega nT}\, d\omega = \delta_{n0} \tag{4.11}$$

A change of variables gives

$$x_n = \frac{1}{2\pi} \sum_k \int_{-\pi/T}^{\pi/T} X\left(\omega + \frac{2k\pi}{T}\right) e^{i\omega T}\, d\omega = \delta_{n0} \tag{4.12}$$

and assuming the integration and summation may be interchanged, Eq. (4.12) may be written in terms of a single equivalent Nyquist bandwidth channel $X_{eq}(\omega)$.

$$x_n = \frac{1}{2\pi} \int_{-\pi/T}^{\pi/T} X_{eq}(\omega)\, e^{i\omega nT}\, d\omega = \delta_{n0} \tag{4.13}$$

where

$$X_{eq}(\omega) = \begin{cases} \sum_k X\left(\omega + \frac{2k\pi}{T}\right) & |\omega| \leq \frac{\pi}{T} \\ 0 & |\omega| > \frac{\pi}{T} \end{cases} \tag{4.14}$$

The equivalent Nyquist bandwidth channel is constructed by slicing the original $X(\omega)$ into segments of width $2\pi/T$ and superimposing all the segments on the interval $[-\pi/T, \pi/T]$.*

We have seen from the sampling theorem that $X_{eq}(\omega)$ may be uniquely reconstructed from its time samples taken at points nT; yet Eq. (4.13) says these must all be zero except at time zero, where a unity value is specified. The only $X_{eq}(\omega)$ meeting these conditions is the rectangular, zero-phase characteristic

*The equivalent Nyquist channel will be recognized by those readers familiar with the analysis of sampled data systems as a form of the sampled spectrum of $X(\omega)$.

$$X_{eq}\,(\text{Nyquist}) = \begin{cases} T & |\omega| \leq \dfrac{\pi}{T} \\ 0 & \text{elsewhere} \end{cases} \tag{4.15}$$

Therefore the equivalent Nyquist bandwidth channel must be "perfect" if $X(\omega)$ is to be free from causing intersymbol interference. This way of stating the Nyquist criterion will be particularly useful later when the evaluation of real channel performance is studied.

In nearly all practical cases of interest the actual bandwidth available does not exceed twice the minimum-required Nyquist bandwidth for the desired symbol rate. If this restriction is made, i.e., if

$$X(\omega) = 0 \qquad |\omega| > \frac{2\pi}{T} \tag{4.16}$$

constructing $X_{eq}(\omega)$ is considerably simplified and results in the original specification given by Nyquist. This is illustrated in Fig. 4.4, which shows a frequency characteristic $X(\omega)$ assumed to be real. The equivalent Nyquist bandwidth channel is obtained by superimposing the three bands labeled X_{-1}, X_0, and X_1. Since $x(t)$ must be real, $X(\omega) = X^*(-\omega)$, and it is only necessary to consider positive frequencies. However, X_1 has no components for positive frequencies when superimposed on X_0. Placing X_{-1} on X_0 is equivalent to "folding" $X(\omega)$ back onto itself about the Nyquist frequency π/T.

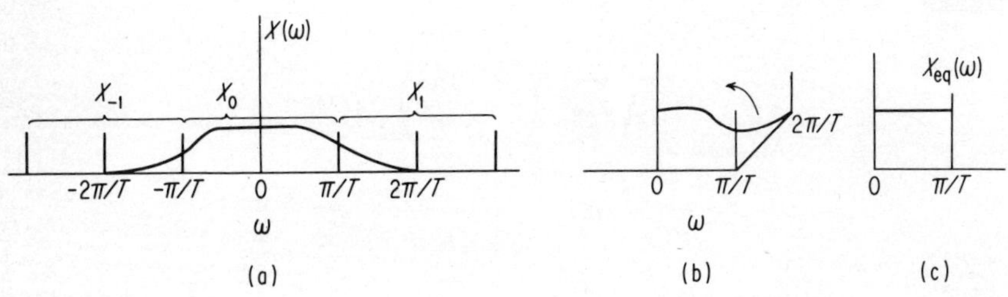

Fig. 4.4. (a) Equivalent Nyquist bandwidth channel is superposition of X_{-1}, X_0, X_1. (b) Folding of the portion of characteristic in excess of Nyquist bandwidth. (c) Equivalent channel must be perfect.

When $X(\omega)$ has a nonzero phase characteristic, care must be exercised in adding the folded portion as a phasor with its imaginary component inverted. Suppose $X(\omega)$ is written in terms of real and imaginary components:

$$X(\omega) = X_x(\omega) + iX_y(\omega) \tag{4.17}$$

Then, for the restricted bandwidth being considered

$$X_{eq}(\omega) = X_x(\omega) + iX_y(\omega) + X_x\left(\omega - \frac{2\pi}{T}\right) + iX_y\left(\omega - \frac{2\pi}{T}\right) \quad \omega \geq 0 \tag{4.18}$$

But since $X(\omega) = X^*(-\omega)$,

$$X_{eq}(\omega) = X_x(\omega) + iX_y(\omega) + \underbrace{X_x\left(\frac{2\pi}{T} - \omega\right) - iX_y\left(\frac{2\pi}{T} - \omega\right)}_{\text{folded portion}} \quad \omega > 0 \tag{4.19}$$

Another way of stating the Nyquist criterion under the restriction of Eq. (4.16) is that $X_x(\omega)$ must consist of the rectangular function plus any arbitrary function odd about π/T, while $X_y(\omega)$ may be any arbitrary even function about π/T.

Obviously the criterion of removal of intersymbol interference does not uniquely specify the pulse spectrum $X(\omega)$ unless the bandwidth is limited to the Nyquist band $[-\pi/T, \pi/T]$. If the available bandwidth is less than this amount, it is impossible to remove intersymbol interference. The ambiguity involved when a greater amount of bandwidth is available must be resolved using other considerations. One such consideration is the rate of decay with time of the pulse $x(t)$. Perturbations in real systems result in nonzero values at the actual pulse sampling times. Generally speaking, the faster $x(t)$ decays with time, the less is the effect of these unavoidable perturbations.

If, for example, $X(\omega)$ is rectangular

$$X(\omega) = \begin{cases} T & |\omega| \leq \dfrac{\pi}{T} \\ 0 & |\omega| > \dfrac{\pi}{T} \end{cases} \tag{4.20}$$

where

$$x(t) = \frac{\sin \pi t/T}{\pi t/T} \tag{4.21}$$

and the pulse response decreases as $1/t$ for large t. The slow decay of the pulse in this case is caused by the discontinuity in $X(\omega)$ at π/T and $-\pi/T$. If the first n derivatives of $X(\omega)$ are defined and bounded over the bandwidth, $t^n x(t)$ must be square-integrable over $(-\infty, +\infty)$ and $x(t)$ must decay asymptotically as $1/t^{n+1}$. In other words, the "smoother" $X(\omega)$ can be made, the faster will be the asymptotic decay of $x(t)$.

Not only is the asymptotic decay of interest, but the immediate decay, the behavior of $x(t)$ between sampling instants and the derivatives of $x(t)$ at the sampling instants are also important considerations. In most practical situations the choice among Nyquist characteristics is influenced by these considerations as well as by relative ease in instrumentation.

One class of Nyquist characteristics which has been extensively used and studied is the so-called *raised-cosine characteristic*. A raised-cosine characteristic consists of a flat amplitude portion and a roll-off portion which has a sinusoidal form. The spectrum $X(\omega)$ is specified in terms of a parameter α which is the amount of bandwidth used in excess of the minimum Nyquist bandwidth divided by the Nyquist bandwidth. Thus an α of unity is referred to as 100 percent roll-off and means that the total bandwidth used is two Nyquist bandwidths or $2\pi/T$, while $\alpha = 0$ indicates that the minimum bandwidth of π/T is employed.

If $X(\omega) = 0$ for $\omega > (1+\alpha)\pi/T$, it can be seen from Fig. 4.4 that the excess "folding" procedure will not affect the portion of $X(\omega)$ on the interval $(1-\alpha)\pi/T$. Since the folded, or equivalent, channel must be rectangular, it is necessary that $X(\omega) = T$ for $0 \le \omega \le (1-\alpha)\pi/T$. Finally, on the interval $[(1-\alpha)\pi/T, (1+\alpha)\pi/T]$ a symmetric, sinusoidal roll-off is used. Thus the raised-cosine characteristics are given as

$$X(\omega) = \begin{cases} T & 0 \le \omega \le \dfrac{\pi}{T}(1-\alpha) \\ \dfrac{T}{2}\left\{1 - \sin\left[\dfrac{T}{2\alpha}\left(\omega - \dfrac{\pi}{T}\right)\right]\right\} & \dfrac{\pi}{T}(1-\alpha) \le \omega \le \dfrac{\pi}{T}(1+\alpha) \end{cases} \tag{4.22}$$

The impulse responses corresponding to these characteristics are

$$x(t) = \frac{\sin \pi t/T}{\pi t/T} \; \frac{\cos \alpha\pi t/T}{1 - 4\alpha^2 t^2/T^2} \tag{4.23}$$

Observe that since the first two derivatives of $X(\omega)$ are bounded and defined, $x(t)$ decreases asymptotically as $1/t^3$.

Some of the raised-cosine characteristics and their impulse responses are shown in Fig. 4.5. The $\alpha = 1$, or 100 percent roll-off, case has a particularly interesting pulse response since $x(t)$ has zeros halfway between sampling times in addition to the usual sampling zeros. As α decreases, the response $x(t)$ becomes more oscillatory and consequently more difficult to deal with in the event of perturbations. However, the smaller α is, the less bandwidth is required for a given rate of transmission. Consequently a given bandwidth is more efficiently utilized by using as small a roll-off as possible; however, the problems of timing and channel equalization become more difficult.

One conceptual difficulty encountered in studying raised-cosine characteristics and most of the other frequency responses mentioned in this book is that none of them is, strictly speaking, physically realizable. A realizable frequency characteristic must have a time response which is zero prior to some time t_0. However, the time t_0 can be made arbitrarily far in the past so that the actual response closely approximates the ideal, unrealizable response.

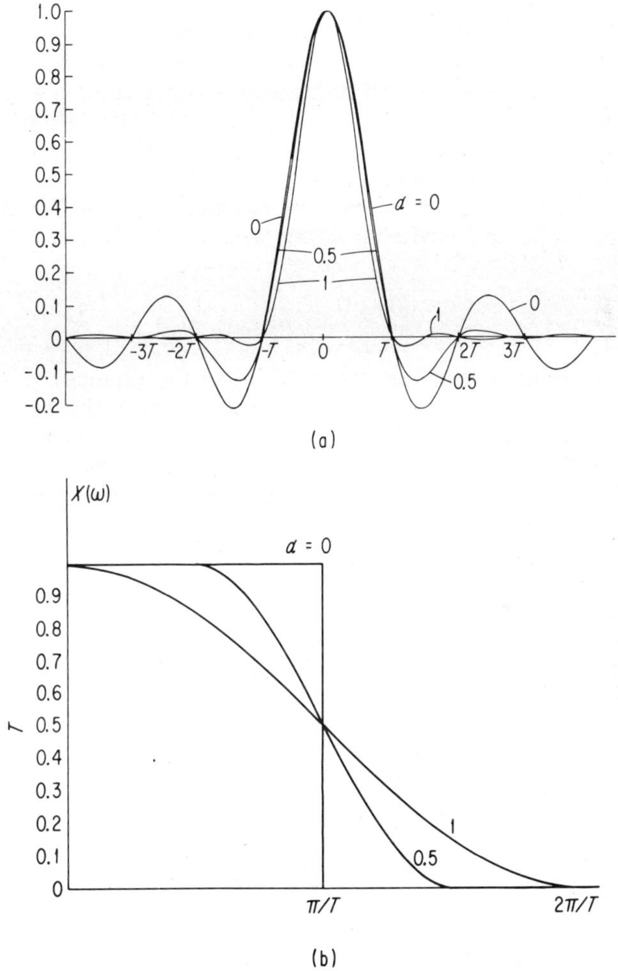

Fig. 4.5. Raised-cosine pulse shaping. (a) Frequency responses. (b) Time responses.

4.1.2 *Transmitting and Receiving Filters for Noise Immunity*

The overall pulse spectrum $X(\omega)$ is chosen to meet the Nyquist condition assuring freedom from intersymbol interference under perfect channel and timing conditions. Usually the particular Nyquist characteristic is picked from the raised-cosine class, depending on the bandwidth allowed and the signaling rate desired. However, this still leaves an arbitrary choise of transmitting or receiving filter since only their product $X(\omega)$ has been specified by intersymbol-interference-removal considerations. The remaining freedom in choice can be resolved by noise-immunity considerations.

The problem to be solved now consists in finding the receiving-filter characteristic $G_R(\omega)$ which maximizes signal-to-noise ratio at the sampling points for a given power input to the channel. Clearly maximizing signal-to-noise ratio will minimize the probability of error as given in Eq. (4.8). Since the overall pulse characteristic must be $X(\omega)$, the transmitting filter will be specified as

$$G_T(\omega) = \frac{X(\omega)}{G_R(\omega)} \tag{4.24}$$

First an expression for power input to the channel is required. Many of these expressions will be required in the remainder of this book; therefore, it will be useful to derive a representation for the power in a data signal here. Recall that the input signal to the channel is

$$s(t) = \sum_{n=-\infty}^{\infty} a_n g(t - nT) \tag{4.25}$$

where $g(t)$ is the inverse Fourier transform of the transmitter filter characteristic $G_T(\omega)$. In finding the average power in the signal described by Eq. (4.25) it is necessary to time-truncate the signal to $2N$ symbols and $2NT$ sec and express the average power as

$$P_s = \left\langle \lim_{N\to\infty} \frac{1}{2NT} \int_{-NT}^{NT} \left[\sum_{n=-N}^{N} a_n g(t - nT) \right]^2 dt \right\rangle \tag{4.26}$$

Interchanging the expectation* and limit gives

$$P_s = \lim_{N\to\infty} \frac{1}{2NT} \sum_{n=-N}^{N} \sum_{m=-N}^{N} \langle a_n a_m \rangle \int_{-NT}^{NT} g(t - nT) g(t - mT)\, dt \tag{4.27}$$

The assumption is made that the input symbols a_n and a_m are uncorrelated for $n \neq m$, so that

---

*The notation $\langle x \rangle$ is used to indicate the ensemble average.

$$\langle a_n a_m \rangle = \begin{cases} 0 & n \neq m \\ \overline{a^2} & n = m \end{cases} \tag{4.28}$$

Then Eq. (4.27) reduces to

$$P_s = \lim_{N \to \infty} \frac{\overline{a^2}}{2NT} \sum_{n=-N}^{N} \int_{-NT}^{NT} g^2(t - nT)\, dt \tag{4.29}$$

$$P_s = \frac{\overline{a^2}}{T} \int_{-\infty}^{\infty} g^2(t)\, dt \tag{4.30}$$

It is frequently convenient to be able to express average signal power in terms of frequency-domain characteristics. This can easily be done through the Parseval relation

$$\int_{-\infty}^{\infty} g^2(t)\, dt = \frac{1}{2\pi} \int_{-\infty}^{\infty} |G(\omega)|^2\, d\omega \tag{4.31}$$

The input power to the channel can therefore be written as

$$P_s = \frac{\overline{a^2}}{2\pi T} \int_{-\infty}^{\infty} |G_T(\omega)|^2\, d\omega \tag{4.32}$$

Since the overall pulse frequency characteristic has been constrained to be $X(\omega)$, the signal power output of the receiver is invariant under changes of the receiving filter, and the signal-to-noise ratio is maximized by minimizing noise power. The noise has spectral density $N(\omega)$ and passes through filter $G_R(\omega)$. The resulting noise power is

$$P_N = \frac{1}{2\pi} \int_{-\infty}^{\infty} N(\omega)\, |G_R(\omega)|^2\, d\omega \tag{4.33}$$

The channel input-power constraint is added to Eq. (4.33) using a Lagrange multiplier λ.

$$V = \frac{1}{2\pi} \int_{-\infty}^{\infty} \left[\frac{\lambda \overline{a^2}}{T} \frac{|X(\omega)|^2}{|G_R(\omega)|^2} + N(\omega)\, |G_R(\omega)|^2 \right] d\omega \tag{4.34}$$

The resulting functional V is minimized by the techniques of the variational calculus. A small variation $\epsilon\, \delta(\omega)$ is added to $|G_R(\omega)|^2$, and the derivative $\partial V/\partial \epsilon |_{\epsilon = 0}$ is set equal to 0. Obviously the phase characteristic of $G_R(\omega)$ is immaterial since only absolute values are involved in Eq. (4.34). The best receiving filter found from the stationary point of Eq. (4.34) is

$$G_R(\omega) = \frac{|X(\omega)|^{\frac{1}{2}}}{N^{\frac{1}{4}}(\omega)} \tag{4.35}$$

while the corresponding transmitting filter becomes

$$G_T(\omega) = \frac{X(\omega)N^{\frac{1}{4}}(\omega)}{|X(\omega)|^{\frac{1}{2}}} \tag{4.36}$$

The two filters may have arbitrary phase characteristics as long as they compensate one another.

If the noise spectral density $N(\omega)$ is flat and $X(\omega)$ is real, the transmitting and receiving filters are identical, each accomplishing "half" of the signal shaping. This is an advantage in system construction since only one filter design must be accomplished.

4.1.3 Performance of Multilevel Baseband with Ideal Channel and Added Gaussian Noise

A family of curves can now be plotted showing probability of error for baseband-pulse-transmission systems under the following conditions:

1. The amplitudes (symbols) a_n are equally likely to assume any of L possible, equally spaced values. Symbols occurring at different times are independent.
2. The additive noise is white gaussian with spectral density N_0 watts per Hz.
3. The pulse shaping $X(\omega)$ is raised-cosine with roll-off parameter α. The bandwidth is therefore $(1 + \alpha)\pi/T$. Since the noise spectrum is flat, the shaping $X(\omega)$ is split equally between transmitting and receiving filters, each of which has characteristic $X^{\frac{1}{2}}(\omega)$.

The power delivered to the channel for this system is

$$P_s = \frac{\overline{a^2}}{2\pi T} \int_{-\infty}^{\infty} |X^{\frac{1}{2}}(\omega)|^2 \, d\omega \tag{4.37}$$

which for the raised-cosine characteristic of Eq. (4.22) becomes

$$P_s = \frac{\overline{a^2}}{T} \tag{4.38}$$

independent of the roll-off parameter α. Since there are L equally likely amplitude values placed at levels $\pm d, \pm 3d, \ldots, \pm(L-1)d$, the average "symbol power" $\overline{a^2}$ is

$$\overline{a^2} = \frac{2}{L} \sum_{i=1}^{L/2} [d(2i-1)]^2 = \frac{d^2}{3}(L^2-1) \tag{4.39}$$

and the channel input power is therefore

$$P_s = \frac{d^2}{T}\,\frac{L^2-1}{3} \tag{4.40}$$

The way the raised-cosine characteristics were defined in Eq. (4.22) leads to unity pulse height at time zero at the output of the receiver filter. Thus the receiver looks for the unattenuated transmitted levels and places slicing levels at $0, \pm 2d, \ldots, \pm(L-2)d$. An error occurs when the noise at a sampling time exceeds in magnitude the distance from the transmitted level to the nearest decision threshold d. However, the outside two levels can only be in error in one direction each; therefore an algebraic factor of $(1-1/L)$ is used to weight the probability of noise magnitude exceeding d. The probability of error is

$$P_e = \left(1 - \frac{1}{L}\right) P[|\eta| > d] \tag{4.41}$$

This probability is easily computed since the gaussian, white noise at the receiver input is simply passed through the receiving filter $X^{\frac{1}{2}}(\omega)$. The output noise voltage is still gaussian with variance

$$\sigma^2 = \frac{N_0}{2\pi} \int_{-\infty}^{\infty} |X^{\frac{1}{2}}(\omega)|^2 \, d\omega = N_0 \tag{4.42}$$

and the probability of error is conveniently expressed in terms of the normal probability integral

$$P_e = 2\left(1 - \frac{1}{L}\right) Q\left[\frac{d}{\sigma}\right] \tag{4.43}$$

$$Q(x) = \frac{1}{\sqrt{2\pi}} \int_x^{\infty} e^{-t^2/2} \, dt \tag{4.44}$$

In terms of transmitted power P_s the probability of error is

$$P_e = 2\left(1 - \frac{1}{L}\right) Q\left[\left(\frac{3}{L^2-1}\,\frac{TP_s}{N_0}\right)^{\frac{1}{2}}\right] \tag{4.45}$$

Finally it is instructive to notice that N_0/T is the input noise power in the Nyquist bandwidth $[-\pi/T, \pi/T]$. Denoting this noise power by P_N, the expression becomes

$$P_e = 2\left(1 - \frac{1}{L}\right)Q\left[\left(\frac{3}{L^2 - 1}\frac{P_s}{P_N}\right)^{\frac{1}{2}}\right] \tag{4.46}$$

Observe that the effective signal-to-noise ratio (the term in parantheses) is modified by a factor of $3/(L^2-1)$ for L-level transmission. Thus a four-level system requires five times, or about 7 db, more power than a binary system to achieve the same probability of error. As L becomes large, about 6 db additional power is required to double the number of levels and retain the same error rate.

A sequence of curves for probability of error P_e as a function of signal-to-noise ratio for $L = 2$, 4, 8, and 16 is shown in Fig. 4.6. The noise power for these curves is measured in the Nyquist band. Although the bandwidth of the PAM system itself will exceed the Nyquist bandwidth by an amount depending on the roll-off used, increasing roll-off α merely increases the bandwidth of the system without changing the received noise power or the probability of error. However, increasing α eases problems of timing and intersymbol interference resulting from channel perturbations. Some compromise value of roll-off is usually chosen to meet specific requirements.

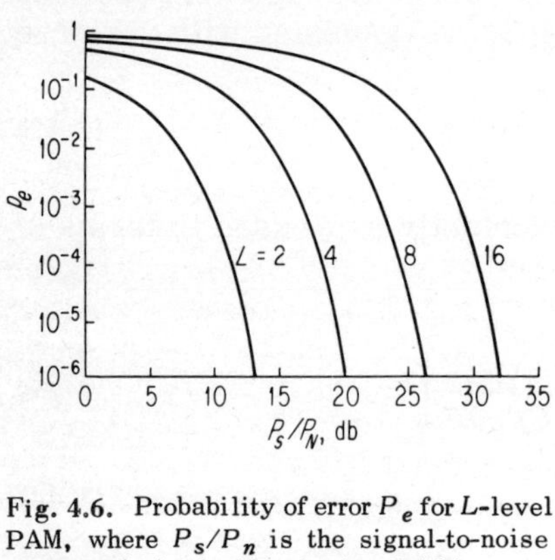

Fig. 4.6. Probability of error P_e for L-level PAM, where P_s/P_n is the signal-to-noise power ratio in the Nyquist bandwidth.

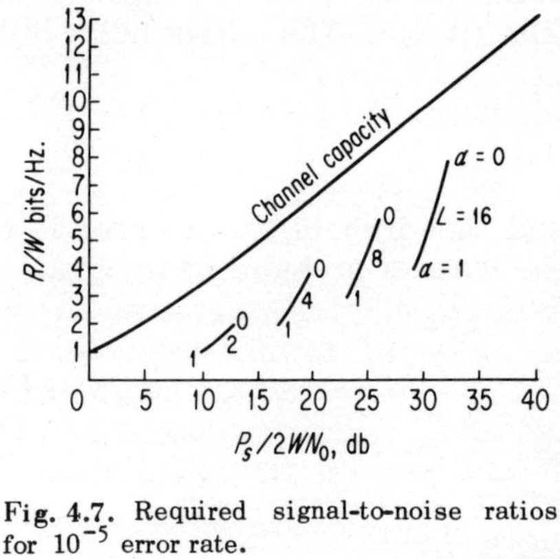

Fig. 4.7. Required signal-to-noise ratios for 10^{-5} error rate.

To bring actual bandwidth and data rate more explicitly into the formulation, the required signal-to-noise ratio for a given error rate and data speed in bits per cycle of bandwidth can be plotted as shown in Fig. 4.7. If a probability of error of 10^{-5} is required, according to Eq. (4.46),

$$\frac{3}{L^2 - 1}\frac{P_s}{P_N} = 4.42^2 \tag{4.47}$$

where a factor of $1-1/L$ has been ignored. Now the data rate of the system is

$$R = \frac{1}{T} \log_2 L \text{ bps} \tag{4.48}$$

while the bandwidth in hertz is

$$W = \frac{1}{2T}(1+\alpha) \tag{4.49}$$

Thus

$$\frac{R}{W} = \frac{2 \log_2 L}{1+\alpha} \tag{4.50}$$

and

$$\frac{P_s}{2N_0 W} = \frac{L^2-1}{3} \frac{4.42^2}{1+\alpha} \tag{4.51}$$

may be used parametrically to plot normalized rate as a function of signal-to-noise ratio in the entire bandwidth for the error rate of 10^{-5}. Also shown in Fig. 4.7 is the rate

$$\frac{R}{W} = \log\left(\frac{P_s}{2N_0 W} + 1\right)$$

at which vanishingly small probability of error can be achieved in theory.

To determine the effectiveness of the PAM technique, we could now compute the bounds discussed in Chap. 3 for the PAM channel as seen by the encoder and decoder. The coder sees a discrete memoryless channel with L input and output symbols transmitted at rate $1/T$ symbols per second with average probability of error P_e. The L^2 transition probabilities between input symbols and output symbols could be easily computed for any specific system configuration, and the performance bounds could be derived from these transition probabilities. To deal with a single figure of merit we shall compute only the channel capacity of the PAM system and compare this figure with the capacity of the original, gaussian modulation channel.

Since we assume L is large, the outer-level asymmetries are relatively unimportant. The best a priori probability weightings of the L input symbols are then equal since no symbol is any different in transition probabilities than any other symbol. The capacity of the digital channel is

$$C = \frac{1}{T}\left[\log_2 L + \sum_i p_i \log_2 p_i\right] \text{ bps} \tag{4.52}$$

where the p_i's are the transition probabilities for any of the symbols. The term $\log_2 L$ is the information conveyed by a symbol, while the summation indicates the information loss, or equivocation, owing to uncertainty in reception.

When P_e is 10^{-3} or less, the capacity C is essentially $1/T \log_2 L$ bps. In the range 10^{-3} to 10^{-1} adjacent level errors become important, and the capacity may be approximated by

$$C = \frac{1}{T}\left[\log_2 L + (1 - P_e) \log (1 - P_e) + P_e \log \frac{P_e}{2}\right] \text{bps} \qquad (4.53)$$

where only the two adjacent level transitions have been considered. For larger P_e, more transition probabilities need to be considered.

After the capacity of the PAM system is computed from Eq. (4.52), the minimum signal-to-noise ratio necessary to attain this capacity may be found from the usual formula $C = W \log (1 + S/N)$ and compared with the actual signal-to-noise ratio necessary to attain this probability of error in the PAM system. The resulting effective loss in signal-to-noise ratio for the most favorable, but unrealistic, case of $\alpha = 0$ is shown in Fig. 4.8 as a function of P_e. The curve shown is an approximation based on large L.

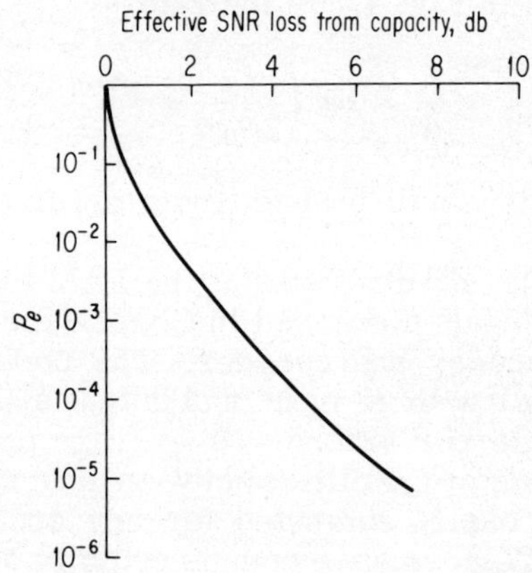

Fig. 4.8. Effective loss in SNR from theoretical channel capacity for PAM as a function of error probability P_e (the curve is an approximation based on large L).

The lesson of Fig. 4.8 is quite apparent. The PAM modulation is inefficient unless the probability of error is large. If the probability of error is small, one might think of the original channel as having been too coarsely quantized. As the number of levels L is increased for a given signal-to-noise ratio, a successively finer channel quantization takes place. The system is more efficient, but the probability of error is also larger. Consequently the encoder and decoder necessary to attain a low output error rate become more complex.

4.2 Effect of Imperfect Channels on the PAM System

With a real channel, intersymbol interference will always be present. This so complicates the analysis of the PAM system that exact results are impossible to obtain; frequently even bounds or approximations become quite tedious to derive.

The received voltage was denoted previously as

$$y(t) = \sum_n a_n x(t - nT) + \eta(t) \tag{4.54}$$

and the voltage at sampling time $kT + t_0$ was written in the form

$$y_k = x_0 \left[a_k + \frac{1}{x_0} \sum_{\substack{n \\ n \neq k}} a_n x_{k-n} + \frac{\eta_k}{x_0} \right] \tag{4.55}$$

In the last section the samples of the impulse response $x(t)$ were 0 except for $x_0 = 1$. Consequently, the middle term in Eq. (4.55) denoting intersymbol interference disappeared, leaving only the desired amplitude a_k and the gaussian noise sample η_k. For an example of the effect of the intersymbol interference term, suppose that the impulse response of the system is as shown in Fig. 4.9. The sample x_1 interferes with the succeeding pulse, while the sample x_{-1} interferes with the previous pulse, etc. The interference with past as well as future symbols is possible because of the time delay t_0 in passage through the channel.

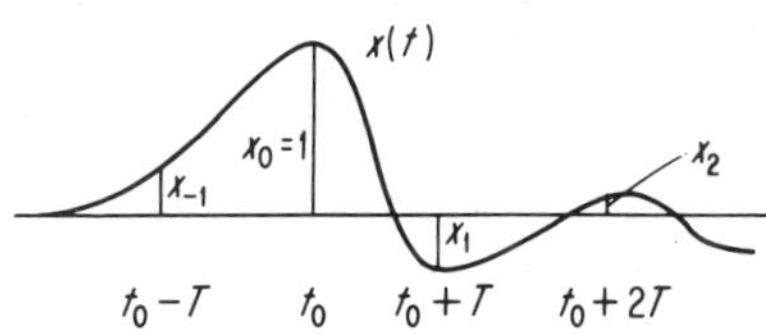

Fig. 4.9. Example impulse response.

Assume that a binary signal $a_k = \pm 1$ is transmitted. Using $x_0 = 1$ in this example (and ignoring, for the moment, the noise sample) gives for y_0‡

$$y_0 = a_0 + \sum_n{}' a_n x_{-n} \tag{4.56}$$

The possible values of this quantity when the transmitted symbol a_0 is +1 are shown diagrammatically in Fig. 4.10. Depending on whether a_{-1} is positive or negative, a voltage $\pm x_1$ is added to the value of +1 for a_0. Similarly $\pm x_{-1}$ is added depending on the sign of

‡A prime on a summation indicates deletion of the zeroth term.

a_1. Continuing to add the effects of more and more symbols results in the treelike diagram of Fig. 4.10.

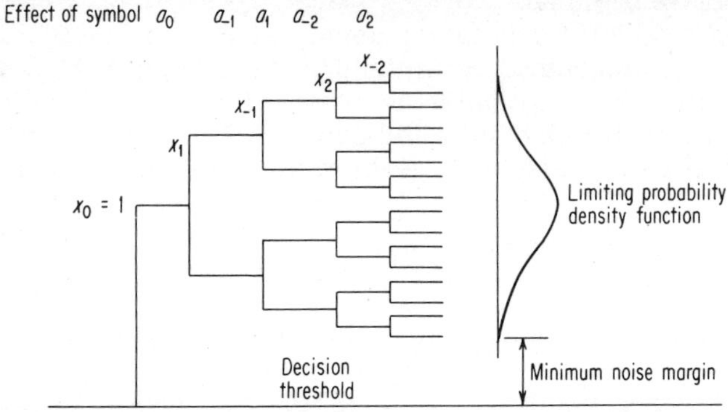

Fig. 4.10. Distribution of intersymbol interference (binary signal).

The limiting distribution of the intersymbol interference as an arbitrarily large number of interfering samples are considered is obviously of great interest. This limiting distribution could be convolved with the gaussian noise distribution to evaluate the probability of error—in this case the probability of crossing the axis into a negative voltage. At first glance it might be thought that the central-limit theorem applies to this situation and that in the limit the distribution of intersymbol interference becomes gaussian. Notice, however, that the maximum amount of intersymbol interference in this situation is

$$D = \sum_n{}' |x_n| \tag{4.57}$$

This maximum interference will be bounded if $x(t)$ decays faster than $1/t$. Since for any real channel impulse response the asymptotic decay must be exponential, it becomes evident that the maximum interference is bounded and the distribution cannot be gaussian.

One question investigated by mathematicians many years ago that seems immediately relevant is whether or not the final distribution becomes smooth enough to possess a probability density. That is, can the spaces between the branches of the tree in Fig. 4.10 become filled as an infinite number of branches is added? If each successive added interfering sample is half the previous sample, it can be shown that all the spaces are eventually filled and a uniform limiting probability density is obtained. On the other hand, if each interference sample is less than half of the preceding sample, no branch of the tree in Fig. 4.10 will ever meet any other branch of the tree.

To be specific, let us assume that the impulse response is zero for negative time and that $x_n = r^n$ for $n \geq 0$ and where $0 < r < \frac{1}{2}$. After

n branches of the tree generated by this impulse response, the distance between closest branches is $2r^n$. However, the total distance each of these branches can be moved toward the other by the cumulative effect of all remaining terms is $r^{n+1}/(1-r)$. For $0 < r < \frac{1}{2}$ this distance is insufficient to bridge the gap between these closest branches.

In this case there is no limiting probability density function. The probability distribution, which always exists, is a staircase from 0 to 1. If N interfering samples are considered, there are generally 2^N steps in the staircase. As $N \to \infty$ the staircase never becomes a smooth incline, but instead become a Cantor-like function which does all its rising on a set of measure zero and is constant almost everywhere. Such mathematical abstractions, of course, do not exist in practice since there must always be some noise to bridge the final gaps between branches—although a large number of separate "packets" may be formed which noise is unable to cover. This is usually the case with relatively good quality communication channels where $x(t)$ decays too fast for a limiting probability density to exist in the absence of noise.

In the laboratory the distribution of intersymbol interference and noise is readily seen by using an oscilloscope to sweep out the received voltage $y(t)$ when the horizontal sweep rate is $1/T$. The resulting oscilloscope display is widely known as an *eye pattern* from its resemblance to the human eye for binary data. To understand and interpret eye patterns examine Fig. 4.11 where two binary

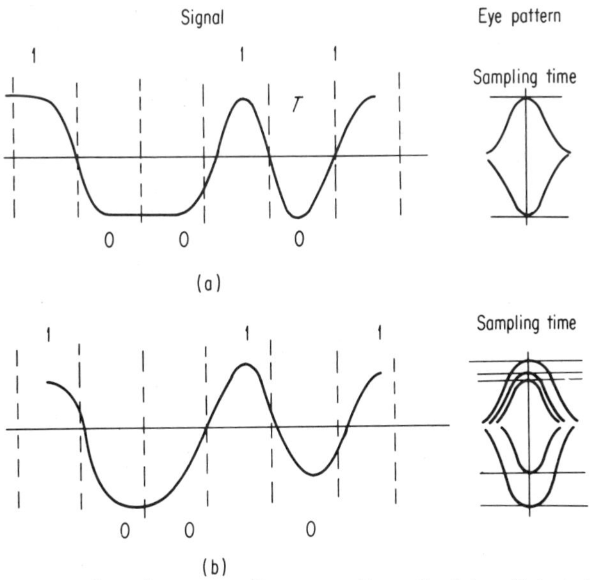

Fig. 4.11. Binary signals and corresponding eye patterns for (a) undistorted and (b) distorted signals. Eye patterns are formed from superposition of dashed segments of received waveforms.

waveforms are shown—one distorted, the other undistorted. In the undistorted waveform when all the T-sec segments of $y(t)$ are superimposed, the "open" eye pattern shown in Fig. 4.11(a) results. A vertical line drawn through the center of the eye pattern shows a superposition of all the received sampled values. When this sampling time is properly adjusted in Fig. 4.11(a), all the sampled values are either +1 or −1.

In Fig. 4.11(b) the waveform is distorted because of the effects of intersymbol interference and noise. Now $y(t)$ no longer passes through the proper values of ± 1 at each sampling point. The distortion is clearly seen in the eye pattern for this waveform; the eye is now partially closed, and detection is obviously more difficult. The distribution of received sample values can be observed along the indicated vertical sampling time.

Eye patterns provide a great deal of information about the performance characteristics of a data system. If the noise is not large, a well-defined pattern, such as that shown in Fig. 4.12, may be studied on an oscilloscope. (1) Notice that the best sampling time must occur approximately where the eye is open widest. (2) The sensitivity of the system to timing error is clearly revealed by the rate of closure of the eye as the sampling time is varied. (3) The maximum signal distortion is indicated by the vertical width of the two branches of the pattern at the sampling time, and the minimum margin against noise error is also shown as the distance from the nearest trace to the threshold at the sampling time.

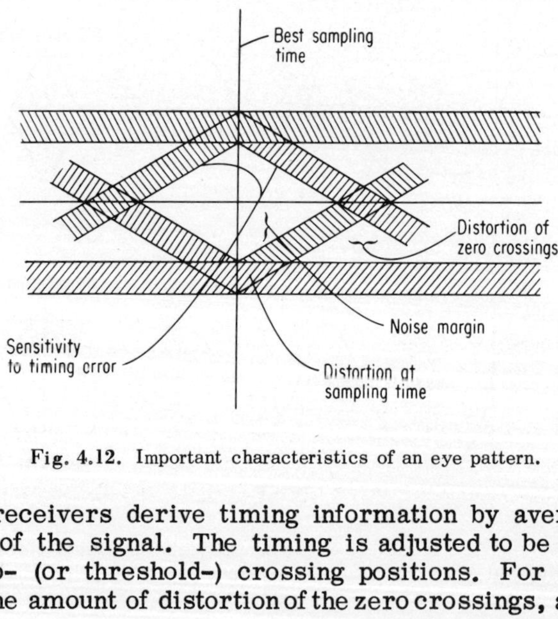

Fig. 4.12. Important characteristics of an eye pattern.

Many receivers derive timing information by averaging zero crossings of the signal. The timing is adjusted to be halfway between zero- (or threshold-) crossing positions. For this type of receiver the amount of distortion of the zero crossings, as indicated in the eye pattern, becomes important. Also the best eye opening should occur in the "middle" of the eye pattern.

For multilevel transmission eye diagrams can be similarly studied. Several multilevel eye diagrams are shown in Fig. 4.13. Frequently in these multilevel eye diagrams it is possible to notice nonlinearities in the transmission channel by asymmetries in the eye pattern. In a strictly linear system with truly random data all eye openings in a multilevel system should be identical.

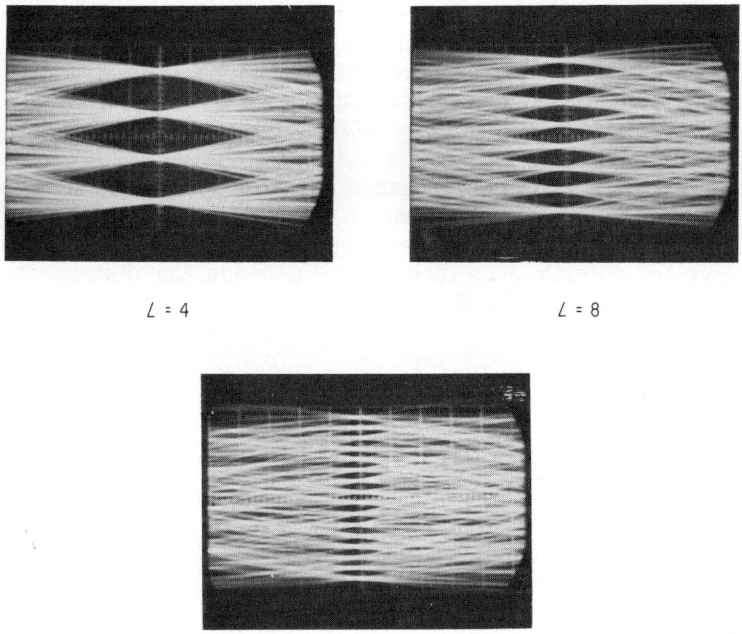

$L = 4$ $\quad$ $L = 8$ $\quad$ $L = 16$

Fig. 4.13. Multilevel eye diagrams.

4.2.1 Probability of Error for PAM with Intersymbol Interference

The previous section has discussed in an heuristic manner some of the mathematical difficulties in obtaining a limiting probability distribution for intersymbol interference in an infinitely long data sequence. Of course, any real message must be time-limited to, say, N symbols; thus intersymbol interference can only assume L^N possible values. A further reduction of this number is possible by assuming that the impulse response $x(t)$ is time-limited to K symbol durations before and after time t_0, and therefore only L^{2K} sequences need be considered to evaluate probability of error. Unfortunately in practice L^{2K} can be a very large number; for example, on a good-quality voice channel K may be 6 or more, while for efficient utilization L will be 8 or 16. The mathematical advantage in treating infinite sequences would be in having a smooth limiting distribution which could be derived analytically in closed form. In the case of intersymbol interference this standard mathematical stratagem fails because the limiting distribution is not the type of function which can be evaluated in a convenient form.

To evaluate the performance of a PAM system operating through a real channel it is necessary to find *probability of error*. No other distortion criterion is as meaningful, inasmuch as the signal can be quite badly distorted by the channel and yet still be received with almost perfect reliability if the distortion is such that decision thresholds are not crossed. The probability of error can always be found to any degree of approximation by considering longer and longer possible data sequences.

Recall that an error occurs in detecting a_0 whenever the sum of intersymbol interference and noise exceeds the distance to the nearest decision threshold.

$$P_e = P\left[\left|\sum_n{}' a_n x_{-n} + \eta_0\right| > x_0 d\right] \tag{4.58}$$

If

$$x_n = 0 \qquad |n| > K$$

$$P_e = P\left[\left|\sum_{n=-K}^{K}{}' a_n x_{-n} + \eta_0\right| > x_0 d\right] \tag{4.59}$$

The probability of error can be found by evaluating the conditional probability of error for each of the L^{2K} data sequences $a_{-K}, a_{-K+1}, a_{-1}, a_1, \ldots, a_K$ and then averaging over all L^{2K} sequences. Let an integer i denote the ith such sequence where the ordering may be accomplished by listing the sequences in order of increasing numerical value—each sequence taken as a number consisting of modulo L digits. The probability of error can be written

$$P_e = \sum_{i=1}^{L^{2K}} P_{ei} P(i) \tag{4.60}$$

where P_{ei} is the probability of error for the ith sequence, and $P(i)$ is the probability of occurrence for the ith sequence.

For each sequence the intersymbol interference $\Sigma_n' a_n x_{-n}$ can be evaluated. Call this quantity $D(i)$; then

$$P_{ei} = P[|D(i) + \eta_0| > x_0 d] \tag{4.61}$$

and for gaussian noise of zero mean and variance σ_N^2,

$$P_{ei} = \frac{1}{\sqrt{2\pi}\,\sigma}\int_{x_0 d - D(i)}^{\infty} e^{-x^2/2\sigma_N^2}\,dx + \frac{1}{\sqrt{2\pi}\,\sigma}\int_{-\infty}^{-x_0 d - D(i)} e^{-x^2/2\sigma_N^2}\,dx \tag{4.62}$$

or

$$P_{ei} = Q\left[\frac{x_0 d - D(i)}{\sigma_N}\right] + Q\left[\frac{x_0 d + D(i)}{\sigma_N}\right] \tag{4.63}$$

Finally if all sequences are equally likely,

$$P_e = \frac{1}{L^{2K}}\sum_{i=1}^{L^{2K}}\left\{Q\left[\frac{x_0 d - D(i)}{\sigma_N}\right] + Q\left[\frac{x_0 d + D(i)}{\sigma_N}\right]\right\} \tag{4.64}$$

Although Eq. (4.64) provides a formal expression for probability of error, it is seldom useful because of the large number of terms involved in the summation. For any specific channel, a digital computer can be used to evaluate Eq. (4.64) if the number of terms L^{2K} is held to several thousand. Unfortunately, this usually leads to a small allowable value for K which poorly approximates the true memory of the channel. Furthermore, although the computation gives a number which is the approximate probability of error, it gives very little insight into the factors determining this probability of error. A basic understanding of the problem is almost impossible to obtain from the computation of Eq. (4.64) since each calculation of P_e takes a great deal of computer time, and many such calculations would have to be made in order to gain any feel for the relationship involved.

These sort of considerations lead one to consider performance criteria other than probability of error. Ideally a performance criterion should enable relative evaluation of data systems and approximate determination of probability of error and yet be easy to manage analytically. Two alternative performance criteria which sometimes have these attributes are the *eye-closure* (or *peak-distortion*) *criterion* and the *mean-square-distortion criterion.*

4.2.2 The Eye Closure Criterion: Peak Distortion

Since engineers in the laboratory examine and measure eye openings to judge the performance of data transmission systems, it would seem appropriate to compute eye openings for mathematical models of communication systems. In the absence of noise (the only case in which the eye opening is meaningfully defined) the maximum opening of the eye pattern shows clearly the minimum margin that the system has against noise at its best sampling point. It may help at this point to reexamine the eye-pattern diagram shown in Fig. 4.12.

As used in this book, the eye opening will always be normalized such that an eye opening of unity is perfect (no intersymbol interference), and an eye opening of zero indicates that the eye is closed. In the latter event errors due to intersymbol interference can be made even in the absence of noise. Typically, an eye opening of 0.5 or better might be required for reliable data transmission. The actual value of opening necessary in any specific case depends on the additive noise, timing jitter, and other factors.

It will be most convenient to compute the eye opening as unity minus the *peak eye closure.* The adjective "peak" is used here since in subsequent sections the rms eye closure will also be considered. The peak eye closure at a given sampling time is defined as the maximum amount that intersymbol interference can perturb a given signal level at the receiver divided by the distance from the level to the nearest decision threshold. For L-level PAM this can be simply found as follows. The received sample y_0 in the absence of noise has been previously given as

$$y_0 = a_0 x_0 + \sum_n{}' a_n x_{-n} \tag{4.65}$$

The maximum value of the summation in Eq. (4.65), indicating intersymbol interference, occurs when that data sequence $\{a_n\}$ is transmitted which utilizes for each symbol a_n the maximum signal level of the same algebraic sign as x_{-n}. The maximum perturbation of the received level is then

$$\text{Max. perturbation} = (L-1)d \sum_n{}' |x_n| \tag{4.66}$$

Since the distance from the received level to the nearest decision threshold is $x_0 d$, the peak eye closure is normalized to

$$\text{Peak eye closure} = (L-1)D \tag{4.67}$$

where

$$D = \frac{1}{x_0} \sum_n{}' |x_n| \tag{4.68}$$

is defined as the *peak distortion* of the system impulse response. In terms of the peak distortion D the eye opening is

$$\text{Eye opening} = 1 - (L-1)D \tag{4.69}$$

When the eye opening is used as a performance criterion, the noise is not directly involved in the calculation. The eye opening simply gives the minimum margin, over all possible data sequences, against noise. It might be argued that the particular data sequence which is used to compute eye opening has a very low probability of occurrence. Sometimes this may be quite true. At other times the most vulnerable data sequence is one which commonly occurs much more frequently than a random-input-sequence assumption would predict. This can happen, for example, when the most vulnerable sequence resembles an all 1s, an all 0s, or a dotting (101010, etc.) pattern. These types of patterns are used for holding a line open, sending black or white areas in facsimile, or cases in which the input redundancy has not been removed.

The eye opening may be used to bound the probability of error if the noise distribution is known. Gaussian noise is always assumed in this text, and for this distribution the bound on probability of error may be a good approximation to its true value. This is because the gaussian distribution weights the nearest traces to the decision threshold much more heavily than those further away in determining probability of error. In terms of Sec. 4.2.1, the bound on probability of error is achieved by selecting the maximum of the conditional probabilities of error P_{ei}. This maximum probability of error occurs for the same "most vulnerable" data sequence used in computing the eye opening. For this sequence, the probability of error is[§]

$$\operatorname*{Max}_{i} P_{ei} = P\{\eta > x_0 d[1 - (L - 1)D]\} \tag{4.70}$$

which for gaussian noise of variance $\sigma_N{}^2$ is

$$\operatorname*{Max}_{i} P_{ei} = Q\left\{\frac{x_0 d[1 - (L - 1)D]}{\sigma_N}\right\} \tag{4.71}$$

The problem of evaluating system performance using the eye-opening criterion reduces to calculation of the peak distortion D for the system impulse response. The response samples x_n which determine D are given by

$$x_n = \frac{1}{2\pi}\int_{-\infty}^{\infty} G_T(\omega)C(\omega)G_R(\omega)e^{i\omega(nT + t_0)}\,d\omega \tag{4.72}$$

Here $G_T(\omega)G_R(\omega)$ is the desired Nyquist pulse characteristic, while $C(\omega)$ is the perturbing channel transfer function. Notice that the peak distortion D is a function of the reference sampling time t_0, and consequently it is necessary to find the time t_0 for which D is minimized. In laboratory measurement this is easy since t_0 is

---

[§]The probability of crossing the opposite threshold has been ignored as being much smaller than the term given.

simply the place where the eye is open widest. However, in calculating D, it becomes necessary to try many values for t_0 to minimize by trial and error. The extra calculations are not worthless since they show the very important *rate* of eye closure with sampling-time variation, which determines the sensitivity of the system to timing jitter. Indeed, so many features of the eye illustrated in Fig. 4.12 are important that it has become common in analysis to have a digital computer calculate and draw eye patterns.

As an example suppose that a 50 percent roll-off ($\alpha = 0.5$) raised-cosine characteristic is chosen for $G_T(\omega) G_R(\omega)$ and that the channel characteristics is flat in its amplitude but quadratic in its delay characteristic (cubic in phase) as illustrated in Fig. (4.14). The delay at the Nyquist frequency of π/T rad is taken to be $\beta_m T$ sec. The delay characteristic for the channel is therefore

$$\beta'(\omega) = \frac{\beta_m T^3}{\pi^2} \omega^2 \tag{4.73}$$

and the phase characteristic is

$$\beta(\omega) = \frac{\beta_m T^3 \omega^3}{3\pi^2} \tag{4.74}$$

The impulse response of this system can be obtained from

$$x(t) = \frac{1}{\pi} \int_0^\infty G_T(\omega) G_R(\omega) \cos[\omega t + \beta(\omega)] \, d\omega \tag{4.75}$$

Unfortunately with $\beta(\omega)$ cubic and $G_T(\omega) G_R(\omega)$ raised-cosine, Eq. (4.75) cannot be evaluated in closed form analytically. This is usually the case in even the simplest of channel models. Using a digital computer to integrate Eq. (4.75) gives the responses shown in Fig. 4.14(b) for various values of normalized Nyquist delay β_m.

It may help in understanding delay distortion to study the responses in Fig. 4.14(b). Notice that since the higher frequencies are delayed relative to lower frequencies, the response tends to oscillate more to the right of the peak, i.e., at a later time. Also the peak of the response drifts to the right as β_m is increased and the magnitude of the oscillation increases.

The peak distortion

$$D = \frac{1}{x_0} \sum_n{}' |x_n|$$

is plotted as a function of β_m for these responses in Fig. 4.15. Several possible strategies for choosing t_0 are shown in this figure. The crudest choice is to use $t_0 = 0$ for all values of β_m. This is optimum for $\beta_m = 0$ but quite inferior elsewhere. A better

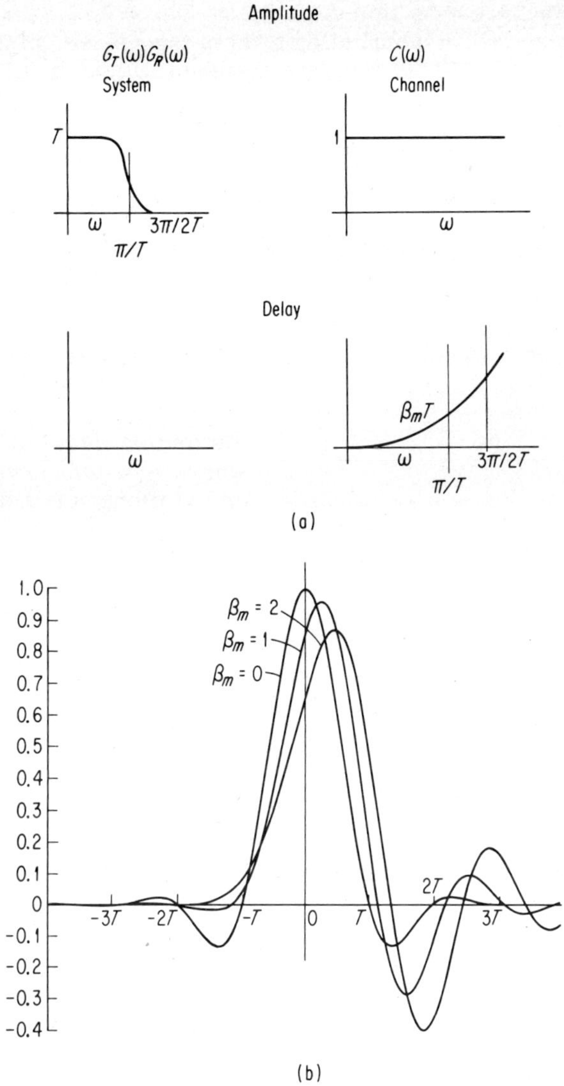

Fig. 4.14. Parabolic delay. (a) Channel and system frequency characteristics. (b) Impulse responses.

choice is to select t_0 at the time of peak of the impulse response. This usually leads to somewhat better results but is sometimes much worse than optimum selection of t_0. Peak distortion in this example is minimized when t_0 is selected such that $x_1 = 0$. As a general rule of thumb, for the best value of t_0 the response sample whose derivative has the largest magnitude (invariably x_1 or x_{-1}) is zero.

What is the effect on this distortion as the roll-off α is changed? If α is decreased, two conflicting effects take place. Since the frequency cutoff in $G_T(\omega)G_R(\omega)$ is steeper, the impulse response tends to oscillate in larger amounts and for longer durations. On the other hand, the bandwidth $(1+\alpha)\pi/T$ rad per sec is decreased, and consequently the system "sees" less of the worsening delay characteristic at the higher frequencies. Generally the latter effect predominates unless α is small, as shown in the distortion curve of Fig. 4.16 where roll-off α is varied while maintaining a constant period T. It would seem advantageous according to these curves to use sharp roll-offs when confronted with delay characteristics which rise near the band edges. Similar curves could be shown for falling amplitude characteristics where again the philosophy would seem to be one of shaping the pulse spectrum to avoid the ranges where the frequency characteristics are worst.

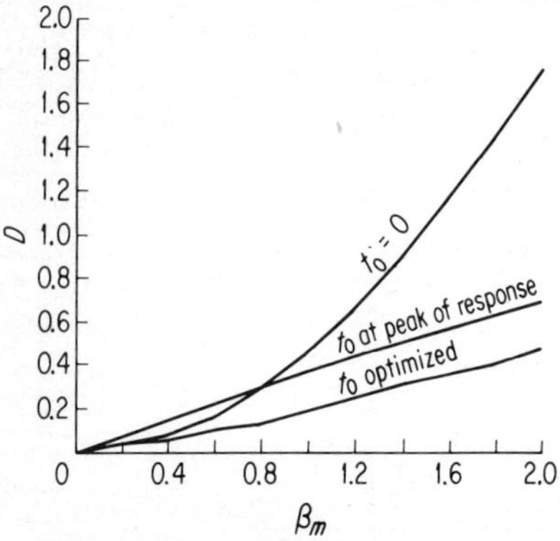

Fig. 4.15. Peak distortion D for parabolic-delay responses. β_m is the delay at Nyquist frequency in terms of pulse intervals.

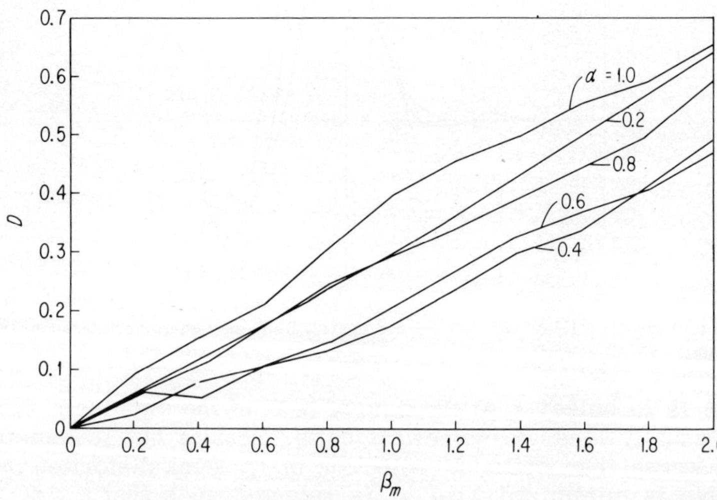

Fig. 4.16. Peak distortion D for parabolic delay for various roll-offs of raised-cosine shaping. β_m is the delay at Nyquist frequency, in terms of pulse intervals, (t_0 optimized).

The penalty exacted by the decrease in roll-off shows up in the rate of closure of the eye with variation in t_0. In Fig. 4.17 the peak distortion is shown as a function of the deviation from optimum t_0 for several values of roll-off and of parabolic delay. The sensitivity to timing jitter can be ascertained from the rate of rise of these curves from their starting points. The horizontal width, or duration, of the eye pattern for an L-level system can be found from the point where $D = 1/(L - 1)$. Thus as α is decreased, the system not only becomes more sensitive to any timing jitter that occurs but the actual occurrence of this jitter becomes more likely because of the distortion of the zero crossings (threshold crossings for $L > 2$) of the received signal.

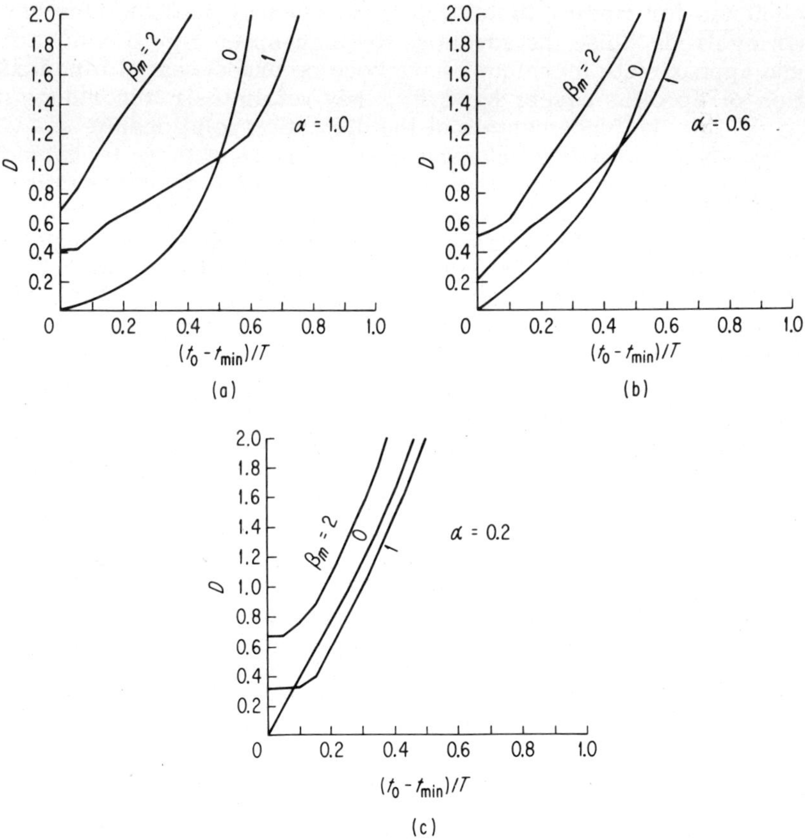

Fig. 4.17. Effect of nonoptimum timing on peak distortion for various roll-offs and parabolic delay.

Notice also that the presence of the delay distortion does not appreciably change the rate of eye closure. This indicates that the sensitivity to timing jetter is a property of the roll-off itself and is only secondarily affected by the delay characteristic of the channel.

Echo theory and approximate distortion methods. The procedure for evaluating data-system performance as given in the preceding material is completely straightforward. The same type of calculations as were applied to parabolic delay could be repeated for many other kinds of channel characteristics, resulting in a catalog of possible distortions and their effect. While this type of detailed analysis is indispensable in the evaluation of specific situations, it lends very little insight into the relationship between channel transfer characteristics on the one hand and distortion, or eye closure, on the other. As an aid to engineering judgment, a number of approximate techniques for finding distortion from channel perturbations have been evolved. Since the advent of the high-speed digital computer, these approximate techniques have become unnecessary for computation of specific cases; however, they retain their importance in illuminating the fundamentals of the distortion relationship.

The *theory of paired echoes* is the best known of these techniques and applies chiefly when the amplitude or phase characteristic of the channel has sinusoidal deviations. Actually, this type of deviation is quite common since filter design techniques tend to fit the desired characteristics only at specific points and leave a residual error which is a sinusoidal ripple. Briefly, the theory states that a ripple in the channel's amplitude characteristic leads to two echoes in the channel's impulse response $x(t)$. The two echoes have the same polarity and occur at times $-T_0$ and T_0, where T_0 is proportional to the number of cycles of ripple in the band. For ripples in the phase characteristic the situation is similar except the two echoes have opposite polarity. If both amplitude and phase ripples are present or if several frequencies of ripple are present simultaneously, interaction terms complicate the simple picture; to a first-order approximation, however, simple addition of the echoes from the separate ripple components suffices.

Suppose initially that the channel has a transfer function which has a cosine-amplitude ripple superimposed on an otherwise perfect characteristic. (Since the amplitude characteristic is even in ω, it is only necessary to consider cosine components. Similarly for the phase characteristic only sine ripples are considered.) Let

$$C(\omega) = 1 + a \cos \omega T_0 \tag{4.76}$$

The impulse response $x(t)$ is then

$$x(t) = \frac{1}{\pi} \int_0^{\infty} G_T(\omega) G_R(\omega) [1 + a \cos \omega T_0] \cos \omega t \, d\omega \tag{4.77}$$

$$x(t) = \frac{1}{\pi}\int_0^\infty G_T(\omega)G_R(\omega)\cos\omega t\,d\omega + \frac{a}{2\pi}\int_0^\infty G_T(\omega)G_R(\omega)\cos[\omega(t+T_0)]\,d\omega + \frac{a}{2\pi}\int_0^\infty G_T(\omega)G_R(\omega)\cos[\omega(t-T_0)]\,d\omega \tag{4.78}$$

If the undistorted response of the system is $g_{TR}(t)$, that is,

$$g_{TR}(t) = \frac{1}{\pi}\int_0^\infty G_T(\omega)G_R(\omega)\cos\omega t\,d\omega \tag{4.79}$$

Eq. (4.78) becomes

$$x(t) = g_{TR}(t) + \frac{a}{2}g_{TR}(t-T_0) + \frac{a}{2}g_{TR}(t+T_0) \tag{4.80}$$

which clearly shows the pair of echoes. Note that no approximation was used in deriving this result (see Fig. 4.18).

To arrive at a similar result for phase ripples of the form

$$\beta(\omega) = b\sin\omega T_0 \tag{4.81}$$

it is more convenient to use the exponential form of the Fourier integral for the impulse response; thus

$$x(t) = \frac{1}{2\pi}\int_{-\infty}^\infty G_T(\omega)G_R(\omega)e^{ib\sin\omega T_0}\,e^{i\omega t}\,d\omega \tag{4.82}$$

If the following expansion in Bessel functions is used:

$$e^{iz\sin\theta} = \sum_{k=-\infty}^{\infty} J_k(z)e^{ik\theta} \tag{4.83}$$

Eq. (4.82) may be expressed as

$$x(t) = \sum_{k=-\infty}^{\infty}\frac{J_k(b)}{2\pi}\int_{-\infty}^\infty G_T(\omega)G_R(\omega)e^{i\omega(t-kT_0)}\,d\omega \tag{4.84}$$

In terms of the undistorted response $g_{TR}(t)$, the response $x(t)$ becomes

$$x(t) = \sum_{k=-\infty}^{\infty} J_k(b)\,g_{TR}(t-kT_0) \tag{4.85}$$

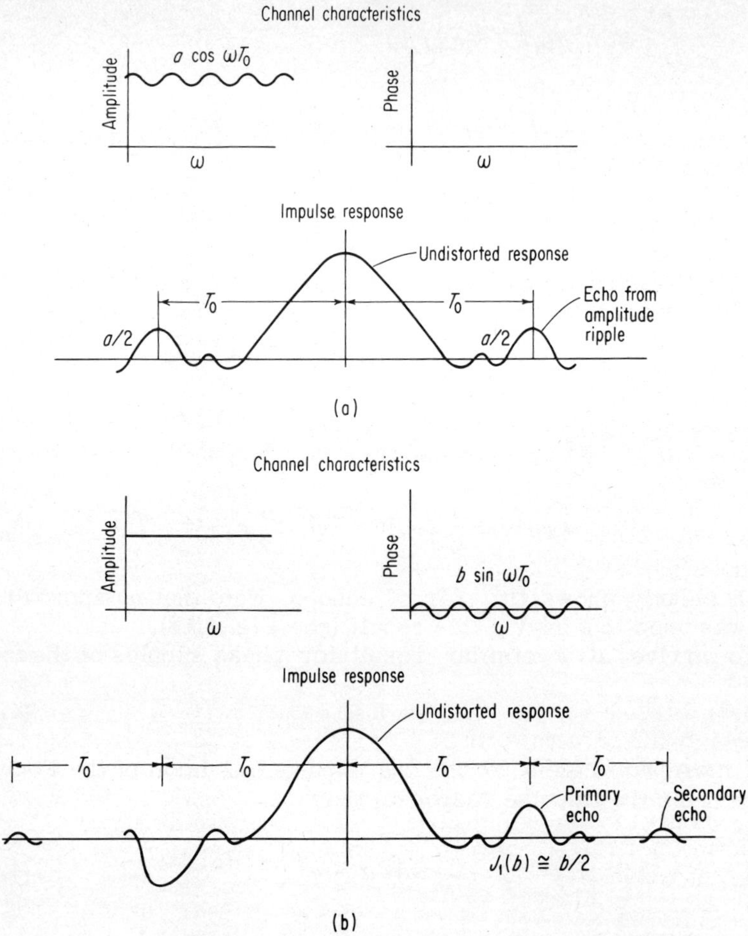

Fig. 4.18. Paired-echo theory. (a) Amplitude ripple. (b) Phase ripple.

Observe that the phase ripple produces an infinite series of echoes. Since $J_{-k}(b) = (-1)^k J_k(b)$, pairs of echoes are alternately of the same and of opposite polarities.

To approximate Eq. (4.85) for small b, the following representation for $J_k(b)$ is employed:

$$J_k(b) = \left(\frac{b}{2}\right)^k \sum_{n=0}^{\infty} \frac{-\frac{1}{4}b^{2n}}{n!\,(k+n)!} \tag{4.86}$$

If only the first term of Eq. (4.86) is significant and if only the adjacent ($k = \pm 1$) echoes are employed,

$$x(t) \cong g_{TR}(t) + \frac{b}{2} g_{TR}(t - T_0) - \frac{b}{2} g_{TR}(t + T_0) \tag{4.87}$$

which expresses the response in a manner analogous to the amplitude echo form of Eq. (4.80). These expressions and their use are illustrated in Fig. (4.18).

The following points should be kept in mind when assessing the effect of amplitude and phase (delay) characteristics:

1. Amplitude fluctuations cause a symmetric distortion of the pulse response. The peak time of the response remains $t = 0$.
2. Phase deviations from linear (delay variations from constant) cause an antisymmetric distortion of the pulse response, which tends to shift the peak time away from $t = 0$.
3. High-frequency, or fine-grain, deviations in either phase or amplitude characteristic cause the distant echoes in the impulse response.
4. Slowly varying, low-frequency deviations lead to nearby echoes. In this case the echoes overlap the main pulse, and the appearance is generally that of a single, distorted pulse.

4.2.3 The RMS Error Criterion

Mean-square-error criteria have achieved a great popularity because of the ease with which they can be handled mathematically. In dealing with intersymbol interference, the *mean-square eye closure* gives some measure of the amount of level perturbation from this cause. The mean-square eye closure is defined much like the peak eye closure, except that the mean-square value of intersymbol interference perturbation, rather than the peak value, is used. For this criterion it is necessary to know the statistics of the input data sequence $\{a_n\}$. For simplicity we shall always consider the input symbols to be independent and equally likely to assume any of the L possible values. Any expression derived under this assumption may be easily generalized by the reader to include other distributions and correlations.

The intersymbol interference for the sample y_0 has been previously written as (in the absence of noise)

$$y_0 - x_0 a_0 = \sum_n{}' a_n x_{-n} \tag{4.88}$$

The mean-square value is

$$\left\langle \left(\sum_n{}' a_n x_{-n} \right)^2 \right\rangle = \overline{a^2} \sum_n{}' x_n^2 \tag{4.89}$$

When this is normalized by the square of the distance to the nearest decision threshold, mean-square eye closure results.

$$\text{Mean-square eye closure} = \frac{\overline{a^2}}{d^2} \frac{1}{x_0^{\,2}} \sum_n{}' x_n^{\,2} \tag{4.90}$$

For L-level PAM the "symbol power" $\overline{a^2}$ was shown to be $d^2(L^2-1)/3$; therefore

$$\text{Mean-square eye closure} = \frac{L^2-1}{3}\,\mathcal{E}^2 \tag{4.91}$$

where $\mathcal{E}^2$ is the mean-square distortion in the system impulse response.

$$\mathcal{E}^2 = \frac{1}{x_0^{\,2}} \sum_n{}' x_n^{\,2} \tag{4.92}$$

Since the noise and intersymbol interference are independent, the total mean-square error is the sum of the mean-square intersymbol interference and the noise variance. Going back to the unnormalized expression Eq. (4.89) and adding the noise variance $\sigma_N^{\,2}$ gives

$$\sigma^2 = \frac{d^2(L^2-1)}{3} \sum_n{}' x_n^{\,2} + \sigma_N^{\,2} \tag{4.93}$$

An error in detecting a symbol occurs whenever the sum of intersymbol interference and noise exceeds $x_0 d$. The noise has been assumed gaussian, but the probability distribution for the intersymbol interference is not known. It is sometimes argued that this distribution is "approximately" gaussian when a large number of significant terms are involved in the summation for intersymbol interference. If the distribution were gaussian, the total error would be gaussian, and the probability of error would be

$$P_e = 2\,Q\left(\frac{x_0 d}{\sigma}\right) \tag{4.94}$$

$$P_e = 2Q\left\{\left[\frac{L^2-1}{3}\,\mathcal{E}^2 + \frac{\sigma_N^{\,2}}{x_0^{\,2} d^2}\right]^{-1/2}\right\} \tag{4.95}$$

The assumption of gaussian intersymbol interference and the resulting probability of error [Eq. (4.95)] is subject to some criticism. The density may indeed be roughly gaussian in its central portion, but it is the "tails" of the density which determine error probability. These tails cannot be even approximately gaussian since the intersymbol interference has an upper bound of only a fraction of the distance to the nearest threshold. Therefore the probability of error [Eq. (4.95)] may be a poor approximation to the true value. In spite of such problems, the total rms error has often been used as a criterion for optimization of digital systems. Unfortunately this criterion treats intersymbol interference and noise as being equivalent in their effect on the error rate, whereas they are actually quite dissimilar processes. It will be the practice here to keep the mean-square eye closure and noise variance as separate quantities to be added only when an approximate probability of error is required.

Mean-square eye closure has been given in terms of the mean-square distortion in the samples x_n of the system impulse response. Unlike the peak distortion D, the mean-square distortion $\mathcal{E}^2$ can be calculated without actually having to compute each of the response samples x_n. The value of $\mathcal{E}^2$ can be derived in the form of a single integral through the concept of the equivalent Nyquist channel which was previously discussed in the section on the Nyquist criterion. It will be recalled that the equivalent Nyquist channel is formed by folding the system transfer function about the Nyquist frequency π/T rad per sec. The resulting transfer function fits into the Nyquist bandwidth and yet has the same values as the original channel at the sampling times nT, that is,

$$x_n = \frac{1}{2\pi}\int_{-\infty}^{\infty} X_{\text{eq}}(\omega)^{i\omega nT}\, d\omega \tag{4.96}$$

$$X_{\text{eq}}(\omega) = \begin{cases} \sum_k X\left(\omega + \dfrac{2k\pi}{T}\right) & |\omega| \le \dfrac{\pi}{T} \\ 0 & |\omega| > \dfrac{\pi}{T} \end{cases} \tag{4.97}$$

The utility of this substitution is that since $X_{\text{eq}}(\omega)$ is limited to the Nyquist band, the sampling theorem may be used to reconstruct $X_{\text{eq}}(\omega)$ uniquely from the samples x_n and, through the Parseval relationship, the energy in $X_{\text{eq}}(\omega)$ may be equated to the energy in the sample values x_n. The reasoning is as follows:

$$\frac{1}{2\pi}\int_{-\pi/T}^{\pi/T} |X_{\text{eq}}(\omega)|^2\, d\omega = \int_{-\infty}^{\infty} x_{\text{eq}}^2(t)\, dt = T\sum_n x_n{}^2 \tag{4.98}$$

Therefore

$$\mathcal{E}^2 = \frac{1}{x_0{}^2}\left[\frac{1}{2\pi T}\int_{-\pi/T}^{\pi/T} |X_{\text{eq}}(\omega)|^2\, d\omega - x_0{}^2\right] \tag{4.99}$$

$$\mathcal{E}^2 = \frac{1}{2\pi x_0{}^2} \int_{-\pi/T}^{\pi/T} |X_{eq}(\omega)|^2 \, d\omega - 1 \qquad (4.100)$$

A good way of visualizing this process is the following: The equivalent channel $X_{eq}(\omega)$ is formed from the actual system response $X(\omega)$ by folding any excess frequency components above π/T back into the basic Nyquist interval. In order to have zero intersymbol interference, $X_{eq}(\omega)$ must be the perfect rectangular characteristic

$$X_{eq}(\omega) = x_0 T \qquad |\omega| \leq \frac{\pi}{T}$$

If there is any deviation from the rectangular characteristic, the integral square error is the energy in the difference between their time responses at the sample values. Thus

$$\sum_n{}' x_n{}^2 = \frac{1}{2\pi T} \int_{-\infty}^{\infty} |X_{eq}(\omega) - x_0 T|^2 \, d\omega = x_0{}^2 \mathcal{E}^2 \qquad (4.101)$$

Since $x_0 T$ is simply the dc value of $X_{eq}(\omega)$, the mean-square distortion $\mathcal{E}^2$ can be thought of as the energy in the ac components of $X_{eq}(\omega)$. Of course, to arrive at an exact expression for $\mathcal{E}^2$ from Eq. (4.101) it is necessary to divide by $x_0{}^2$.

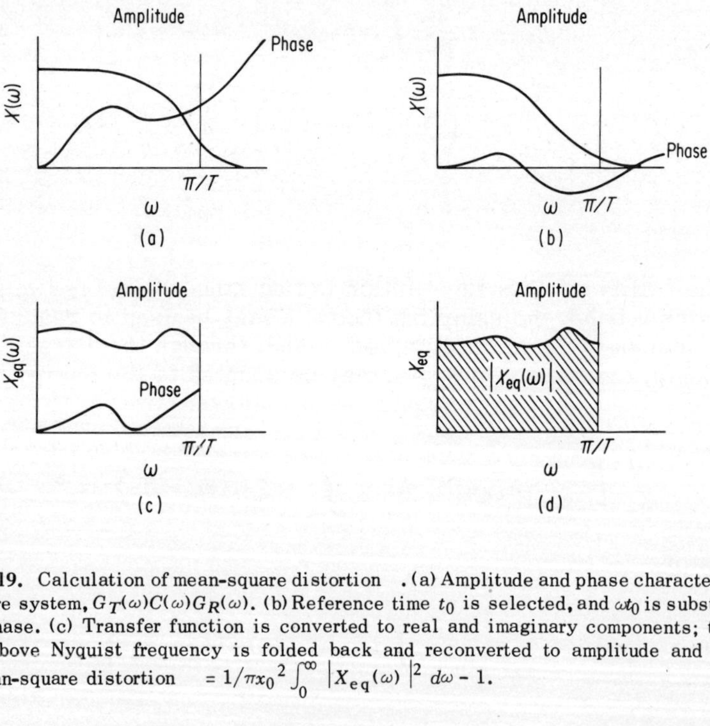

Fig. 4.19. Calculation of mean-square distortion . (a) Amplitude and phase characteristics of entire system, $G_T(\omega)C(\omega)G_R(\omega)$. (b) Reference time t_0 is selected, and ωt_0 is substracted from phase. (c) Transfer function is converted to real and imaginary components; the excess above Nyquist frequency is folded back and reconverted to amplitude and phase. (d) Mean-square distortion $= 1/\pi x_0{}^2 \int_0^{\infty} |X_{eq}(\omega)|^2 \, d\omega - 1$.

As in the previous two criteria, probability of error and peak eye closure, it is necessary to optimize the mean-square criterion with respect to the reference sampling time t_0. Since $X_{eq}(\omega)$ has an impulse response which corresponds with the actual system response $x(t)$ *only* at the sampling times nT, for all integers n, the phase characteristic ωt_0 must be subtracted from the phase of $X(\omega) = G_T(\omega)C(\omega)G_R(\omega)$ in order to shift the sampling times of $x(t)$ from $(nT + t_0)$ to nT. Thus $X_{eq}(\omega)$ depends on t_0 and a trial-and-error or search routine must be used to minimize $\mathcal{E}^2$ with respect to t_0. The procedure of calculating $\mathcal{E}^2$ is illustrated in Fig. 4.19(d).

4.2.4 Summary of Criteria: Numerical Example

The preceding sections have discussed three performance criteria: probability of error, peak eye closure, and rms eye closure. The advantages and drawbacks of each have been pointed out in these sections. Briefly, the probability of error criterion is too difficult to compute, but it is the most meaningful; rms closure is easiest to compute, but it is perhaps the least meaningful. The peak eye-closure, or peak-distortion, criterion is not as simple to calculate as rms closure but leads to a definite upper bound on error rate and relates well to measurement and intuition.

If the channel response $x(t)$ is assumed to be zero for $|t| > kT$, there are $2K$ terms in the summation for intersymbol interference. Computing the probability of error would require L^{2K} calculations of conditional probabilities of error, while peak distortion requires $2K$ calculations of impulse response samples, and rms distortion needs only one integration after some manipulation of the frequency characteristics for the system. The principal definitions and formulas for the peak and mean-square eye-closure criteria are summarized in Table 4-1.

Table 4-1. Definitions for Peak- and Rms-distortion Criteria,

Criterion	Formula: Peak	Formula: Rms
Impulse response, distortion	$D = \frac{1}{x_0} \sum_n{}' \lvert x_n \rvert$	$\mathcal{E} = \frac{1}{x_0}\left(\sum_n{}' x_n^2\right)^{1/2}$
L-level PAM, eye closure	$(L-1)D$	$\left(\frac{L^2-1}{3}\right)^{1/2} \mathcal{E}$
Approximate probability of error for gaussian noise, var σ_N^2	$Q\left(\frac{x_0 d}{\sigma_N}\right)[1-(L-1)D]$ (upper bound)	$2Q\left(\frac{L^2-1}{3}\mathcal{E}^2 + \frac{\sigma_N^2}{x_0^2 d^2}\right)^{-1/2}$

For a specific numerical example giving insight into the three performance criteria a channel with a perfect phase characteristic, but an attenuation characteristic which falls off linearly in decibels, has been selected. The channel characteristic is taken as

$$C(\omega) = 10^{-TA_m\omega/20\pi} \tag{4.102}$$

where A_m has been defined as the attenuation in decibels at the Nyquist frequency.

The system filtering $G_T(\omega)G_R(\omega)$ is raised-cosine with roll-off parameter α. For the channel characteristic [Eq. (4.102)] the resulting impulse response is symmetrical since no phase distortion is present, and in all cases $t_0 = 0$. First the peak distortion D and the rms distortion $\mathcal{E}$ were calculated for various values of A_m and α. The results of these computer runs are shown in Figs. (4.20) and (4.21). Next it is informative to determine the ratio of peak-to-rms distortion for the intersymbol interference in these examples. Curves of $D/\mathcal{E}$ are shown in Fig. 4.22 as a function of roll-off α with the amount of attenuation A_m as a parameter. The typical behavior as seen in these curves is that $D/\mathcal{E}$ is somewhat independent of the roll-off α but increases with the attenuation of A_m. The latter effect results from the more prolonged oscillations

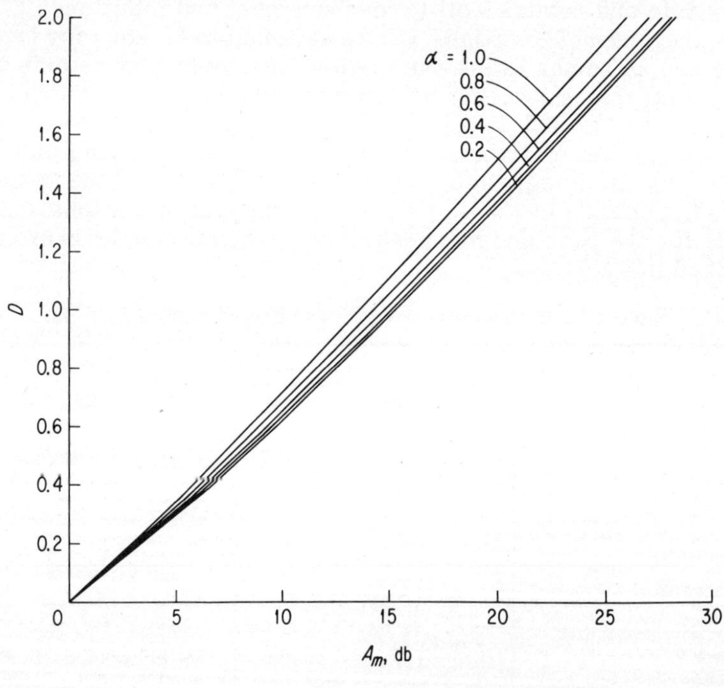

Fig. 4.20. Peak distortion D for linear attenuation at Nyquist frequency (see Table 4-1).

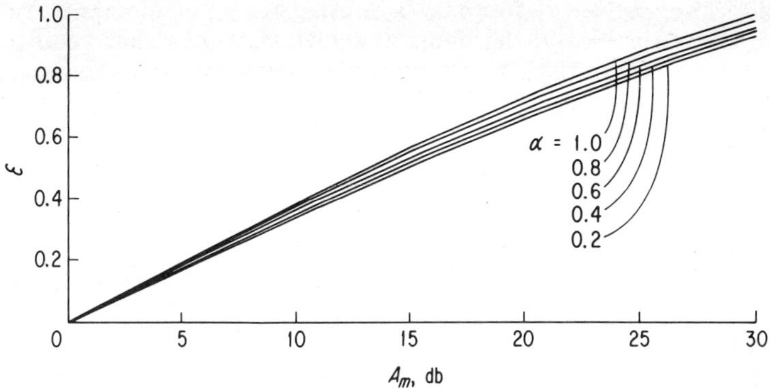

Fig. 4.21. Rms distortion ϵ for linear db attenuation at Nyquist frequency (see Table 4-1).

of the impulse response when A_m is large. Many small samples x_n contributing to the intersymbol interference in this case result in a large peak-to-rms ratio.

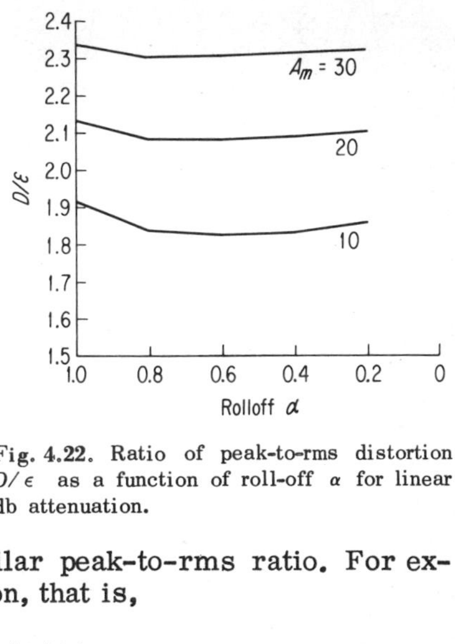

Fig. 4.22. Ratio of peak-to-rms distortion D/ϵ as a function of roll-off α for linear db attenuation.

The peak-to-rms distortion ratio $D/\mathcal{E}$ gives some indication of the shape of the probability density for intersymbol interference. The value of this ratio in the present example is approximately 2. This value is typical of many simple examples, and it leads to the temptation to approximate the density function of intersymbol interference by some mathematically simple function which has a similar peak-to-rms ratio. For example, a full raised-cosine function, that is,

$$1 + \cos \omega \qquad |\omega| \leq \pi$$

suitable scaled and normalized is sometimes used to approximate the probability density for intersymbol interference. For the raised cosine, $D/\mathcal{E} = 2.76$, and the general shape of the function is similar to what one expects from intersymbol interference. However, this type of reasoning is not generally applicable and sometimes leads to poor approximation of error rate.

The actual density of the intersymbol interference for the present example may be approximated by computing all possible values of $\Sigma_{n=-K}^{K} a_n x_n$ for some sequence length K. For a specific case it was assumed that binary transmission was employed and that $\alpha = 0.6$ and $A_m = 10$ db. When all binary sequences were considered equally likely, the density shown in Fig. (4.23) resulted for $K = 5$. Remember that, mathematically speaking, this density does not actually exist as the smooth function in Fig. 4.23. In practice, however, this density would lead to a very good approximation for probability of error if convolved with gaussian noise.

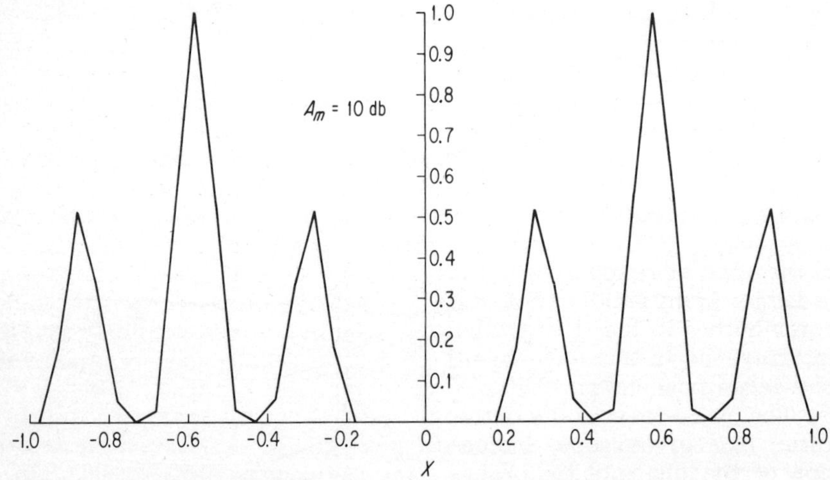

Fig. 4.23. Approximate probability density for intersymbol interference for linear db attenuation; A_m = 10 db, α = 0.6.

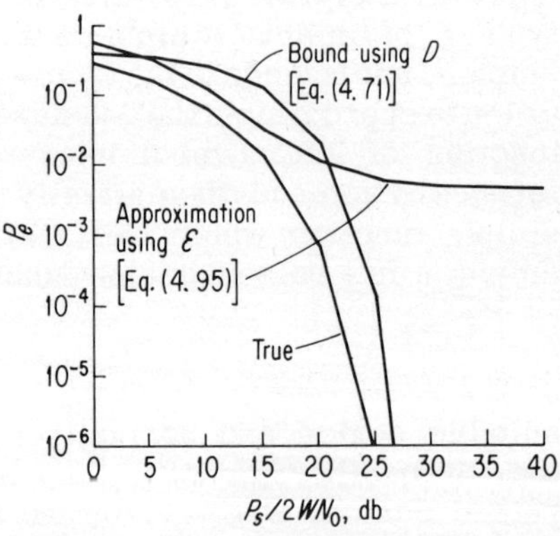

Fig. 4.24. Probability of error P_e for binary PAM for linear db attenuation; A_m = 10 db, $\alpha = 0.6$.

Finally the probability of error was approximated in three ways: first by computing the conditional probability of error for each of 2^{11} equally likely sequences of length 11 from Eq. (4.64), second by bounding the probability of error using D and Eq. (4.71), and third by assuming gaussian intersymbol interference and applying Eq. (4.95). The technique of computing conditional probabilities of error is clearly the best approximation; in this instance, for $K = 5$, it would be extremely close to the to the true value. Consequently the corresponding curve in Fig. 4.24 is labeled "True" probability of error. The closeness of the other approximations may be judged from their relative positions in this same figure.

4.3 Controlled Intersymbol Interference in PAM: Duobinary and Partial-response Techniques

Thus far intersymbol interference has been treated as an undesirable phenomenon. Indeed, its very name connotes a nuisance effect. In the last few years, however, engineers have been becoming increasingly aware of the advantages of certain signaling systems, called variously *duobinary*, *polybinary*, *partial response*, or *biternary*, which utilize a controlled amount of intersymbol interference to achieve certain beneficial effects in data transmission. To explain the basic idea in the simplest manner, the duobinary scheme for binary data will first be described and then somewhat generalized.

Suppose that a sequence $\{a_n\}$ of binary digits is to be transmitted at the ideal Nyquist rate of $2W$ symbols (bits) per second over an ideal channel of bandwidth W Hz. According to what has been said in the previous sections of this chapter, the only possible way this could be accomplished *without intersymbol interference* would be to use perfect rectangular filters ($\alpha = 0$) at the transmitter and receiver. This system is obviously unrealizable; any real system which would closely approximate it would be extremely sensitive to any perturbations in rate, timing, or channel characteristics.

If, instead of the ideal low-pass characteristic, the cosine (not the *raised-cosine*) filter shown in Fig. 4.25(a) is used for the composite signal-shaping characteristic $G_T(\omega)G_R(\omega) = X(\omega)$, intersymbol interference must be expected. In fact, the system impulse

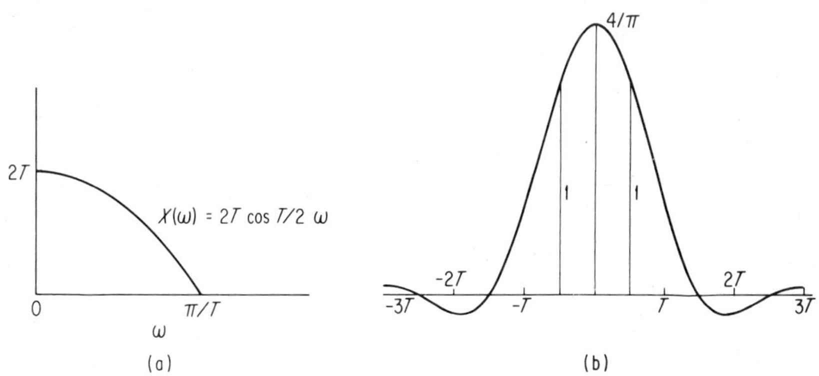

Fig. 4.25. (a) Duobinary filter and (b) its impulse response.

response looks as depicted in Fig. 4.25(b). The transfer function and its response are the following:

$$X(\omega) = \begin{cases} 2T \cos \frac{T}{2}\omega & |\omega| \leq \frac{\pi}{T} \\ 0 & |\omega| > \frac{\pi}{T} \end{cases} \tag{4.103}$$

$$x(t) = \frac{4}{\pi}\left[\frac{\cos \pi t/T}{1 - 4t^2/T^2}\right] \tag{4.104}$$

Now it may be observed that if the response sampling time t_0 is chosen to be $-T/2$,

$$x_n = x(nT + t_0) = \begin{cases} 1 & n = 0,1 \\ 0 & \text{otherwise} \end{cases} \tag{4.105}$$

and the intersymbol interference comes only from the preceding symbol; i.e., in the absence of noise

$$y_k = a_k + a_{k-1} \tag{4.106}$$

If a_k can assume the two possible values $\pm d$, y_k has the three values $\pm 2d$ and 0. Thus the binary input to the system has been converted to a three-level output. However, the data can still be decoded correctly using Eq. (4.106). If a_{k-1} is decided, its effect on y_k can be eliminated by subtraction, and a_k can be decided, etc.

One immediately apparent drawback of this system is that errors, once made, tend to propagate. If a_{k-1} is in error, then its effect is incorrectly compensated for when deciding a_k; consequently a_k is likely to be in error also. A means for avoiding this error propagation was introduced by Lender.[3] Lender's scheme eliminates the error propagation by removing the necessity for the compensation [Eq. (4.106)] at the receiver by *precoding* at the transmitter. The precoding operation is performed on the sequence of binary input digits $\{a_k\}$. This sequence is converted to another binary sequence $\{b_k\}$ before transmission using the following rule:

$$b_k = a_k \oplus b_{k-1} \tag{4.107}$$

(The symbol $\oplus$ represents modulo-2 summation of the binary digits.) The sequence $\{b_k\}$ of bits is then transmitted using the levels $\pm d$ for 1 or 0, respectively.

Suppose that a_k is 0; therefore $b_k = b_{k-1}$, and according to Eq. (4.106) y_k will be either $2d$ or $-2d$. On the other hand if a_k is 1, b_k is the reverse of b_{k-1}, and the received level y_k is always 0. Thus a_k can always be decoded by the rule

$$\begin{aligned} a_k &= 0 \quad \text{if} \quad y_k = \pm 2d \\ a_k &= 1 \quad \text{if} \quad y_k = 0 \end{aligned} \tag{4.108}$$

The rule obviously involves rectification of y_k and a simple threshold binary decision. Since no knowledge of any symbol or sample other than y_k is involved in deciding a_k, error propagation cannot occur.

The accomplishment of this duobinary scheme has been to transmit binary digits at the Nyquist rate $2W$ using realizable and perturbation-tolerant filters. In fact, it has been shown[4] that the rate insensitivity of the system is such that the sampling rate $(1/T)$ of the duobinary system can be varied by 43 percent for the given filter configuration without closing the (three-level) eye pattern. Remember that the only comparable binary conventional PAM system must use an abrupt-cutoff ($\alpha = 0$) rectangular-filter characteristic and consequently has no tolerance to variations in rate whatsoever. However, the duobinary signal has three levels in the channel which must be distinguished; thus it requires a somewhat higher signal-to-noise ratio for equal performance than the ideal $\alpha = 0$ binary system. The exact penalty for the privilege of relative insensitivity will now be calculated.

4.3.1 Performance of Duobinary

If the input symbols a_n are equally likely and independent, so, too, are the precoded symbols b_n. The average "symbol power" input to the transmitting filter is $\overline{b_n^2} = d^2$. The power input to the channel is

$$P_S = \frac{d^2}{2\pi T} \int_{-\pi/T}^{\pi/T} |G_T(\omega)|^2 \, d\omega \tag{4.109}$$

As with ordinary PAM, it is advantageous from a noise-power standpoint to split the overall duobinary signal shaping equally between the transmitter and the receiver. Therefore, we use

$$G_T(\omega) = G_R(\omega) = \left(2T \cos \frac{T}{2}\omega\right)^{1/2} \qquad |\omega| \leq \frac{\pi}{T} \tag{4.110}$$

and, performing the integration in (4.109), we obtain

$$P_S = \frac{4d^2}{\pi T} \tag{4.111}$$

At the receiver the noise output $G_R(\omega)$ is similarly evaluated.

$$\sigma_N^2 = \frac{4N_0}{\pi} \tag{4.112}$$

The received levels in the absence of noise are $2d$, 0, and $-2d$, but the probabilities of receiving these are 1/4, 1/2, and 1/4, respectively. Therefore the probability of error is

$$P_e = \frac{3}{2} P[\eta > d] \tag{4.113}$$

$$P_e = \frac{3}{2} Q\left[\frac{\pi}{4}\left(\frac{TP_S}{N_0}\right)^{1/2}\right] \tag{4.114}$$

In terms of the noise power in the Nyquist bandwidth P_N, the probability of error is

$$P_e = \frac{3}{2} Q\left[\frac{\pi}{4}\left(\frac{P_S}{P_N}\right)^{1/2}\right] \tag{4.115}$$

which when compared with the equivalent expression for binary transmission with ideal filters, Eq. (4.46), is poorer by a factor of $\pi/4$ or 2.1 db.

When channel distortion is present, the criteria of peak and rms eye closure can be applied to duobinary in a fashion similar to that previously described for L-level PAM. However, now the problems of proper definition and normalization become quite confusing for the following reason. Consider the received sample y_0:

$$y_0 = a_0 x_0 + a_{-1} x_1 + \sum_{n \neq 0, -1} a_n x_{-n} \tag{4.116}$$

If the pulse $x(t)$ is ideal then $x_0 = x_1 = 1$, and the other samples are zero. When $a_n = \pm d$, y_0 takes on only the three values $+2d$, 0, and $-2d$ in the absence of noise.

If the pulse $x(t)$ is distorted, the summation in Eq. (4.116) represents the intersymbol interference of the same type as has been handled previously, but the unbalance in x_0 and x_1 splits the first two terms into *four* (rather than just three) possible values $\pm(x_0 + x_1)d$ and $\pm(x_0 - x_1)d$. When the intersymbol interference is added to these four levels, the eye diagram of Fig. 4.26 results. The troublesome feature of this diagram, from a theoretical viewpoint, is that the probability distributions about the three desired received levels are not equivalent, and it is not at all obvious where the slicing (decision) thresholds should be placed. In fact, to minimize average probability of error these threshold placements will depend on the noise and intersymbol-interference distributions in a complicated fashion.

A practical receiver has no such difficulties, of course, since it has simply been instructed to slice at a level such as $d(x_0 + x_1)/2$.

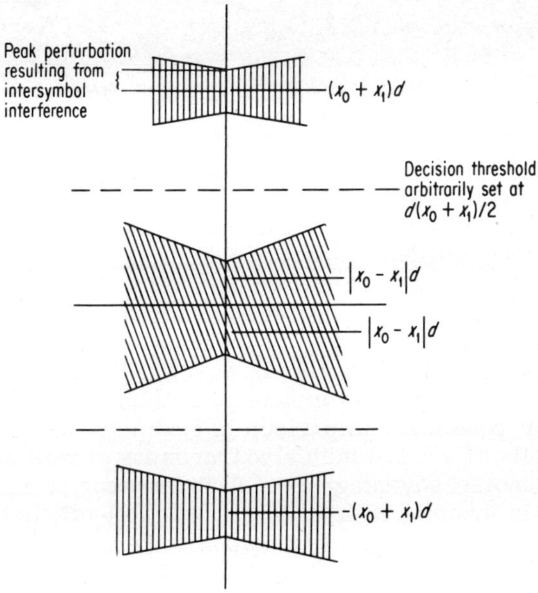

Fig. 4.26. Duobinary eye pattern.

Although there are other choices for this threshold (depending on the gain-control circuitry), this particular threshold will be employed here for definiteness. With this threshold the probabilities of error for the three received levels will be unequal, the zero level ($a_k = 1$) more likely to be in error than the outer levels ($a_k = 0$). Thus the eye closure could be defined in a number of ways. In probably the most consistent definition the peak eye closure would be based on the worst-case zero level. The eye closure would then be

$$\underset{\text{(zero level)}}{\text{Peak eye closure}} = \frac{|x_0 - x_1| + \sum_{n \neq 0,1} |x_n|}{(x_0 + x_1)/2} \tag{4.117}$$

A simpler expression which is practically little different from Eq. (4.117) results if the eye is based on the outer levels.

$$\underset{\text{(outer levels)}}{\text{Peak eye closure}} = \frac{\sum_{n \neq 0,1} |x_n|}{(x_0 + x_1)/2} \tag{4.118}$$

For the mean-square eye closures, the corresponding expressions are

$$\underset{\text{(zero level)}}{\text{Mean-square eye closure}} = \frac{(x_0 - x_1)^2 + \sum_{n \neq 0,1} x_n^{\,2}}{[(x_0 + x_1)/2]^2} \tag{4.119}$$

$$\begin{matrix}\text{Mean-square eye closure} \\ \text{(outer levels)}\end{matrix} = \frac{\sum_{n\neq 0,1} x_n{}^2}{[(x_0 + x_1)/2]^2} \tag{4.120}$$

These expressions are most easily calculated using Eq. (4.100) and the fact that $X_{eq}(\omega) = X(\omega)$ for duobinary since the system bandwidth is exactly equal to the Nyquist bandwidth.

The inevitable question of comparing duobinary and PAM occurs. The most difficult aspect of this as in many other comparisons is to find similar systems which can be fairly compared. The duobinary system sends only at rate $2W$ bps. It has been shown that this system is, under ideal conditions, 2.1 db inferior to binary PAM with abrupt roll-off $\alpha = 0$. However, with any realistic channel, the PAM eye closes completely; duobinary is thus preferred.

The other possible comparison is duobinary against four-level PAM with roll-off $\alpha = 1$, which also transmits at rate $2W$ bps. Here the inherent noise advantage is 4.9 db in favor of duobinary; this time the PAM system, with its slow $\alpha = 1$ roll-off, is less susceptible to channel distortion. It is helpful to bear in mind that in this comparison the pulse interval of the PAM is twice that of the duobinary system. Thus for a given range of pulse distortion, the PAM pulse tends to have fewer samples x_n disturbed. Which system performs better depends on whether or not the amount of difference in intersymbol-interference susceptibilities is enough to compensate for the 4.9-db noise advantage of duobinary. The timing-jitter sensitivities are also different and must be taken into consideration in any realistic appraisal of the two systems.

4.3.2 An Extension and a Generalization of the Duobinary Technique

Duobinary can easily be extended to a multilevel input. If the input symbols a_n are chosen from an L-level alphabet, the received samples y_k, under ideal conditions, can be all possible sums of the a_n values taken two at a time, which comprise $(2L-1)$ possible values. Thus when a four-level input is transmitted through the duobinary filter, a seven-level output results. For equally probable input symbols the output levels have a triangular probability distribution, i.e., the probability that y_k is $2md$ is

$$P(y_k = 2md) = \frac{L - |m|}{L^2} \tag{4.121}$$

for $m = \pm 1, \pm 2, \ldots, \pm(L-1)$.

A suitable precoding can be found such that the detection of symbols is accomplished without having to compensate for past decisions and consequently risking error propagation. In this precoding

$$b_n = a_n - b_{n-1} \pmod{L} \tag{4.122}$$

and the receiver decodes the received levels modulo L to recover the original symbols a_n.

These multilevel duobinary systems send data at a rate R of $2W \log_2 L$ bps in a bandwidth W Hz. Naturally they require a higher signal-to-noise ratio for equivalent error performance. When their noise performance is assessed, the familiar degrading factor of $3/(L^2-1)$ crops up in signal-to-noise ratio. The expression for probability of error is

$$P_e = 2\left(1 - \frac{1}{L^2}\right) Q\left[\frac{\pi}{4}\left(\frac{3}{L^2-1}\,\frac{P_S}{P_N}\right)^{1/2}\right] \tag{4.123}$$

A convenient graphical representation of the performance of duobinary is a chart similar to that already shown for PAM in Fig. 4.7 where the *required signal-to-noise ratio* (noise measured in the entire band where data energy is present) for a 10^{-5} error rate is shown for various systems as a function of the normalized rate of R/W bits per cycle. This chart is repeated in Fig. 4.27 with the duobinary information included. It can be seen that the duobinary systems fit neatly into the scheme of things by providing the missing links between the regions of availability of conventional PAM systems.

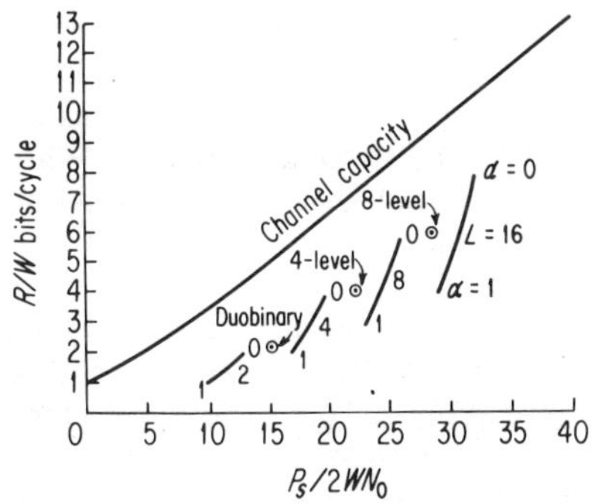

Fig. 4.27. Performance curves for duobinary and PAM. $P_s/2WN_o$ is the required signal-to-noise ratio for a 10^{-5} error rate.

When the channel is imperfect, the peak and mean-square eye closures are simply $(L-1)$ and $(L^2-1)/3$ times the eye closures computed for the $L = 2$ duobinary system. As L becomes relatively large, the probability of receiving the zero level $(1/L)$ is small enough that the anomaly involved in the eye-pattern definition is unimportant and the outer-level eye closures, Eqs. (4.118) and (4.120) should be used.

The duo-binary technique can also be generalized to what is known as *partial-response signaling.* The duobinary response is but one of a great variety of responses which have intersymbol interference in only a limited number of positions. The intersymbol interference can be removed at the receiver by proper compensation; sometimes a precoding scheme is available which eliminates any risk of error propagation. In any event, if a signal spectrum is desired which is limited to the ideal Nyquist band of $1/2T$ Hz, the impulse response may be expanded in terms of the samples

$$x(t) = \sum_{n=-k}^{k} x_n \frac{\sin \pi(t - nT)}{\pi(t - nT)} \tag{4.124}$$

and the frequency characteristic can be written in the form

$$X(\omega) = \begin{cases} T \displaystyle\sum_{n=-k}^{k} x_n e^{-i\omega nT} & |\omega| \leq \dfrac{\pi}{T} \\ 0 & |\omega| > \dfrac{\pi}{T} \end{cases} \tag{4.125}$$

If various combinations of integer values for the samples x_n are chosen, the frequency characteristic $X(\omega)$ can be made to take on "useful" shapes. A useful shape generally means that the characteristic itself would be continuous and perhaps one or more of its derivatives would also be continuous. The smoothness of $|X(\omega)|^2$ would help provide protection against the inevitable perturbations encountered in practice. Kretzmer[5] has tabulated and classified a number of these partial-response systems which appear to have useful properties. The responses—their frequency characteristics, their signal-to-noise ratio degradations over ideal binary, and their speed tolerances before peak eye closure is unity—are shown in Table 4-2. Notice particularly the bottom two frequency characteristics. These functions go to zero at zero frequency (in addition to at the Nyquist frequency) and thus become attractive for the many occasions in which frequencies near dc are prohibited or are severely attenuated. In Chap. 7 we discuss an example of this use in conjunction with single-sideband modulation.

4.4 Summary

In this chapter we have discussed the design and performance evaluation of simple pulse-amplitude-modulated transmission systems at baseband. The design criteria discussed were confined to elimination of intersymbol interference and proper division of the overall shaping characteristic between transmitting and receiving

Table 4-2. Partial-response Systems. (From Kretzmer[5])

Impulse response $x(t)$	Frequency characteristic $X(\omega)$	$X(\omega)$	No. of received levels	Speed tolerance, %	SNR degradation over ideal binary, db
	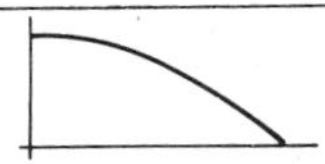	$2T \cos \frac{\omega T}{2}$	3	43	2.1
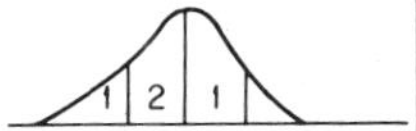	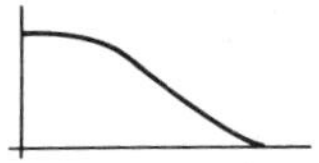	$4T \cos^2 \frac{\omega T}{2}$	5	40	6.0
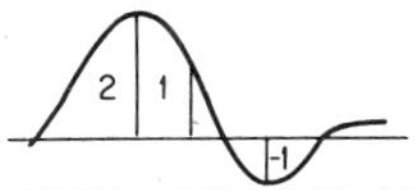		$T[2 + \cos \omega T - \cos 2\omega T] + iT[\sin \omega T - \sin 2\omega T]$	5	38	1.2 (7.2 precoded)
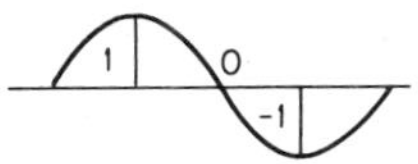	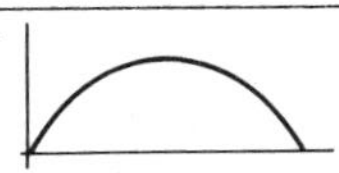	$2T \sin \omega T$	3	15	2.1
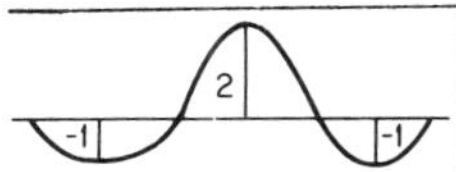	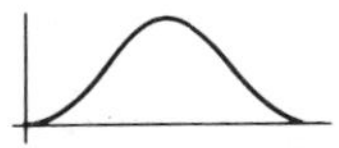	$4T \sin^2 \omega T$	5	8	6.0

filters for noise immunity. These considerations did not completely specify the filters, and we were left with the engineering considerations of ease of instrumentation and sensitivity to various perturbations.

The performance of the PAM system was first evaluated for ideal channels and added gaussian noise. The effect of an imperfect channel on the PAM system was then considered in some detail. Although the mathematical model of the PAM system seemed entirely straightforward, the difficulty in computing probability of error was nearly insurmountable. Thus several alternative formulations of the problem involving distortion criteria rather than error-rate calculations had to be studied. Example calculations of peak and rms distortion and approximate error rate were performed for particular channels to lend some insight into their interrelationship.

Throughout this chapter the viewpoint has been that the system was to be designed without knowledge of the channel characteristics. When the channel was not perfect, the actual characteristics degraded the system performance by an amount which we attempted to evaluate mathematically. In the following chapter we shall discuss the design of the PAM system when the channel characteristic and noise statistics are known exactly by the system designer. Thus the present chapter reflects current engineering practice, while the next chapter will present an attainable ideal. The actual approach of practical application to the ideal will become the subject of the subsequent chapter on equalization.

Notes

The version of the Nyquist criterion presented in this chapter is from Gibby and Smith.[6] For more information on baseband pulse transmission the reader is referred to the text by Bennett and Davy[4] or to the excellent article by Sunde.[7] This latter reference also includes much material on the transmission of pulses at passband. The theory of paired echoes, which is also covered by Sunde, is originally from Wheeler.[8] Finally, the concept of multilevel duobinary was given by Gerrish and Howson.[9]

5

Optimization of the Baseband System

In the previous chapter the designer was placed in the realistic position of having no detailed knowledge of the actual characteristics of the transmission channel. Therefore, the design procedure concentrated on achieving systems which were relatively immune to expected channel perturbations. Obviously such systems could not be expected to perform well over an ensemble of diverse channels. In fact, we found that a small amount of pulse distortion in the channel was sufficient to cause a large increase in error rate in data systems designed without taking channel characteristics into account. This error rate, a result of intersymbol interference, persisted in some cases even when the signal-to-noise ratio was made arbitrarily large.

Now we suppose that the designer knows, or is capable of learning, the exact transmission characteristics of the particular channel to be used. The reason we should be interested in what seems to be such an unrealistic situation are twofold. First, by evaluating the performance of optimized systems of this kind, we find bounds which can be used for comparison purposes with more practical systems so that we know when a system cannot be substantially improved by further design. Second, by studying the form of the optimum system within certain practically contrained frameworks, we can often invent systems which are capable of approaching optimum performance asymptotically as they learn the channel characteristics. These adaptive systems will be discussed in the following chapter, where their motivation and evaluation will depend on material from the present chapter.

The optimizations in this chapter will apply to the PAM system described in Chap. 4 rather than to the general, unconstrained communication problem for which bounds were obtained in Chap. 3. The questions to be considered are how well a PAM system can perform through a dispersive channel with gaussian noise and how should this system be designed, i.e.: What are the best transmitting and receiving filters? While deriving these results, we shall also consider, where possible, the broader optimization problems that are only partially within the PAM framework. For example, given a PAM transmitter, what is the best possible receiver and how much better does it perform than the best PAM receiver?

The chapter will be divided into three main sections: receiver optimization, transmitter optimization, and joint optimization of transmitter and receiver. Within each section several optimization criteria, such as probability of error or mean-square error, may be used.

5.1 Receiver Optimization

The transmitter will now be specified as a conventional PAM transmitter whose filtering $G_T(\omega)$ has been chosen previously such that the transmitted waveform is

$$s(t) = \sum_n a_n g(t - nT) \tag{5.1}$$

where

$$g(t) = \frac{1}{2\pi} \int_{-\infty}^{\infty} G_T(\omega) e^{i\omega t}\, d\omega \tag{5.2}$$

This waveform is passed through a known channel $C(\omega)$ and is corrupted by additive gaussian noise $\xi(t)$. The received waveform is designated $z(t)$ and can be written

$$z(t) = y(t) + \xi(t) \tag{5.3}$$

where

$$y(t) = \sum_n a_n h(t - nT)$$

and

$$h(t) = \frac{1}{2\pi} \int_{-\infty}^{\infty} G_T(\omega) C(\omega) e^{i\omega t}\, d\omega \tag{5.4}$$

The receiver must decide the coefficients a_n in some suitably defined fashion. Our main interest in this chapter will be in the conventional PAM receiver consisting only of a filter $G_R(\omega)$ and a threshold detector. In this instance we shall be concerned with optimizing $G_R(\omega)$ against the intersymbol interference and noise according to some criterion or combination of criteria discussed in the previous chapter. However, before delving into this problem, we shall mention briefly what is known about optimum reception of the waveform described by Eq. (5.3). We shall then be able to compare the performance of the optimized PAM receiver with theoretical bounds given by decision theoretic considerations. Fortunately the simple receiver often performs so well that little would be gained from attempting further improvement by more complicated signal processing.

5.1.1 Matched-filter Reception

A message must always be limited to some finite number, say N, of symbols. Since each symbol a_n has L possible values, only L^N messages are possible during this interval. Because of the action of the encoder, not all of these L^N messages are actually allowed to be transmitted, but rather only about L^{RN}, where R is the rate (efficiency) of the code. The best possible receiver for the PAM waveform [Eq. (5.3)] would examine the entire received signal, which is possibly of infinite duration, and then make a single choice as to which of the L^{RN} messages was transmitted. However, we have drawn a sharp line of demarcation between the duties of the decoder and the demodulator, which precludes the consideration of this extremely complex type of receiver. Instead, the mathematical model for the modem specifies *independent* input symbols a_n which are each to be decided at the receiver with no knowledge of the coding structure. Therefore, according to the receiver's ground rules, L^N equally likely messages are to be decided among on the basis of the entire received waveform.

Since the channel is presumed known, each possible message waveform $y_k(t)$, where $k = 1, \ldots, L^N$, can be computed through Eq. (5.4):

$$y_k(t) = \sum_{n=1}^{N} a_{kn} h(t - nT) \tag{5.5}$$

where a_{kn} is the nth symbol in the kth message. The receiver processing consists of a selection among L^N-known waveforms in the presence of gaussian noise. This is a familiar problem in statistical decision theory which is amply treated elsewhere;[1] accordingly only a brief and simplified sketch of the optimum procedure is given here.

The received signal is denoted

$$z(t) = y_k(t) + \xi(t) \tag{5.6}$$

and for simplicity the noise is taken to be gaussian noise of uniform spectral density. Both signal and noise are assumed to be bandlimited to W Hz. The receiver observes $z(t)$ over a finite interval of $M/2W$ sec, during which time the signal may be approximated using the expansion

$$z(t) = \sum_{m=1}^{M} z\left(\frac{m}{2W}\right) \frac{\sin \pi(2Wt - m)}{\pi(2Wt - m)} \tag{5.7}$$

Thus $z(t)$, $y_k(t)$, and $\xi(t)$ may be envisioned as vectors in an M-dimensional space. The *decision rule* consists in partitioning this space into L^N regions whereby the entire message is decided according to region into which the received vector z falls.

Let $\Phi_k(z)$ be a characteristic function that denotes in which region of the signal space z will be classified as message y_k; i.e., let

$$\Phi_k(z) = \begin{cases} 0 & [z: \text{message } y_k \text{ is declared}] \\ 1 & \text{elsewhere} \end{cases} \tag{5.8}$$

The probability of error may then be written

$$P_e = \frac{1}{L^N} \sum_{k=1}^{L^N} \int_Z p(z \mid y_k) \Phi_k(z)\, dz \tag{5.9}$$

Here all messages are equally likely, and the probability of error for each is simply the probability that z falls outside the proper decision region. The regions $\Phi_k(z)$ will now be defined so as to minimize P_e.

When the summation and integration are interchanged, it will be recognized that the Φ_k's can be selected so as to minimize each point of the integrand. Since the probability $p(z \mid y_k)$ is always positive, and one of the Φ_k's must always be unity while the rest are zero at each point of the space, it becomes obvious that P_e is minimized by the rule

$$\Phi_k(z) = \begin{cases} 0 & [z:\ p(z \mid y_k) \geq p(z \mid y_j) \quad \text{for all} \quad j \neq k] \\ 1 & [z:\ p(z \mid y_j) > p(z \mid y_k) \quad \text{for some} \quad j \neq k] \end{cases} \tag{5.10}$$

In other words the message y_k is decided when $p(z \mid y_k)$ is greater than $p(z \mid y_j)$ for all $j \neq k$. Therefore the receiver needs only to compute the conditional probabilities $p(z \mid y_k)$ and to select the largest. This is known as a *maximum likelihood receiver.*

Because the noise samples are gaussian and independent with variance σ^2, the conditional probabilities may be written simply in terms of the Nyquist samples z_m and y_{km}.

$$p(z \mid y_k) = \prod_{m=1}^{M} \frac{1}{\sigma\sqrt{2\pi}} e^{-(z_m - y_{km})^2/2\sigma^2} \tag{5.11}$$

Usually it is more convenient to make the decision-rule formulation in terms of the logarithm of the probabilities in Eq. (5.11). Inasmuch as the logarithm is a monotone function of the positive argument, this change does affect the decision rule. Thus we use

$$\log p(z \mid y_k) = -\frac{1}{2\sigma^2} \sum_{m=1}^{M} (z_m - y_{km})^2 - M \log(\sigma\sqrt{2\pi}) \tag{5.12}$$

The constant term does not enter into the decision; therefore the rule simply says to select the message y_k for which the distance in the M-dimensional euclidean space is least to the received vector. This rule will hardly come as a surprise to the reader since it is quite evident from basic considerations.

The test statistic in Eq. (5.12) can be further simplified by removing a z_m^2 term which is common to the distance calculation for each of the y_k. This results in the test statistics d_k, where

$$d_k = \sum_{m=1}^{M} z_m y_{km} - \frac{1}{2} \sum_{m=1}^{M} y_{km}^2 \tag{5.13}$$

and the message y_k is chosen for which $d_k \geq d_j$, all $j \neq k$. It can be shown that, under proper conditions, for an infinite observation interval $(M \to \infty)$ the sums in Eq. (5.13) go over to integrals giving[1]

$$d_k = \int_{-\infty}^{\infty} z(t) y_k(t)\, dt - \frac{1}{2} \int_{-\infty}^{\infty} y_k^2(t)\, dt \tag{5.14}$$

Thus the only values the receiver needs to compute are the cross correlations between the received signal $z(t)$ and each of the possible messages $y_k(t)$. Each cross correlation is adjusted by a bias which is half the message energy and the largest selected as corresponding to the most likely transmitted signal. These cross correlations are most easily implemented as *matched filters*; i.e., if the signal $z(t)$ is passed through a filter whose impulse response is $y_k(T - t)$, the output of the filter at time T is the desired cross correlation. A block diagram of the optimum receiver is shown in Fig. 5.1.

When the noise is not white, but has some arbitrary correlation function $R_{NN}(\tau)$, the same general considerations still hold, except

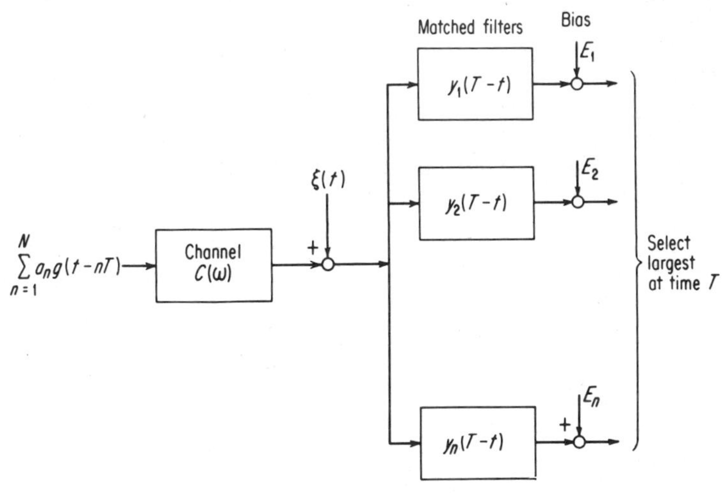

Fig. 5.1. Optimum receiver for a message of N symbols.

that expansion (5.7) becomes inappropriate because the noise samples are then correlated. However, an appropriate basis for expansion of the received signal can be found in which the noise samples are independent. In this expansion the basis functions $\psi_m(t)$ are a set of orthonormal functions satisfying the integral equation

$$\int_0^T R_{NN}(t - s)\psi_m(s)\, ds = \sigma_k^2 \psi_k(t) \qquad 0 < t < T \tag{5.15}$$

On the interval $[0, T]$ the noise $\xi(t)$ can be written

$$\xi(t) = \sum_m \xi_m \psi_m(t) \tag{5.16}$$

where

$$\xi_m = \int_0^T \xi(t)\psi_m^*(t)\, dt \tag{5.17}$$

and

$$\langle \xi_m \xi_n \rangle = \sigma_n^2 \delta_{mn} \tag{5.18}$$

It can be shown that the decision rule becomes: Decide s_k if

$$\sum_{m=1}^{M} \frac{(z_m - y_{km})^2}{\sigma_m^2} \le \sum_{m=1}^{M} \frac{(z_m - y_{jm})^2}{\sigma_m^2} \qquad \text{all } j \neq k \tag{5.19}$$

The coefficients z_m and y_{km} are relative to the expansion of $z(t)$ and $y_k(t)$ in terms of the basis functions $\{\psi_m\}$. The decision rule again is simply the selection of the closest allowed vector $\mathbf{y}_k$ according to the metric of the particular M-dimensional space. In this space the contours of equal noise probability are not spheres since the noise components ξ_m have different variances.

Decision rule (5.19) can also be given an integral formulation and implemented by means of matched filters. The form of the matched filter in this case has an interesting interpretation, which is illustrated in Fig. 5.2. In Fig. 5.2(a) the white-noise matched filter is shown for a signal whose Fourier transform is $G_T(\omega)$ which has been transmitted through a channel $C(\omega)$. The frequency characteristic of the matched filter described earlier (for $T = 0$) is the complex conjugate of the frequency characteristic of the received signal without noise, or $G_T^*(\omega)C^*(\omega)$. The interpretation of the corresponding colored-noise matched filter for noise of spectral density $N(\omega)$ is shown in Fig. 5.2(c). Here we have inserted a "whitening" filter $N^{-\frac{1}{2}}(\omega)$ at the input to the receiver. The noise output of this filter has a uniform spectral density.

Now if $N^{-\frac{1}{2}}(\omega)$ and $N^{\frac{1}{2}}(\omega)$ exist, no irreparable harm has been done to the received signal; furthermore, the proper matched filter from this point on is seen to correspond to the white noise example in Fig. 5.2(a), except that the signal transform is $G_T(\omega)C(\omega)N^{-\frac{1}{2}}(\omega)$. The composite matched filter for the colored noise is therefore $G_T(\omega)C(\omega)N^{-1}(\omega)$ as shown in Fig. 52(b).

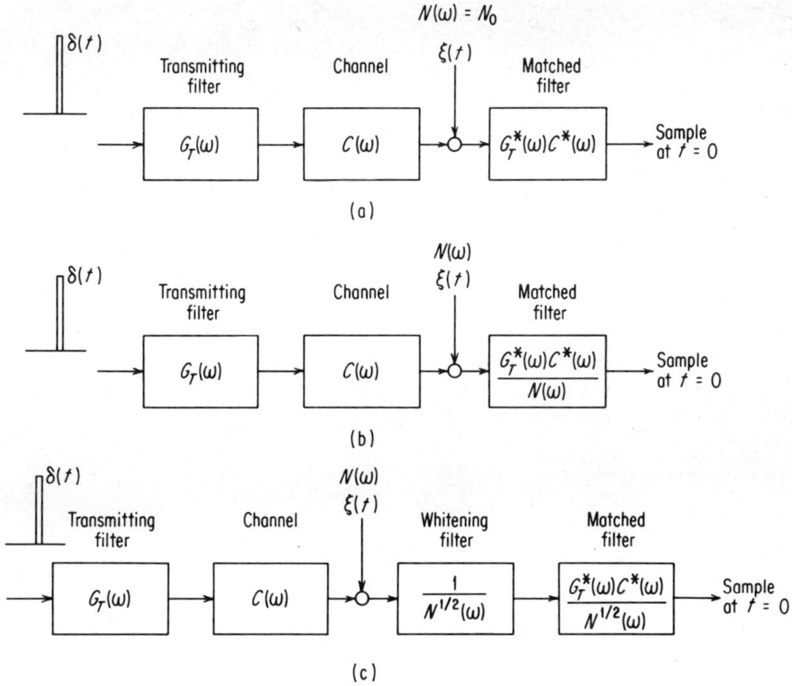

Fig. 5.2. Matched filters, (a) White noise, (b) Colored noise, (c) Interpretation of colored-noise filter.

Optimum reception for the ideal channel. Suppose now that the channel is perfect, that is, $C(\omega) = 1$. In Chap. 4 we considered the design of transmitting and receiving filters to eliminate intersymbol interference and concluded that an infinite class of possible characteristics could be used for the overall shaping. However, to minimize the probability of error for the PAM system, we found that the shaping should be evenly distributed between transmitter and receiver; i.e., if a real composite characteristic is $X(\omega)$, $G_T(\omega) = G_R(\omega) = X^{\frac{1}{2}}(\omega)$. The performance of this system was then calculated for $X(\omega)$ equal to a raised cosine with roll-off α.

We shall now show that for white gaussian noise and an ideal channel the simple PAM receiver consisting of a filter $X^{\frac{1}{2}}(\omega)$ and a sampler and threshold detector is optimum in the decision theoretic sense. This fact is sometimes overlooked in a search for more complicated and efficient receivers. Also, the optimality of the simple receiver may be doubted in view of the curve in Fig. 4.8, which shows an effective loss from theoretical capacity for the PAM system as a function of probability of error. This loss in fact results from the inefficient selection of the PAM waveforms (the signal space has been too coarsely quantized), not to any fault in the receiver design.

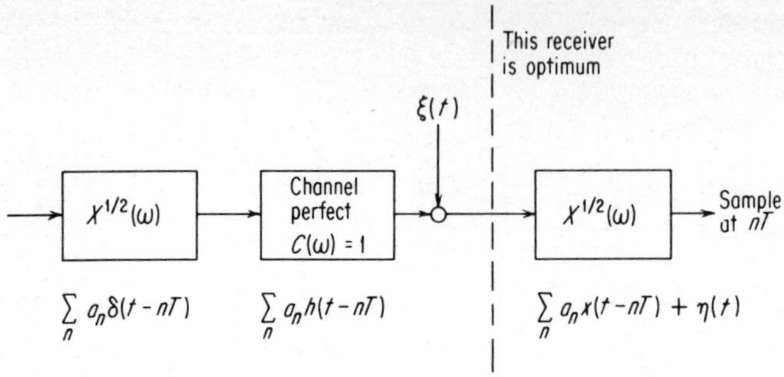

Fig. 5.3. PAM system.

For the PAM system shown in Fig. 5.3 the received waveform is

$$z(t) = \sum_{n=1}^{N} a_n h(t - nT) + \xi(t) \tag{5.20}$$

where
$$h(t) = \frac{1}{2\pi} \int_{-\infty}^{\infty} X^{\frac{1}{2}}(\omega) e^{i\omega t} \, d\omega \tag{5.21}$$

The optimum receiver has been shown to choose that data sequence $\{b_{kn}\}$ such that the corresponding test statistic d_k is maximized. We choose

$$y_k(t) = \sum_{n=1}^{N} b_{kn} h(t - nT) \tag{5.22}$$

where $k = 1, \ldots, L^N$. If we use these particular expressions for $y_k(t)$ and $z(t)$ in determining the d_k from Eq. (5.14), we arrive at

$$\begin{aligned} d_k = & \sum_{n-1}^{N} \sum_{m=1}^{N} b_{kn} a_m \int_{-\infty}^{\infty} h(t - nT) h(t - mT) \, dt \\ & + \sum_{n=1}^{N} b_{kn} \int_{-\infty}^{\infty} \xi(t) h(t - nT) \, dt \\ & - \frac{1}{2} \sum_{n=1}^{N} \sum_{m=1}^{N} b_{kn} b_{km} \int_{-\infty}^{\infty} h(t - nT) h(t - mT) \, dt \end{aligned} \tag{5.23}$$

The salient points of this equation are that $h(t - nT)$ and $h(t - mT)$ are orthonormal on $(-\infty,\infty)$ and that

$$\eta_n = \int_{-\infty}^{\infty} \xi(t) h(t - nT) \, dt \tag{5.24}$$

is the noise output of the receiving filter in Fig. 5.3 at time nT. The former point comes about because the frequency characteristic of $h(t)$ was chosen to be the *square root* of a Nyquist characteristic. Therefore, from the Parseval relationship,

$$\int_{-\infty}^{\infty} h(t - nT)h(t - mT)\,dt = \frac{1}{2\pi}\int_{-\infty}^{\infty} X(\omega)e^{i\omega(n-m)T}\,d\omega \tag{5.25}$$

and $X(\omega)$ is a Nyquist shape such that its response has the property $x(nT) = \delta_{n0}$. The latter point, Eq. (5.24), arises since $h(t) = h(-t)$ as a consequence of $X^{\frac{1}{2}}(\omega)$ being real.

When these observations are used in Eq. (5.23), we arrive at

$$d_k = \sum_{n=1}^{N}\left\{b_{kn}a_n + b_{kn}\eta_n - \frac{1}{2}b_{kn}^{\;2}\right\} \tag{5.26}$$

which can be put in the form

$$d_k = \frac{1}{2}\sum_{n=1}^{N}\left\{(a_n + \eta_n)^2 - [b_{kn} - (a_n + \eta_n)]^2\right\} \tag{5.27}$$

To choose the sequence b_{kn} which maximizes d_k it is only necessary to choose *each* coefficient b_n such that it is as close as possible to $(a_n + \eta_n)$, which is simply the output voltage of the receiving filter at time nT. Therefore, common symbol-by-symbol threshold detection following the filter $X^{\frac{1}{2}}(\omega)$ is as good as one can possibly do with this particular PAM transmitted sequence. Note, however, that this would not have been true had the noise been nonwhite or the shaping $X(\omega)$ not a real Nyquist characteristic.

A matched-filter bound for the nonideal channel. If a *single* pulse $a_0 g(t)$ were to be transmitted through a channel $C(\omega)$, ideal reception would call for a matched filter $G_T^*(\omega)C^*(\omega)$ at the receiver, followed by a threshold detector. When a sequence of pulses which overlap in time is transmitted, the simple receiver is no longer optimum unless, as in the ideal channel situation just described, the individual pulses do not interfere with each other in the process of matched-filter detection. When the channel is imperfect, there will be intersymbol interference at the output of a matched filter and the system is obviously no longer optimum.

It is possible to arrive at a quite useful bound on probability of error by assuming that one can detect *each* pulse $a_n g(t - nT)$ optimally by a matched filter without running into interference from other symbols. This would be equivalent to transmitting each of the N pulses through a separate, identical channel and using N perfect, matched filters. Clearly, it is impossible for an ideal receiver to perform better than this fictional system, but it will be interesting to see later on that even relatively simply instrumented receivers can perform nearly as well under certain conditions.

The probability of error for the fictional parallel channel system illustrated in Fig. 5.4 can be easily arrived at through the techniques described in Chap. 4. We find for an L-level PAM system

$$P_e \geq 2\left(1 - \frac{1}{L}\right) Q\left[\left(\frac{3}{L^2 - 1}\, \beta\, \frac{P_S}{P_N}\right)^{\frac{1}{2}}\right] \tag{5.28}$$

where

$$\beta = \frac{\int_{-\infty}^{\infty} |C(\omega)|^2\, |G_T(\omega)|^2\, d\omega}{\int_{-\infty}^{\infty} |G_T(\omega)|^2\, d\omega} \tag{5.29}$$

represents a signal-to-noise degradation (or enhancement) caused by the channel $C(\omega)$.

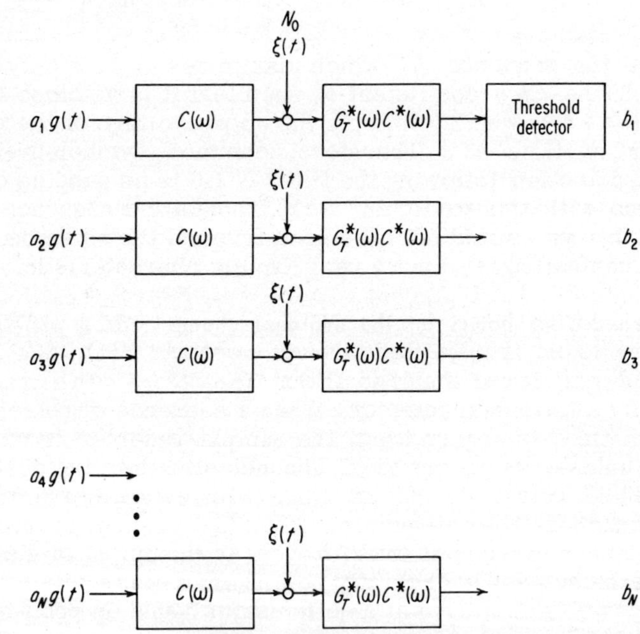

Fig. 5.4. Matched-filter bound on PAM reception.

Iterative, sequential detection. We have seen that the optimum detection of an N-symbol message requires a *joint* decision concerning the data sequence $\{a_n\}$, where $n = 1, \ldots, N$, after processing the entire received signal. Since L^N messages are possible, the size of the optimum receiver grows exponentially with the message length N. In practice such a receiver is an impossibility. A basic requirement any practical receiver must meet is that the number of computations required to process a message of N symbols must grow only linearly with N. A signal-detection system based on an iterative procedure which meets this requirement (and under certain conditions is optimal), has been proposed by Chang and Hancock.[2]

In describing briefly this procedure we shall suppose that the symbol a_n causes intersymbol interference only with the $(K-1)$ succeeding symbols $a_{n+1}, \ldots, a_{n+K-1}$. The symbols are then detected in blocks of K symbols labeled A_n representing the K-symbol block $a_{n-K+1}, a_{n-K+2}, \ldots, a_n$. Thus each block has L^K-possible values from which a decision must be made. As an example, the following sequence has been grouped into blocks for the case of $K = 3$, where intersymbol interference from any symbol extends into the two succeeding symbols.

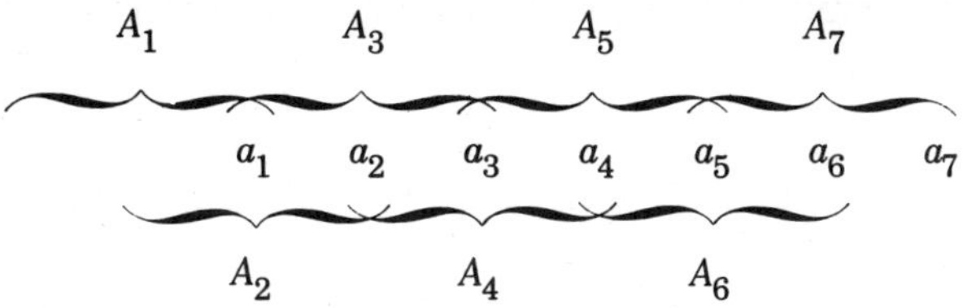

Whereas optimum detection would require a joint decision on all symbols, $a_1 \ldots, a_N$, this procedure is a sequential detection system in that decisions are made on a block-by-block basis. Each block is classified independently of other decisions on the basis of maximum a posteriori probability. For convenience the L^K-possible values each clock can assume will be denoted by the integers $1, \ldots, L^K$. Then the decision rule is to choose the value j for block A_k for which $P(A_k = j \mid z)$ is rendered greatest. It is easily shown that this procedure will lead to the same classification of all N symbols as the optimum procedure, provided that, given the received vector z, the probability of having transmitted the "true" message (the data sequence actually transmitted) is greater than 1/2. This provision is more severe than requiring that the optimum receiver would have classified the entire message $a_1, \ldots, a_N$ correctly, given the received waveform z. Thus it could not possibly be satisfied unless the probability of error and message length were both small. Nevertheless the procedure is quite a reasonable approach to the problem of limiting the growth of computation with message length.

The advantages of blocking the data into the groups of K symbols will not be apparent to the reader without his studying the original paper by Chang and Hancock.[2] Basically this blocking enables the probabilities $p(A_k = j \mid z)$ to be computed iteratively much in the manner of probabilities of a Markov chain. In fact it is quite evident that the input blocks A_n form a Markov chain since

$$P(A_n | A_{n-1}, A_{n-2}, \ldots, A_1) = P(A_n | A_{n-1}) \tag{5.30}$$

as can be seen from the previous example of the blocking for the case $K = 3$. The detection problem is then exactly that of observing a Markov source through a noisy channel.

The probability $P(A_k = j | \mathbf{z})$ can be formulated as the product of two probabilities—one of which involves all of the past symbols and the present symbol a_k and the other of which involves the "future" symbols a_{k+1} through a_N. This can be written

$$P(A_k = j | \mathbf{z}) = P(A_k = j | \mathbf{v}_k) P(A_k = j | \mathbf{u}_N) \tag{5.31}$$

where $\mathbf{v}_k$ indicates the received waveform from the beginning through the symbol interval corresponding to a_k, and $\mathbf{u}_N$ denotes the received waveform from this time on. It can be shown further that each of these component probabilities can be computed in an iterative fashion. For example, the L^K a posteriori probabilities $P(A_k | \mathbf{v}_k)$, where $j = 1, \ldots, L^K$, can be found in the form of a column vector by a matrix operation involving the product of the previous a posteriori probability vector, the transition matrix for the input Markov process, and a diagonal matrix representing the a posteriori probabilities for the current block. In other words, the probabilities $P(A_k = j | \mathbf{v}_k)$ are computed as if they had arisen from a Markov chain with known a priori transition matrix $P_{ij} = P(A_k = i | A_{k-1} = j)$, except they need to be modified by the current a posteriori probabilities for the block in question.

In all the computations involved in this procedure L^K sets of numbers are involved. Furthermore, each current a posteriori probability, of which there are L^K, requires about L^K calculations itself. Therefore, in many applications the cost of implementation would be prohibitive.

5.1.2 Optimization of the Receiving Filter

The optimization of the conventional PAM receiver will now be considered. In this case the form of the receiver is constrained to consist of a filter $G_R(\omega)$, followed by a sampler and threshold detector. We seek the optimum form of the receiving filter G_R for a given transmitting filter G_T. The development will parallel the discussion of intersymbol-interference effects of Chap. 4 in that we shall first try to minimize the probability of error for a finite data sequence. As might be expected, a general form for $G_R(\omega)$ can be found; in any specific case, however, the exact result is nearly impossible to compute. Therefore we shall turn to easier criteria, such as eye opening and mean-square error for minimization.

Minimization of probability of error. For simplicity it is assumed that a finite message of $(2N+1)$ symbols is to be passed through the transmitting filter $G_T(\omega)$, the channel $C(\omega)$, and the receiving filter $G_R(\omega)$. White, gaussian noise of spectral density N_0 watts per Hz is added to the signal in the channel. The receiving filter $G_R(\omega)$ is to be determined so as to minimize the probability of error in detecting symbol a_0, the center symbol of the data sequence.

At the sampling time the received voltage can be written

$$y_0 = \sum_{n=-N}^{N} a_n x_{-n} + \eta_0 \tag{5.32}$$

where
$$x_n = \frac{1}{2\pi} \int_{-\infty}^{\infty} G_T(\omega) C(\omega) G_R(\omega) e^{in\omega T} \, d\omega \tag{5.33}$$

An error occurs in detecting a_0 whenever the sum of intersymbol interference and noise exceeds the distance dx_0 to the nearest threshold; that is,

$$P_e = P\left[\left|\sum_{n=-N}^{N}{}' a_n x_{-n} + \eta_0\right| > dx_0\right] \tag{5.34}$$

To find P_e, the conditional probability for each of the L^{2N} possible intersymbol interference values can be evaluated and weighted by the probability of occurrence for that particular data sequence. It was observed by Aaron and Tufts[3] that given the $(2N+1)$ values x_n, the conditional probability of error for each possible data sequence is minimized by minimizing the noise variance $\sigma^2 = \langle \eta_0{}^2 \rangle$, provided the eye is not completely closed in the absence of noise. The provision of an open eye is necessary since otherwise some particular data sequence would always cause an error in the absence of noise and increasing σ^2 would then *decrease* the probability of error for this sequence. For all sequences which do not cause errors in the absence of noise, the conditional probability of error is a monotonically increasing function of σ^2. We therefore restrict consideration to only those values of $\{x_n\}$, where $n = -N, \ldots, N$ for which the eye is open in the absence of noise.

Minimizing P_e is then accomplished by the following procedure:

1. Constrain the $(2N+1)$ values of x_n, where $n = -N, \ldots, +N$.
2. Subject to this constraint, select $G_R(\omega)$ so as to minimize the noise variance σ^2, where

$$\sigma^2 = \frac{N_0}{2\pi} \int_{-\infty}^{\infty} |G_R(\omega)|^2 \, d\omega \tag{5.35}$$

3. Repeat steps 1 and 2 while varying the response sequence $\{x_n\}$ over all possible values in its $(2N+1)$-dimensional space until a minimum probability of error is obtained for the data system.

In approaching the minimization problem outlines in steps 1 and 2 and many of the other optimization questions of a similar nature, it is perhaps most convenient to use the same type of development employed in deriving the Nyquist criterion in terms of the equivalent Nyquist bandwidth channel. This procedure, initiated by Gibby and Smith,[4] often greatly simplifies the analysis and optimization of synchronous, baseband data-transmission systems.

The equivalent channel is formed by superimposing selections of the original characteristic of width $2\pi/T$ on the central Nyquist band $[-\pi/T,\pi/T]$. In the case we are considering, the overall response $X(\omega)$ is the product $G_T(\omega)C(\omega)G_R(\omega)$ and the equivalent channel $X_{\text{eq}}(\omega)$ will be written

$$X_{\text{eq}}(\omega) = \begin{cases} \sum_k X^{(k)}(\omega) & |\omega| \le \dfrac{\pi}{T} \\ 0 & |\omega| > \dfrac{\pi}{T} \end{cases} \tag{5.36}$$

where we set

$$X^{(k)}(\omega) = X\left(\omega + \frac{2k\pi}{T}\right) \tag{5.37}$$

It will be recalled that although the equivalent channel is band-limited to the Nyquist interval, it retains the same sample values as the original channel response. Thus the $(2N+1)$ constraint equations are written

$$x_n = \frac{1}{2\pi}\int_{-\pi/T}^{\pi/T} \sum_k X^{(k)}(\omega)\, e^{i\omega nT}\, d\omega \tag{5.38}$$

where $n = -N, \ldots, +N$. The noise variance, Eq. (5.35) may be put into the Nyquist band similarly.

$$\sigma^2 = \frac{N_0}{2\pi}\int_{-\pi/T}^{\pi/T} \sum_k |G_R{}^{(k)}(\omega)|^2\, d\omega \tag{5.39}$$

Now the optimum amplitude and phase characteristics for each of the segments $G_R{}^{(k)}(\omega)$ may be found through a straightforward application of the calculus of variations. Adding the contraint equations with Lagrange multipliers gives the functional V where

$$V = \frac{1}{2\pi}\int_{-\pi/T}^{\pi/T} \sum_k \Bigg\{ N_0 |G_R{}^{(k)}(\omega)|^2 + \sum_{n=-N}^{N} \lambda_n G_T{}^{(k)}(\omega)\, C^{(k)}(\omega)\, G_R{}^{(k)}(\omega)\, e^{i\omega nT} \Bigg\}\, d\omega \tag{5.40}$$

whose stationary point may easily be shown to be given by the segments

$$G_R^{(k)}(\omega) = [G_T^{(k)}(\omega)C^{(k)}(\omega)]^* \sum_{n=-N}^{N} \frac{-\lambda_n}{2N_0} e^{i\omega nT} \tag{5.41}$$

Equation (5.41) is typical of all optimizations of this type and should be closely examined. The terms in brackets indicate a filter which is *matched* to the received pulse. Remember that this type of filtering is optimum if the channel is ideal; however, since the channel may not be ideal, this filter must be cascaded with another filter indicated by the summation. The most important characteristic of this filter is its *independence of the segment index k.* Therefore, the response of this filter is repetitive in the frequency domain with period $2\pi/T$ rad.

The optimum linear receiver is illustrated in Fig. 5.5(a). The receiver filtering is decomposed into the aforementioned two filters: the matched filter $G_T^*(\omega)C^*(\omega)$ and the periodic filter $T(\omega)$ of the form

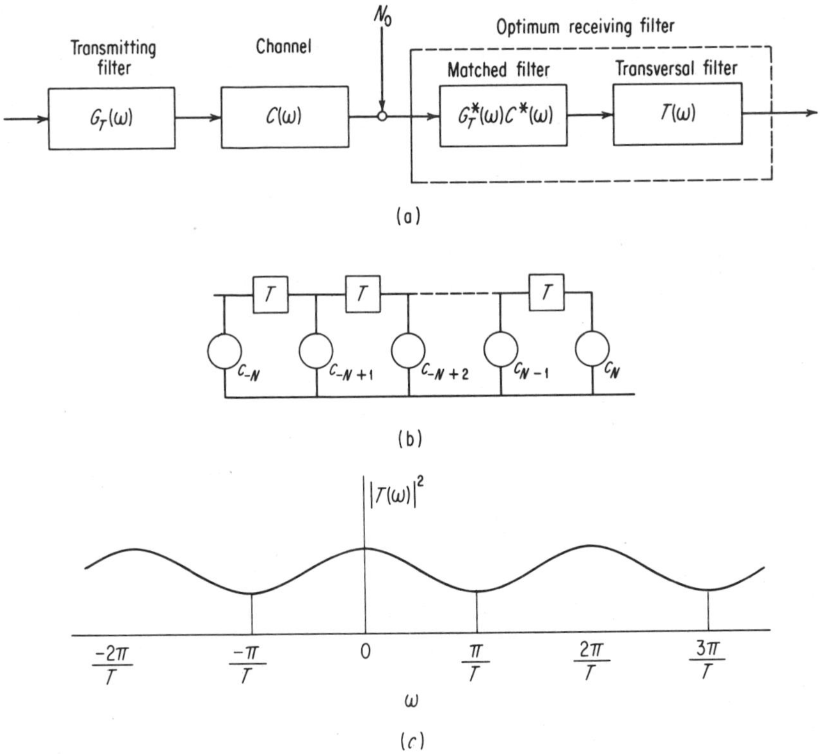

Fig. 5.5. Composition of the optimum linear receiver. (a) Optimum linear receiver. (b) Transversal filter $T(\omega)$. (c) Frequency response of transversal filter.

$$T(\omega) = \sum_{n=-N}^{N} c_n e^{-i\omega nT} \tag{5.42}$$

which will be called a *transversal filter* since this is the nomenclature for its implementation shown in Fig. 5.5(b). The transversal filter consists of a delay line tapped at T-sec intervals. Each tap along the delay line is connected through an amplifier (and possibly an inverter to enable negative gains) to a summing bus which provides the output. The frequency response for the transversal filter is repetitive with points of symmetry in real and imaginary components located at $2n\pi/T$ rad as shown in the example response of Fig. 5.5(c).

Now that the form of the optimum linear receiver has been found, it remains to find the $(2N+1)$ values of the impulse response constraints x_n which lead to a system with minimum probability of error. Equivalently, we must find the $(2N+1)$ settings of the transversal-filter tap gains c_n, where $n = -N, \ldots, +N$ that minimize probability of error in the system of Fig. 5.5(a). Since the overall pulse response at the output of the matched filter is $|G_T(\omega)C(\omega)|^2$, the time-domain response must be symmetrical. Clearly no improvement could be achieved in the transversal filter by introducing any asymmetry in the tap-gain settings; therefore $c_n = c_{-n}$. Also, the overall gain factor $T(\omega)$ is unimportant; thus c_0 can be arbitrarily selected at, say, unity. The minimization must therefore be carried out over the N variables c_n, where $n = 1, \ldots, N$.

In principle a minimum vector of N tap settings can be shown to exist. However, the minimization requires the solution of N simultaneous nonlinear equations which are arrived at by writing the expression for probability of error and differentiating with respect to the N free tap gains. Since $(2N+1)$ is the message length, N is almost always a very large number. Once again we have arrived at an intuitively interesting solution but one which we cannot in practice compute for any specific case. Therefore it is necessary to turn to the alternative criteria of peak and mean-square eye closure.

Here the peak eye opening in the absence of noise does not in itself serve as a criterion for minimization since there is usually an infinite set of possible filters $G_R(\omega)$ for which intersymbol interference can be completely eliminated making peak eye closure zero. To eliminate intersymbol interference it is necessary and sufficient that the equivalent channel frequency response be a constant over the Nyquist band. Thus any receiver filter $G_R(\omega)$ for which

$$\sum_k G_T^{(k)}(\omega) C^{(k)}(\omega) G_R^{(k)}(\omega) = \text{const} \qquad |\omega| \leq \frac{\pi}{T} \tag{5.43}$$

will result in a completely open eye pattern. In particular, the response

$$G_R(\omega) = \begin{cases} \dfrac{\text{const}}{G_T(\omega)C(\omega)} & |\omega| \le \dfrac{\pi}{T} \\ 0 & |\omega| > \dfrac{\pi}{T} \end{cases} \tag{5.44}$$

is the only frequency response limited to the Nyquist band which will satisfy Eq. (5.43); however, allowing more bandwidth for $G_R(\omega)$ results in the infinite class of possible characteristics.

Obviously $G_R(\omega)$ cannot be allowed to have infinite gain; therefore it is necessary to require that $|G_T(\omega)C(\omega)|$ not be zero on the Nyquist interval $[-\pi/T,\pi/T]$ in order for $G_R(\omega)$ to exist in Eq. (5.44). More generally, a necessary and sufficient condition for the existence of a filter $G_R(\omega)$ which eliminates intersymbol interference is that

$$\sum_k |G_T^{(k)}(\omega)C^{(k)}(\omega)|^2 \neq 0 \quad |\omega| \le \frac{\pi}{T} \tag{5.45}$$

This simply states that there must be energy transmission at every frequency in the Nyquist band or at corresponding points in the translated Nyquist bands $(2k-1)\pi/T \le \omega \le (2k+1)\pi/T$. It is now of interest to find the particular filter $G_R(\omega)$ of the class which eliminates intersymbol interference leading to minimum probability of error.

Minimization of probability of error in the absence of intersymbol interference. If intersymbol interference is constrained to be zero, the probability of error is minimized by simply minimizing the noise variance σ^2. Thus we minimize

$$\sigma^2 = \frac{N_0}{2\pi} \int_{-\pi/T}^{\pi/T} \sum_k |G_R^{(k)}(\omega)|^2 \, d\omega \tag{5.46}$$

subject to the constraint

$$\sum_k G_T^{(k)}(\omega)C^{(k)}(\omega)G_R^{(k)}(\omega) = Tx_0 \quad |\omega| \le \frac{\pi}{T} \tag{5.47}$$

The value of the central pulse response sample x_0 can be arbitrarily assigned without affecting the performance.

The function $G_R^{(k)}(\omega)$ in Eq. (5.46) can be minimized simply point-by-point for each value of ω since constraint (5.47) is not in integral form. Thus for each value of ω, $|\omega| \le \pi/T$, we look for the stationary point of

$$V = \sum_k [|G_R^{(k)}(\omega)|^2 + \lambda(\omega)G_T^{(k)}(\omega)C^{(k)}(\omega)G_R^{(k)}(\omega)] \tag{5.48}$$

This gives

$$G_R^{(k)}(\omega) = \frac{-\lambda(\omega)}{2}[G_T^{(k)}(\omega)C^{(k)}(\omega)]^* \tag{5.49}$$

and solving Eqs. (5.47) and (5.49) simultaneously to eliminate the constraint yields

$$G_R^{(k)}(\omega) = [G_T^{(k)}(\omega)C^{(k)}(\omega)]^* \frac{Tx_0}{\sum_k |G_T^{(k)}(\omega)C^{(k)}(\omega)|^2} \tag{5.50}$$

Thus the optimum receiver filter which minimizes probability of error while eliminating intersymbol interference has the same interpretation as the previous filter shown in Fig. 5.5. It consists of a filter matched to the receiver signal followed by a transversal filter, i.e.,

$$G_R(\omega) = [G_T^*(\omega)C^*(\omega)]\,T(\omega) \tag{5.51}$$

This time the transversal filter is infinite in extent since we are no longer dealing with the finite input message of the previous section. Because the transversal filter has a response which is repetitive in the frequence domain with a period equal to the Nyquist interval, only one setting of the filter can eliminate intersymbol interference. This setting can be obtained from

$$T(\omega) = \sum_n c_n e^{-in\omega T} = \frac{Tx_0}{\sum_k |G_T^{(k)}(\omega)C^{(k)}(\omega)|^2} \tag{5.52}$$

In other words the transversal-filter settings are the coefficients in a Fourier expansion of the inverse of the equivalent channel up to its input. In fact, since $T(\omega + 2k\pi/T) = T(\omega)$, we have the useful relationship for the transversal filter

$$[T(\omega)X(\omega)]_{\text{eq}} = X_{\text{eq}}(\omega)\,T(\omega) \tag{5.53}$$

Since to eliminate intersymbol interference the equivalent channel must be the constant Tx_0, Eq. (5.52) can be derived by inspection of Fig. 5.5(a).

The probability of error for this system can be easily shown to be

$$P_e = 2\left(1 - \frac{1}{L}\right) Q\left[\left(\frac{3}{L^2 - 1}\,\beta_1 \frac{P_S}{P_N}\right)^{1/2}\right] \tag{5.54}$$

where, as usual, P_N is the noise power in the Nyquist bandwidth and the signal-to-noise attenuation factor is

$$\beta_1 = \left\{\frac{T}{2\pi}\int_{-\infty}^{\infty} |G_T(\omega)|^2\, d\omega \; \frac{T}{2\pi}\int_{-\pi/T}^{\pi/T} \frac{1}{\sum_k |G_T^{(k)}(\omega)\, C^{(k)}(\omega)|^2}\, d\omega\right\}^{-1} \tag{5.55}$$

The form of the attenuation factor is somewhat inconvenient but is easily evaluated in any specific case. Of particular interest is the ratio β/β_1 [see Eq. (5.29)], which gives the degradation in signal-to-noise ratio over the matched-filter bound. This degradation is the price which must be paid for the use of a PAM receiver and elimination of intersymbol interference. In Sec. 5.4 of this chapter we shall compute this ratio for a particular numerical example.

Minimization of mean-square error. The last possibility for optimization of $G_R(\omega)$ which we shall examine here is the minimization of mean-square error resulting from both the intersymbol interference and noise. Some caution must be entered here in that although the intersymbol interference and noise are dissimilar statistical processes, the minimization of total mean-square error attaches equal significance to each process. In a practical sense, however, the solution arrived at here is nearly the same as in the previous two cases if the signal-to-noise ratio is relatively large. If the signal-to-noise ratio is small there may be merit in seeking a simultaneous minimization of noise and intersymbol interference in spite of the philosophical difficulties.

This case, as well as the previous two cases, was originally treated in the literature by Tufts,[5] who assumed a finite message of N symbols. Here the optimum $G_R(\omega)$ depends on the symbol being detected; hence the optimum linear receiver is time-varying. However, most actual messages are so long that the assumption of an infinite message length is quite good. This assumption leads to a simpler analytical problem and to an optimum filter which is time-invariant.

The sum of mean-square noise and intersymbol interference for an L-level PAM system has previously been shown to be

$$\sigma^2 = d^2 x_0^2 \frac{L^2 - 1}{3} \mathcal{E}^2 + \sigma_N^2$$

where σ_N^2 is the noise variance and

$$\mathcal{E}^2 = \frac{1}{2\pi x_0^2 T}\int_{-\pi/T}^{\pi/T} \left|\sum_k G_T^{(k)}(\omega)\, C^{(k)}(\omega)\, G_R^{(k)}(\omega)\right|^2 d\omega - 1 \tag{5.56}$$

To effect the minimization the pulse height x_0 is constrained using

$$x_0 = \frac{1}{2\pi} \int_{-\pi/T}^{\pi/T} \sum_k G_T^{(k)}(\omega) C^{(k)}(\omega) G_R^{(k)}(\omega)\, d\omega \tag{5.57}$$

and the usual expression

$$\sigma_N^2 = \frac{N_0}{2\pi} \int_{-\pi/T}^{\pi/T} \sum_k |G_R^{(k)}(\omega)|^2\, d\omega \tag{5.58}$$

is used for the noise variance. Smith[6] has shown that the optimum filter for this situation is given by

$$G_R^{(k)}(\omega) = \frac{\lambda [G_T^{(k)}(\omega) C^{(k)}(\omega)]^*}{1 + \dfrac{d^2(L^2 - 1)}{3N_0} \sum_k |G_T^{(k)}(\omega) C^{(k)}(\omega)|^2} \tag{5.59}$$

The constant λ is adjusted to make the output pulse the desired height x_0. The explicit solution for λ is a little lengthy and will not be given here.

Notice again that the optimum filter $G_R(\omega)$ is a matched filter in tandem with a transversal filter, i.e., a filter whose frequency response is repetitive at Nyquist intervals. When the noise density N_0 approaches zero, this filter approaches the previous result for minimizing P_e while completely removing intersymbol interference. On the other hand, for large N_0, the filter approaches a matched filter which would be optimum in the case of no intersymbol interference. In other words we generally have two filters: a matched filter and a transversal filter. The matched filter might be interpreted as treating the noise, while the transversal filter corrects intersymbol interference. When the signal-to-noise ratio is low, the principal correction must come from the matched filter; for high signal-to-noise ratios intersymbol interference must be almost completely suppressed and the principal correction effected by the transversal filter.

5.2 Transmitter Optimization

The problem of optimizing the transmitting filter $G_T(\omega)$ for a given receiving filter is treated identically to the receiver optimization problems discussed in Sec. 5.1.2. The results are nearly identical also, each consisting of a filter matched to the channel—receiving-filter combination followed by a transversal filter. After summarizing the principal results of this nature, we shall consider the selection of transmitted signals to accomplish certain special purposes, such as causing a time-limited output pulse or accomplishing maximum energy transfer in a given time-bandwidth region.

5.2.1 Optimization of the Transmitting Filter

For a fixed channel $C(\omega)$ and receiving filter $G_R(\omega)$ the transmitting filter $G_T(\omega)$ may be selected to eliminate intersymbol interference provided the condition

$$\sum_k |C^{(k)}(\omega) G_R^{(k)}(\omega)|^2 \neq 0 \qquad |\omega| \leq \frac{\pi}{T} \tag{5.60}$$

is satisfied. Of this class, the filter $G_T(\omega)$ which minimizes probability of error subject to a constraint on average signal power can be shown to be given by[6]

$$G_T(\omega) = [C(\omega) G_R(\omega)]^* \frac{x_0 T}{\lambda_2 \sum_k |C^{(k)}(\omega) G_R^{(k)}(\omega)|^2} \tag{5.61}$$

which is exactly equivalent to expression (5.50).

To minimize mean-square error from both noise and intersymbol interference for a given average transmitted power the following expression can be used for $G_T(\omega)$:

$$G_T(\omega) = \frac{\lambda_1 [C(\omega) G_R(\omega)]^*}{1 + \lambda_2 \sum_k |C^{(k)}(\omega) G_R^{(k)}(\omega)|^2} \tag{5.62}$$

The constants λ_1 and λ_2 are used to satisfy constraints on transmitted power and pulse height.

Generally speaking, there is little difference in performance between optimization of the transmitting filter for a fixed receiving filter and the alternative of optimization of the receiving filter for a fixed transmitting filter. In practice where the optimization must be carried out by adaptive circuitry certain advantages and disadvantages are involved with each of these choices. If the transmitting filter is to be optimized, a feedback path is required so that the channel characteristics can be learned at the transmitting point. Since this feedback path is not always available and since the mathematical results are nearly equivalent, we shall concentrate the numerical results and equalization techniques to be discussed later upon the problem of receiver optimization for a fixed transmitter.

5.2.2 Time and Bandwidth Properties of Transmitted Waveforms

In 1961 Gerst and Diamond published a paper which showed that input pulses $g(t)$ can be selected for any linear, lumped-parameter system such that the output pulse $x(t)$ is time-limited to any duration T desired. Such pulses could then be used for data transmission at arbitrarily high rates without intersymbol interference. The seeming contradiction of the Nyquist theorem results from the fact that any real, lumped-parameter system must have transmission to infinite frequencies rather than having a point of cutoff at W Hz where the frequency response drops identically to zero.

Requiring the output pulse $x(t)$ to be time-limited to the interval $[0, T]$ seems like a difficult way to eliminate intersymbol interference compared with the usual technique of asking that $x(t)$ be zero at points $t_0 + nT$ for all $n \neq 0$. Indeed, it is usually true that the requirement of a time-limited output pulse is so severe that it results in a class of allowable input pulses having very high attenuation in passage throughout the channel and hence being very susceptible to noise. For this reason the Gerst and Diamond approach cannot in general be recommended for the type of channels being considered here. However, a brief look at the Gerst and Diamond theorems is useful in the study of transmitter pulse shaping.

Since these theorems will apply to lumped-parameter linear networks, the Laplace transform is more easily applicable than the Fourier transform we have been using thus far. Let the Laplace transforms of the input pulse, the channel and the output pulse be $G(s)$, $C(s)$, and $X(s)$, respectively. If the output pulse $x(t)$ is time-limited to the interval $[0, T]$,

$$X(s) = \int_0^T x(t)e^{-st}\, dt \tag{5.63}$$

which, because of the finite range of integration, must converge for all complex s. Thus $X(s)$ has no singularities in the finite plane, i.e., $X(s)$ is an entire function. Although this condition is necessary on $X(s)$ if its pulse is to be time-limited, it is not a sufficient condition. The theorem, quoted by Gerst and Diamond and attributed to Doetsch,[8] that gives both necessary and sufficient conditions is as follows:

Theorem 5-1: Necessary and sufficient conditions that $x(t)$ be time-limited to the interval $[0, T]$ are that

1. $X(s)$ *be an entire function*
2. $X(s)$ *be bounded in the right-half plane.*
3. $e^{Ts}X(s)$ *be bounded in the left-half plane*

Since the output pulse $X(s)$ is the product of $G(s)$ and $C(s)$, condition (1) requires that $G(s)$ have zeros at the points in the complex plane where the channel $C(s)$ has transmission poles. Simply constructing the input pulse $G(s)$ from these zeros is not an acceptable solution since this makes $g(t)$ a sum of higher-order impulses.

There are, of course, an infinite number of $G(s)$'s for which $X(s) = G(s)C(s)$ will satisfy the conditions of the theorem. One of the problems Gerst and Diamond approached was to find a convenient way of expressing some of the more useful, admissible input pulses $G(s)$. They gave us two techniques by which the pulses $G(s)$ could be constructed.

Method 1: Let channel $C(s)$ have the form

$$C(s) = \frac{N(s)}{D(s)} = \frac{\sum_{i=0}^{m} k_i s^i}{\sum_{i=0}^{n} h_i s^i} = \frac{N(s)}{\prod_{i=1}^{n} (s - \gamma_i)} \tag{5.64}$$

where $N(s)$ and $D(s)$ are real polynomials with no common factor, $m \leq n$, $n \geq 1$, and $h_n = 1$. Then the input pulse whose Laplace transform is

$$G(s) = \left[\frac{1 - e^{-Ts/(n+1)}}{s}\right] \prod_{j=1}^{n} \left[1 - e^{-T(s - \gamma_j)/(n+1)}\right] \tag{5.65}$$

when passed through the channel results in an output pulse $x(t)$ time-limited to the interval $[0, T]$.

Notice in Eq. (5.65) that the input pulse $G(s)$ can be interpreted as a narrow rectangular pulse of width $T/(n+1)$ being passed through an $(n+1)$-tap transversal filter. The transversal-filter portion is indicated by the product in Eq. (5.65) since

$$\begin{aligned} \prod_{j=1}^{n} \left[1 - e^{-T(s - \gamma_j)/(n+1)}\right] &= 1 - e^{-Ts/(n+1)} \left[\sum_{j=1}^{n} e^{T\gamma_j/(n+1)}\right] \\ &\quad + e^{-2Ts/(n+1)} \left[\cdots\right] \\ &\quad - \cdots + (-1)^n e^{-nTs/(n+1)} \left[e^{T \sum_{j=1}^{n} \gamma_j/(n+1)}\right] \end{aligned} \tag{5.66}$$

which is the standard representation of a tapped delay line. An example of the Gerst and Diamond technique of pulse shaping for a channel with two poles is shown in Fig. 5.6.

Method 1 can be generalized somewhat by allowing other than rectangular pulse inputs to the transversal filter and other than uniform spacing between taps on the delay line. The basic idea, however, is this. An $(n+1)$-tap delay line has a Laplace transform which, as shown in Eq. (5.66), can be put in the form of an n-fold product showing the locations of complex zeros at points γ_j, where $j = 1, \ldots, n$. These zeros can, by proper adjustment of the tap weights, be located so as to coincide with the n poles of the channel. Thus if the input to the transversal filter is an entire function, the output of the channel will be also.

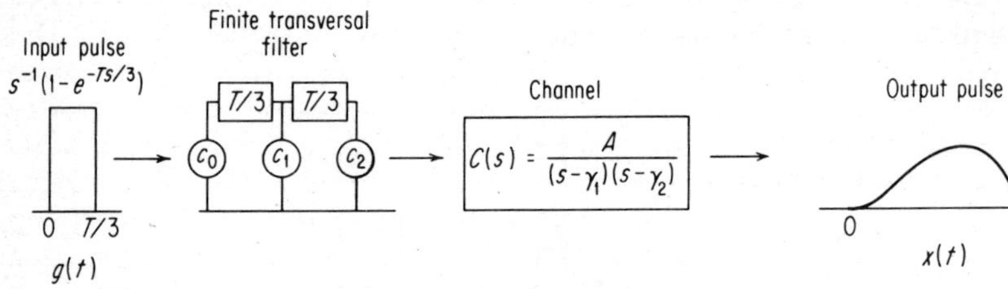

Fig. 5.6. Gerst and Diamond technique.

Method 2: The input pulse $G(s)$ must be representable as a product of an entire function $E(s)$ and the function $D(s)$ from Eq. (5.64) which cancels the channel poles. Thus

$$G(s) = E(s) \sum_{i=0}^{n} h_i s^i \tag{5.67}$$

and, inverting,

$$g(t) = h_0 e(t) + h_1 e^{(1)}(t) + h_2 e^{(2)}(t) + \cdots + h_n e^{(n)}(t) \tag{5.68}$$

It can be shown that the output pulse corresponding to the input pulse $g(t)$ in Eq. (5.68) will be time-limited to the interval $[0, T]$ as long as $e(t)$ itself is time-limited to this same interval and is differentiable the required n times.

If the Gerst and Diamond pulses are used for data transmission, matched-filter reception of each pulse can be employed at the receiver. The probability of error will then be determined by the ratio of received signal energy to noise power. Since transmitter power is limited, the designer of a transmission system is concerned with maximizing the energy transfer through the channel. There is reason to believe that the Gerst and Diamond pulses have poor energy transfer ratios because of the severe constraint on the output pulse. There is a large class of possible waveforms which do result in time-limited output pulses, and in fairness one should find the particular Gerst and Diamond input pulse which has maximum energy transfer through the channel. This problem has been considered by Hancock and Schwarzlander,[9] but the optimization requires the solution of a high-order differential equation with boundary conditions at both $t = 0$ and $t = T$. Thus the solution can be found only for particularly simple problems.

Most practical channels exhibit rather sharp cutoffs in their frequency characteristics. This behavior is indicative of a many-pole transfer function. In the Gerst and Diamond pulses of Eqs. (5.65) and (5.68) the input pulse transform must have zeros at the points where the channel transform has poles. This would ordinarily indicate poor energy transmission efficiency, although this effect can be mitigated by other freedom in the choice of the input waveform. A real channel may appear to be band-limited. If it actually were band-limited, it would be impossible to obtain a time-limited output pulse. However, by receiving energy outside the apparent passband the output pulse may be caused to be time-limited. The central question is this: What percentage of the pulse energy can be contained in the apparent passband? Such questions dealing with time- and band-limited pulses have been dealt with by Slepian and Pollak[10] and Landau and Pollak.[10,11] Of particular relevance here is a theorem proved by Landau and Pollak.

Theorem 5-2: There is a function g, with unit total energy, with energy α^2 in the time interval $[-T/2, T/2]$, and with energy β^2 in the frequency interval $(-W, W)$ Hz, under the following conditions and only under the following conditions:

1. *If* $\alpha = 0$, *when* $0 \le \beta < 1$
2. *If* $0 < \alpha < \sqrt{\lambda_0}$, *when* $0 \le \beta \le 1$
3. *If* $\sqrt{\lambda_0} \le \alpha < 1$. *when* $\cos^{-1}\alpha + \cos^{-1}\beta \le \cos^{-1}\sqrt{\lambda_0}$
4. *If* $\alpha = 1$, *when* $0 < \beta \le \sqrt{\lambda_0}$

The constant λ_0 is a function of the time-bandwidth product WT. Specifically, it is the largest eigenvalue of the integral equation

$$\lambda\Psi(t) = \frac{1}{\pi}\int_{-T/2}^{T/2} \Psi(\tau)\,\frac{\sin 2\pi W(t-\tau)}{(t-\tau)}\,d\tau \tag{5.69}$$

The right-hand side of Eq. (5.69) will be recognized as the response of an ideal low-pass filter to the time-truncated portion of $\Psi(t)$. The eigenfunctions of this equation were recognized by Slepian and Pollak[10] to be normalized versions of the prolate spheroidal wave functions. Many of the interesting properties of these functions are related in the original papers by Slepian and Pollak[10] and Landau and Pollak[11] to which the reader is referred for further detail.

As a result of Theorem 5-2 we are able to show in Fig. 5.7 permissible regions in the (α^2, β^2) plane for the pulse $\Psi(t)$ for various values of the time-bandwidth product WT. In the Gerst and Diamond case previously discussed, α^2 must equal 1 since all the pulse energy must be contained in a T-sec interval. Now if the bandwidth W Hz is taken to be the Nyquist band $1/2T$, we see from Fig. 5.7 that not more than about 75 percent of the pulse energy can be contained in $[-W, W]$ if the pulse is to be time-limited to a

T-sec interval. The actual pulse $x(t)$ which achieves this highest concentration of energy is the eigenfunction $\psi_0(t)$ corresponding to the largest eigenvalue λ_0 of integral equation Eq. (5.69). The necessary energy outside $[-W,W]$ which must be present in the output pulse must have passed through a region of high attenuation in the channel. Therefore the input pulse must have contained a much greater percentage of energy outside $[-W,W]$; hence the inefficient energy transfer.

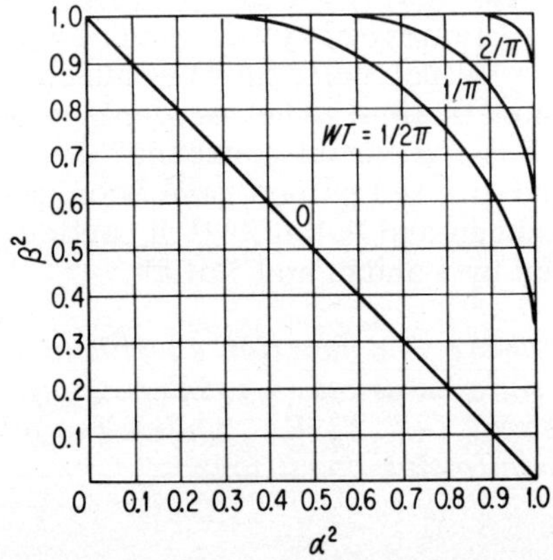

Fig. 5.7. Possible combinations of α^2 (percent pulse energy in I-sec time slot) and β^2 (percent pulse energy in W-Hz frequency slot) for different WT. (From Landau and Pollak[11])

While it is conceivable that situations requiring time-limited output pulses can occur, most data-transmission systems operate under the much less stringent requirement of zeros at sampling times only. Therefore the topic of obtaining time-limited pulses will not be discussed further here.

5.3 Joint Transmitter and Receiver Optimization

The joint optimization of transmitter and receiver is the central pulse-transmission problem. We have seen that this optimization would require selection of 2^{RNT} waveforms and detection using 2^{RNT} matched filters. However, by backing off considerably from this ideal, a tractable problem can be made using a signal which is the superposition of time translates of a single pulse $g(t)$. The receiver is then specified as consisting only of a filter $G_R(\omega)$ and of a sampler and threshold detector. In other words we have restricted consideration to only the simple PAM system and the only optimization required is that of the filters $G_T(\omega)$ and $G_R(\omega)$.

Even with the problem thus restricted, the minimization of probability of error for gaussian noise does not appear to be solvable. However, if intersymbol interference is constrained to be zero, an easy solution can be obtained. For high signal-to-noise ratios the optimum filters to minimize probability of error will be very close to those obtained when intersymbol interference is constrained to zero. This constraint requires

$$\sum_k G_T^{(k)}(\omega) G_R^{(k)}(\omega) = T x_0 \qquad |\omega| \le \frac{\pi}{T} \tag{5.70}$$

We have previously shown in Eq. (5.49) that the best receiving filter for a given transmitted signal subject to this constraint is of the form

$$G_R^{(k)}(\omega) = \frac{-\lambda_1(\omega)}{2} [G_T^{(k)}(\omega) C^{(k)}(\omega)]^* \tag{5.71}$$

Similarly the best transmitting filter for a given receiver filter and allowed transmitter power is

$$G_T^{(k)}(\omega) = \frac{-\lambda_1(\omega)}{2\lambda_2} [C^{(k)}(\omega) G_R^{(k)}(\omega)]^* \tag{5.72}$$

Simultaneous solution of Eq. (5.71) and (5.72) requires that either $G_R^{(k)}(\omega) = G_T^{(k)}(\omega) = 0$ or that

$$1 - \frac{\lambda_1^2(\omega)}{4\lambda_2} |C^{(k)}(\omega)|^2 = 0 \tag{5.73}$$

Thus

$$\lambda_1(\omega) = \frac{\pm 2\lambda_2^{\frac{1}{2}}}{|C^{(k)}(\omega)|} \tag{5.74}$$

Now Eq. (5.74) can only be satisfied for one value of k at each point ω. For all other k at that same frequency ω we must use the trivial solution $G_T^{(k)}(\omega) = G_R^{(k)}(\omega) = 0$. The trivial solution cannot be used for all integers k because of constraint (5.70). Thus for each value of ω there is one and only one integer k for which $G_T^{(k)}(\omega)$ and $G_R^{(k)}(\omega)$ are nonzero. For this particular value of k, we obtain, by combining Eqs. (5.70) through (5.72) and (5.74)

$$|G_R^{(k)}(\omega)| = \frac{\lambda [T x_0]^{\frac{1}{2}}}{|C^{(k)}(\omega)|^{\frac{1}{2}}} \tag{5.75}$$

$$|G_T^{(k)}(\omega)| = \frac{[T x_0]^{\frac{1}{2}}}{\lambda |C^{(k)}(\omega)|^{\frac{1}{2}}} \tag{5.76}$$

and $$\arg G_R^{(k)}(\omega) + \arg G_T^{(k)}(\omega) = -\arg C^{(k)}(\omega)$$

The remaining constant λ is determined by the desired transmitter power. As indicated in Eq. (5.76), an arbitrary phase characteristic may be added to $G_T^{(k)}(\omega)$ and compensated for in $G_R^{(k)}(\omega)$ without affecting the solution. Basically then, aside from constants, the transmitting and receiving filters have amplitude characteristics which vary as $|C^k(\omega)|^{-\frac{1}{2}}$ and have arbitrary phase characteristics which must add such that they exactly cancel the channel phase characteristic.

One would intuitively expect that the value of k for which $G_T^{(k)}(\omega)$ and $G_R^{(k)}(\omega)$ are nonzero would be the value for which the transmission $|C^{(k)}(\omega)|$ is greatest; indeed this is easily shown to be true. Let the set L_k be defined as that set of ω for which $|C^{(k)}(\omega)| \geq |C^{(j)}(\omega)|$ for all $j \neq k$. The complete definitions for transmitting and receiving filters are then

$$|G_R^{(k)}(\omega)| = \begin{cases} \dfrac{\lambda[Tx_0]^{\frac{1}{2}}}{|C^{(k)}(\omega)|^{\frac{1}{2}}} & \omega \subset L_k \\ 0 & \text{elsewhere} \end{cases} \tag{5.77}$$

$$|G_T^{(k)}(\omega)| = \begin{cases} \dfrac{[Tx_0]^{\frac{1}{2}}}{\lambda|C^{(k)}(\omega)|^{\frac{1}{2}}} & \omega \subset L_k \\ 0 & \text{elsewhere} \end{cases} \tag{5.78}$$

$$\arg G_R^{(k)}(\omega) + \arg G_T^{(k)}(\omega) = -\arg C^{(k)}(\omega)$$

An example of the use of Eqs. (5.77) and (5.78) is shown in Fig. 5.8. This section emphasizes an important consideration in data communications: that energies at points ω and $\omega + (2k\pi/T)$ are equivalent for synchronous baseband transmission at rate $1/T$. When we are confronted with selecting optimum signal shaping and receiver filtering, we might tend to use only the Nyquist band $[-\pi/T, \pi/T]$ rad per sec. However, should the channel have greater transmission at some frequency of the form

$$\omega + \frac{2k\pi}{T} \qquad k \neq 0$$

$$|\omega| < \frac{\pi}{T}$$

than it does at ω, the frequency $\omega + (2k\pi/T)$ should be used for transmission instead of frequency ω. The total measure of the frequency support for $G_T(\omega)$ and $G_R(\omega)$ should be the Nyquist range of $2\pi/T$.

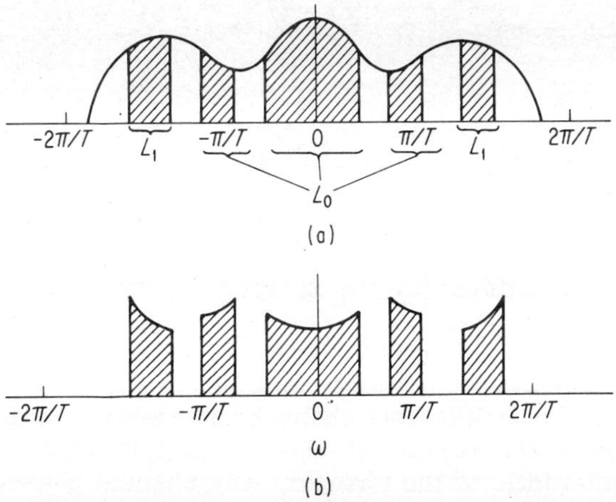

Fig. 5.8. Optimum transmitting and receiving filters which eliminate intersymbol interference. (a) Channel response $|C(\omega)|$. (b) Transmitting- and receiving-filter responses $|G_T(\omega)|$ and $|G_R(\omega)|$.

One bothersome facet of the filters described by Eqs. (5.77) and (5.78) is that the signal power spectrum varies as $|C|^{-1}$; thus where the channel has poorest transmission, we use greatest signal energy. This reasoning is completely opposite that of the water-pouring theorem given in Chap. 3 for the most efficient use of the given channel. However, the requirement of zero intersymbol interference causes the overall spectrum to be flat. If the transmitter filter deemphasized the frequencies with poor transmission, the receiving filter would be forced to amplify these frequencies at the expense of considerably increasing the noise.

The probability of error for an L-level PAM system employing the filters described by Eqs. (5.77) and (5.78) may be shown to be

$$P_e = 2\left(1 - \frac{1}{L}\right) Q\left[\left(\frac{3}{L^2 - 1}\,\beta_2\,\frac{P_S}{P_N}\right)^{1/2}\right] \tag{5.79}$$

where P_N is the noise power in the Nyquist band $[-\pi/T, \pi/T]$ and

$$\beta_2 = \left[\frac{T}{2\pi} \sum_k \int_{L_k} \frac{1}{|C(\omega)|}\, d\omega\right]^{-2} \tag{5.80}$$

The joint optimization of transmitter and receiver using a criterion of total mean-square error has also been carried out by Smith[6] and by Berger and Tufts.[14] The resulting filters are similar to Eqs. (5.77) and (5.78), particularly for high signal-to-noise ratios. Since considerable detail is involved in this optimization, the reader is referred to the original papers for this result.

5.4 Numerical Example

For an example of the computation of optimum performances we use a channel with slope-attenuation distortion and an arbitrary (assumed-zero) phase characteristics, i.e.,

$$C(\omega) = 10^{-TA_m\omega/20\pi} \tag{5.81}$$

where A_m is the attenuation in decibels at the Nyquist frequency. This example was previously used in Sec. 4.2.4 in which the performance of the ordinary raised-cosine PAM system was assessed in the face of this particular channel characterization.

We do not use an example channel with delay (phase) distortion because none of the results of this chapter is dependent upon the phase characteristic of the channel. Any channel phase characteristic can, in theory, be perfectly compensated by an all-pass network which changes neither signal- nor noise-power levels. Thus all equations for optimum system performance levels depend only upon the channel transfer function magnitude $|C(\omega)|$. Observe also that all of the optimum filters have a stage which involves a channel-matching filter $C^{k*}(\omega)$. The effect of this stage is to square the amplitude characteristic of the channel while eliminating (making zero) the phase characteristic. For the same kind of reasons we have just given the reference sampling time t_0 can be assumed zero without loss of generality in our results.

We shall assume now that a transmitter filter which accomplishes half of a 60 percent roll-off raised-cosine shaping is used. In the conventional configuration an identical receiver filter would be employed so that the signal shaping would be a 60 percent roll-off raised-cosine characteristic. Because of the distortion present in the channel, the conventional configuration is degraded in performance by the presence of intersymbol interference. The amount of this degradation and the resulting error rates were computed in the previous chapter. Now we shall compute error rates for this system when optimum receiver and then optimum transmitter filters (which eliminate intersymbol interference) are substituted for the conventional-cosine roll-off filters.

The formation of the optimum receiver filter for this example is illustrated in Fig. 5.9. Here the transmitting filter is the specified square-root-of-raised-cosine, 60 percent roll-off characteristic. The channel characteristic is pictured on a linear scale and shows the exponential falloff with increasing frequency. For this situation the best receiving filter (which eliminates intersymbol interference) consists of a matched filter followed by a transversal filter. The matched filter contains, first, the other half of the 60 percent roll-off shaping filter, which is precisely the filter (and the only filter) that would be employed in the conventional system. Secondly, the matched filter contains a duplicate channel characteristic. (If the channel had a nonzero phase characteristic, this filter would have needed a compensating phase characteristic.)

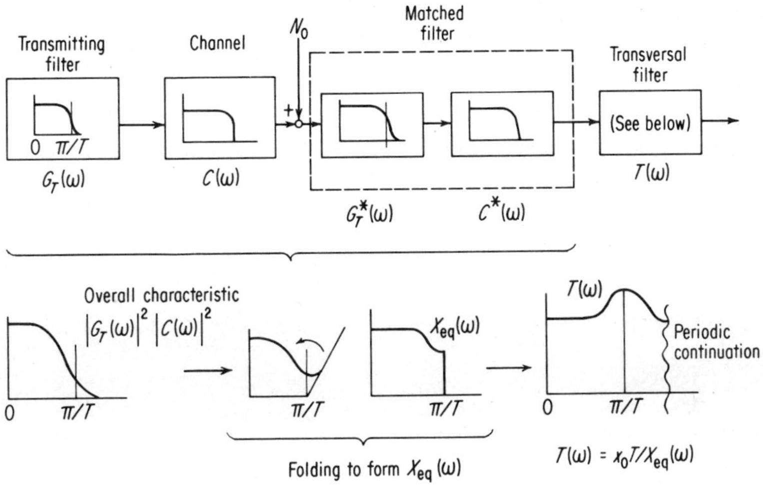

Fig. 5.9. Formation of the optimum linear receiver for elimination of intersymbol interference.

The transversal filter following this matched filter has the assignment of eliminating intersymbol interference. The response of the transversal filter in the frequency domain can be found as illustrated in the bottom half of Fig. 5.9. First, the overall transfer function up to the input of the transversal filter is formed. Second, the equivalent Nyquist channel is formed by folding the overall characteristic about the Nyquist frequency π/T. Now the transversal filter must cause this equivalent channel to be the (arbitrary) constant x_0T. Therefore the transversal-filter response is the function $x_0T/X_{eq}(\omega)$ illustrated in the figure. This response then repeats itself during every Nyquist interval. Finally, the actual tap gains of the transversal filter can be computed by calculating the Fourier coefficients in the expansion of this repetitive frequency response.

The performance in terms of probability of error as a function of signal-to-noise ratio for this optimized receiver has been calculated for the specific case in which A_m = 10 db and plotted in Fig. 5.10 along with the corresponding curve for the conventional receiver. Also plotted in this figure are the bound on P_e for this particular transmitted signal and the performance of the system when both transmitter and receiver filters are optimized. Observe that the bound on performance is close to the performance of the optimum linear receiver system. Therefore very little benefit could be achieved by the very much more complicated signal processing which might be envisioned for this particular dispersive channel.

The curve showing performance of the linear system when both transmitter and receiver are optimized shows only a slight improvement is attainable in this example by designing the transmitting filter for this particular channel characteristic. Two notes of caution must be interjected here. First the matched filter bound is not generally an attainable one and the optimum receiver performance may lie anywhere between curves A and B in Figs. 5.10 and 5.11. Second since the bound applies only to a given transmitter configuration, the joint optimization of curve C can occasionally prove better than the bound of curve B.

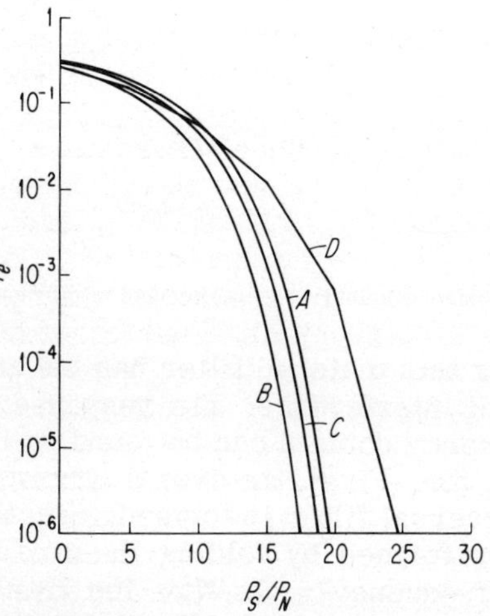

Fig. 5.10. Probability of error P_e for various systems operating through 10 db; linear attenuation distortion; $\alpha = 0.6$, $A_m =$ 10 db. Curve *A*: using best $G_R(\omega)$, no intersymbol interference. Curve *B*: bound for given $G_T(\omega)$. Curve *C*: using best $G_R(\omega)$ and $G_T(\omega)$, no intersymbol interference. Curve *D*: raised-cosine filtering.

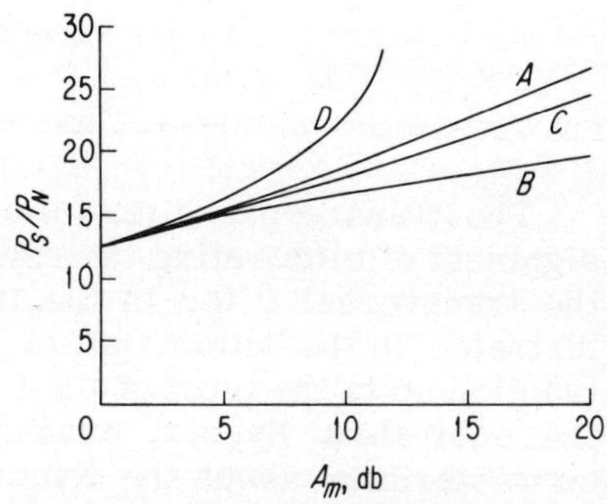

Fig. 5.11. Required SNR for 10^{-5} error rate for various systems operating through linear attenuation distortion. Curve *A*: using best $G_R(\omega)$, no intersymbol interference. Curve *B*: bound for given $G_T(\omega)$. Curve *C*: using best $G_R(\omega)$ and $G_T(\omega)$, no intersymbol interference. Curve *D*: raised-cosine filtering.

The relative performances of the systems we have been discussing are very much a function of the severity of the linear attenuation distortion. To illustrate this dependence the signal-to-noise ratio required by each system to achieve a 10^{-5} error rate is shown as a function of A_m in Fig. 5.11. Notice that increasing the slope of the attenuation increases the disparity between the various systems; however, at points where channel distortion is not inordinately severe the optimum linear PAM receiver performs close to the theoretical bound, and little improvement is possible by the additional optimization of the transmitter.

The results given thus far have applied to the system where the roll-off was 60 percent so that the pulse transmission rate was five-eighths of the Nyquist limit for the total system bandwidth. The effect of using sharper roll-offs in an effort to achieve greater bandwidth utilization usually results in a greater disparity between the various theoretical bounds. As they approach the Nyquist limit, the simpler systems tend to perform more poorly relative to the theoretically achievable levels.

5.5 Summary

This chapter began by considering the optimal reception of a PAM message. Although the optimal receiver required L^N-matched filters to detect N symbols from an L-level alphabet, it was shown that the simple PAM receiver using a single filter and threshold detector provided equivalent performance when the channel was ideal and the signal shaping was a Nyquist characteristic. In practice, channels may be in some sense "close" to ideal, thus one might feel that some modification of the PAM receiver filter would result in performance which would closely approximate the performance of the optimal matched-filter receiver. Thus best forms of the receiver filter were derived under the criteria of probability of error, probability of error with intersymbol interference constrained to zero, and total rms error.

The three optimizations of the receiver filter were illuminating in that all three resulted in filters of the same general form—a matched filter followed by a transversal filter. The tap coefficients (gains) of the transversal filter were slightly different in each case, and in the direct minimization of probability of error the tap gains could not be evaluated in closed form. For relatively high signal-to-noise ratios the difference between filters is usually not great, and the expression for probability of error for the system in which intersymbol interference was eliminated [Eq. (5.54)] provides a good indication of performance for the optimum linear receiver.

To compare the performance of this best, linear receiver with what might be achieved with more elaborate instrumentation, an upper bound was derived on the performance of the matched-filter receiver. In most cases where the channel has reasonably "good" transfer characteristics, this bound, Eq. (5.28), will be very close to the aforementioned probability of error for the best linear receiver.

Also considered in this chapter was the optimization of the transmitter and receiver filters. The most useful result from these sections is the probability of error, Eq. (5.79), for the best filter combination under the constraint of zero intersymbol interference. This expression might be compared with Eq. (5.54) where the best receiver filter (under a zero intersymbol-inter- used at the transmitter. Again, very often, in practice the additional gain realized by transmitter optimization instead of by the usage of a conventional filter might prove to be small. Another consideration here is that the best transmitter filter uses sharp corners in its frequency characteristic; consequently it is relatively sensitive to any perturbations in timing or in system parameters.

Each of the three most useful equations for probability of error for the optimized systems was given in the form

$$P_e = 2\left(1 - \frac{1}{L}\right) Q\left[\left(\frac{3}{L^2 - 1} \beta \frac{P_S}{P_N}\right)^{1/2}\right]$$

where L = number of levels

$$Q(x) = \int_x^\infty e^{-t^2/2}/\sqrt{2\pi}\, dx$$

P_S = transmitted signal power

$P_N = N_0/T$ = noise in Nyquist band

For convenience we repeat in Table 5.1 the expressions derived for the signal-to-noise enhancement factor β.

Table 5-1. Probability of Error for Optimized Systems

System	β
Lower bound on P_e for matched filter receiver, given $G_T(\omega)$	$\dfrac{\int_{-\infty}^{\infty} \lvert C(\omega)\rvert^2 \lvert G_T(\omega)\rvert^2\, d\omega}{\int_{-\infty}^{\infty} \lvert G_T(\omega)\rvert^2\, d\omega}$
Best linear receiver which eliminates intersymbol interference, given $G_T(\omega)$	$\left[\dfrac{T}{2\pi}\int_{-\infty}^{\infty} \lvert G_T(\omega)\rvert^2\, d\omega \, \dfrac{T}{2\pi}\int_{-\pi/T}^{\pi/T} \dfrac{1}{\sum_k \lvert G_T^{(k)}(\omega)\, C^{(k)}(\omega)\rvert^2}\, d\omega\right]^{-1}$
Best combination of transmitting and receiving filters which eliminates intersymbol interference	$\left[\dfrac{T}{2\pi}\sum_k \int_{L_k} \dfrac{1}{\lvert C^{(k)}(\omega)\rvert}\, d\omega\right]^{-2}$ $L_k = [\omega: \lvert C^{(k)}(\omega)\rvert \geq \lvert C^{(j)}(\omega)\rvert \quad \text{all } j \neq k]$

Notes

The books by Helstrom[1] and Hancock and Wintz[12] are basic references on signal detection theory. For PAM systems with intersymbol interference see Aaron and Tufts,[3] Tufts,[5], Smith,[6] George,[13] and Berger and Tufts.[14] Berger and Tufts[14] contains additional material which includes the effect of timing on the filter-optimization problem.

6

Equalization of the Baseband System

In Chaps. 4 and 5 we first discussed and evaluated the detrimental effects of intersymbol interference on conventional PAM systems and then formulated certain optimal PAM-type systems for dispersive channels. These optimal systems are in a sense chimeras since the designer seldom has certain foreknowledge of the exact channel characteristics and never has the capability of exactly producing the necessary optimal filter characteristics. This last point in fact needs some emphasis. In the laboratory, the imprecision in design of the transmitting and receiving filters prohibits transmission of more than about eight levels per symbol, even when the transmitter is connected directly to the receiver without the intervention of a transmission medium. In any real system intersymbol interference from this and other sources is always present, and, if the signal-to-noise ratio is sufficiently large, this intersymbol interference will limit the data rate of the system to some figure which is much less than theoretical capacity.

The only recourse of the system designer to mitigate the effects of the intrinsic residual distortion is to include within the data system an adjustable filter or filters which can be trimmed to fit closely the required characteristics for any individual data call. Traditionally this kind of filter has been called an *equalizer* since this was the term applied to early filters designed to compensate the amplitude characteristics of transmission cables. Now any filter whose purpose is some sort of compensation is known as an equalizer. Any real transmission system usually has a great many such filters. The telephone system has a complicated hierarchy of equalizers to provide flat attenuation characteristics over the various transmission channels.

For synchronous data transmission we require a different sort of equalization. Instead of the traditional aim of flattening the amplitude characteristic and linearizing the phase characteristic of the channel, we have the very specific goal of achieving minimum probability of error in a given data call. We have discussed in

Chap. 5 the kinds of filters which in theory approach this goal. At the receiver, for example (in addition to the usual receiver filter of perhaps the raised-cosine form), we require (1) a filter capable of being adjusted to match the exact channel characteristic and (2) an infinite transversal filter whose tap gains are adjusted to eliminate intersymbol interference.

In reality any adjustable filter can only have a finite number of variable parameters, and no characteristic can be fitted exactly. Furthermore, the cost of the adjustable filter (equalizer) will usually be directly proportional to the number of variable parameters. Now given only, say, N adjustable coefficients in an equalizer, it is by no means clear that the form of the equalizer should approximate the form of the optimum filters discussed in the previous chapter. The problem becomes one of finding the type of adjustable filter best able to compensate for the ensemble of possible channel characteristics having the least number of variable parameters.

Without better statistical characterization of the ensemble of channels than is usually available, the problem of defining optimum, finite filters seems unapproachable. Therefore the field of adjustable equalization for data transmission has been based on the availability of certain types of adjustable networks. Some equalizers use a sequence of bandpass filters, each tuned to a different portion of the data band and adjustable in gain. These are called *bump equalizers* since the attenuation characteristic for these equalizers displays a number of adjustable bumps. The most useful form of a variable network, however, has been the transversal filter. The transversal filter was invented by Wiener and Lee in 1935 and first described in the literature by Kallman in 1940.

The application of the transversal filter to adjustable equalization might seem a natural outgrowth of the optimization results of Chap. 5, but, of course, the practical application of transversal filters to data transmission predates the availability of the theoretical results. Also, it must be emphasized that the theoretical results require an infinite transversal filter and that no optimality claim may be made for a finite transversal filter as an equalizer. However, the theory does provide a certain insight into the requirements for good equalization. It was shown that optimum filtering required two filters in addition to the usual receiver filter. One filter matched the channel and had the basic job of minimizing the effect of noise. The other filter, a transversal filter, had the job of eliminating (or in some sense minimizing) intersymbol interference.

Since cost is proportional to the number of adjustable parameters, the available number of adjustable parameters must be carefully divided between matched filtering (noise) and transversal filtering (intersymbol interference). In a high-noise environment, for example, with small intersymbol interference it might be best to concentrate on adjustable matched filtering to the sacrifice of transversal filtering. On the other hand, in low-noise environments, such as telephone channels, the most important consideration by far is minimization of intersymbol interference. For a given amount of equipment here it usually pays to forget adjustable channel matching completely and put all the effort into intersymbol interference reduction. We shall show that such a philosophy results in performance only very slightly worse than is theoretically possible with ideal filtering.

Practically speaking, since the transversal filter is so versatile, it can be used to accomplish matched filtering also, except that the taps on the delay line are spaced at Nyquist intervals rather than at symbol (T-sec) intervals. DiToro[2] has constructed an automatically adjusted equalizer using transversal filters for both channel matching and intersymbol-interference removal. Schreiner, Funk, and Hopner[3] and Becker, Holzman, Lucky, and Port[4] have constructed equalizers using transversal filters for intersymbol-interference removal. Earlier references relate the use of transversal filters as adjustable matched filters.[5]

The need for precise, automatically adjustable equalization has been a phenomenon of the data explosion. A great many instrumentations for automatic equalizers are presently being investigated—and all these instrumentations employ transversal filters as the adjustable element. One reason this type of filter is so popular is that the channels to which it is being applied exhibit time dispersion over only a small number of symbols duration. Therefore a time-domain device such as the transversal filter is able to correct the small time dispersion with only a few adjustable parameters (tap gains). It is quite conceivable that future requirements will involve the equalization of channels with extremely long time dispersion, but with a simple characterization in terms of some other viewpoint (such as location of transmission poles and zeros). In this chapter, however, we shall consider solely the transversal-filter equalizer. We begin with a basic description of the action of a transversal filter on a data signal. Subsequently, we shall consider strategies for automatically setting the equalizer before or during data transmission.

6.1 Transversal-filter Equalization

The transversal filter was described in Chap. 5. Basically the device consists of a delay line tapped at T-sec intervals. Each tap is connected through a variable gain (which can be negative) to a summing bus. For convenience we shall always assume that there are $(2N+1)$ taps on the transversal filter with gains $c_{-N}, c_{-N+1}, \ldots, c_N$. The impulse response of this equalizer is then

$$e(t) = \sum_{n=-N}^{N} c_n \,\delta(t - nT) \tag{6.1}$$

while the frequency characteristic is

$$E(\omega) = \sum_{n=-N}^{N} c_n \, e^{-in\omega T} \tag{6.2}$$

If the equalizer, shown in Fig. 6.1, is connected in tandem with a separate system whose impulse response is $x(t)$, the impulse response of the equalized system $x(t) * e(t)$ becomes (see Fig. 6.1)

$$h(t) = \sum_{n=-N}^{N} c_n x(t - nT) \tag{6.3}$$

At the sampling times $(kT + t_0)$ we have

$$h(kT + t_0) = \sum_{n=-N}^{N} c_n x[(k - n)T + t_0] \tag{6.4}$$

or

$$h_k = \sum_{n=-N}^{N} c_n x_{k-n} \tag{6.5}$$

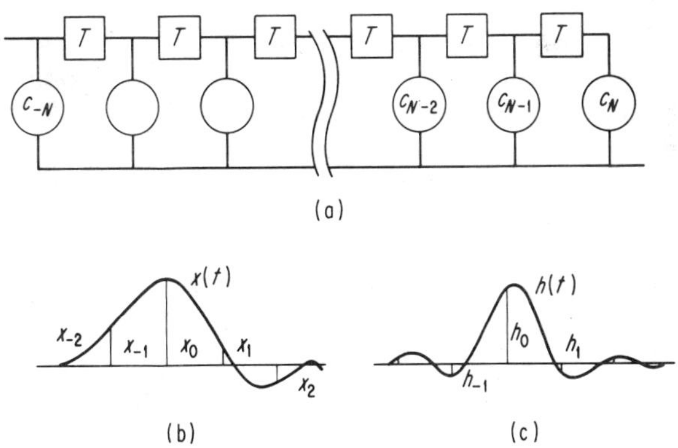

Fig. 6.1. Transversal-filter equalizer [see Eq. (6.3)]. (a) Transversal filter. (b) Input pulse. (c) Output pulse.

Equation (6.5) indicates that the output sample sequence is formed from the input sequence by polynomial multiplication. For example, if the input sequence is $x_{-1} = 1/4$, $x_0 = 1$, and $x_1 = 1/2$ and if the tap gains are $c_{-1} = -1/4$, $c_0 = 1$, and $c_1 = -1/2$, the output sequence is the product

$$
\begin{array}{ccccc}
\frac{1}{4} & 1 & \frac{1}{2} & & \\
-\frac{1}{4} & 1 & -\frac{1}{2} & & \\
\hline
-\frac{1}{16} & -\frac{1}{4} & -\frac{1}{8} & & \\
 & \frac{1}{4} & 1 & \frac{1}{2} & \\
 & & -\frac{1}{8} & -\frac{1}{2} & -\frac{1}{4} \\
\hline
-\frac{1}{16} & 0 & \frac{3}{4} & 0 & -\frac{1}{4}
\end{array}
$$

Thus $h_{-2} = -1/16$, $h_{-1} = 0$, $h_0 = 3/4$, $h_1 = 0$, and $h_2 = -1/4$.

Observe that the output sequence is longer than the input sequence. In this example there are three taps on the equalizer and five output samples. No adjustment of the three taps in this example can eliminate intersymbol interference. Except in special cases, it will be true in general that a finite transversal filter cannot eliminate intersymbol interference. However, the reader should not be confused by the Gerst and Diamond technique described in Sec. 5.2.2, where an $(n+1)$-tap transversal filter with tap spacing $T/(n+1)$ is used to eliminate intersymbol interference in a system characterized by n poles.

In theory as the length N of the transversal filter becomes large, intersymbol interference can usually be made arbitrarily small. We have previously laid down guidelines for the determination of tap settings for an infinite transversal filter. Recall that in order to eliminate intersymbol interference the equivalent Nyquist channel must have the characteristic $h_0 T$. Therefore

$$H_{\mathrm{eq}}(\omega) = h_0 T = \sum_k H\left(\omega + \frac{2k\pi}{T}\right) \qquad |\omega| \leq \frac{\pi}{T} \tag{6.6}$$

But the frequency response of the transversal filter, given by Eq. (6.2), repeats itself every $2\pi/T$ rads. The equivalent channel for the equalized system becomes

$$H_{\mathrm{eq}}(\omega) = \sum_k X\left(\omega + \frac{2k\pi}{T}\right) E\left(\omega + \frac{2k\pi}{T}\right) \tag{6.7}$$

$$H_{\mathrm{eq}}(\omega) = E(\omega) \sum_k X\left(\omega + \frac{2k\pi}{T}\right) = E(\omega) X_{\mathrm{eq}}(\omega) \tag{6.8}$$

Thus

$$E(\omega) = \frac{h_0 T}{X_{eq}(\omega)} \qquad |\omega| < \frac{\pi}{T} \tag{6.9}$$

uniquely determines the frequency characteristic of the transversal equalizer.

As $N \to \infty$ the transversal-filter characteristic becomes a Fourier series with real coefficients and can approximate (in an integral-square sense) any function on the interval $[-\pi/T, \pi/T]$ which has an even real part and an odd imaginary part. From Eq. (6.9) we see that any channel can be equalized perfectly *with a transversal equalizer* as long as $X_{eq}(\omega) \neq 0, |\omega| \leq \pi/T$. This is somewhat more stringent than the requirement of Eq. (5.45) that

$$\sum_k \left| X\left(\omega + \frac{2k\pi}{T}\right)\right|^2 \neq 0 \qquad |\omega| \leq \frac{\pi}{T}$$

which determined the existence of *any* linear filter to eliminate intersymbol interference.

Starting with the sample value sequence $\{x_n\}$ of the uncorrected pulse response, we can find the tap-gain settings for the infinite transversal filter as follows:

1. Form the equivalent Nyquist channel frequency response.

$$X_{eq}(\omega) = T \sum_n x_n e^{-in\omega T} \qquad |\omega| \leq \frac{\pi}{T} \tag{6.10}$$

2. Find the frequency response of the transversal equalizer.

$$E(\omega) = \frac{h_0 T}{X_{eq}(\omega)} \tag{6.11}$$

3. Invert $E(\omega)$ to find the tap-gain settings.

$$c_n = \frac{T}{2\pi} \int_{-\pi/T}^{\pi/T} E(\omega)\, e^{in\omega T}\, d\omega \tag{6.12}$$

The resultant output response will have $h_n = 0,\ n \neq 0$.

For particularly simple pulses the tap settings can be obtained directly from the input sequence $\{x_n\}$. For example, assume that the input pulse has $x_0 = 1$, $x_1 = 1/2$, and elsewhere $x_n = 0$. The polynomial product of (1, 1/2) with the equalizer tap-gain sequence has to be simply unity (assuming $h_0 = 1$); therefore by direct division we can arrive at the proper gain settings.

$$
\begin{array}{rllll}
 & c_0 & c_1 & c_2 & \\
 & 1, & -\frac{1}{2}, & \frac{1}{4} & \cdots \\
\hline
1,\frac{1}{2}\,\big) & 1 & & & \\
 & 1 & \frac{1}{2} & & \\
\cline{2-3}
 & & -\frac{1}{2} & & \\
 & & -\frac{1}{2} & -\frac{1}{4} & \\
\cline{3-4}
 & & & \frac{1}{4} & \\
 & & & \frac{1}{4} & \frac{1}{8} \\
\cline{4-5}
 & & & & -\frac{1}{8} \ \cdots
\end{array}
$$

This technique is useful only when the input pulse has no distortion before the reference sample x_0 (no precursors). A similar procedure can be used when the input pulse has no tail distortion.

An examination in depth of the properties of the input and output time sequences is facilitated by the use of z transforms. Letting $z = e^{i\omega T}$ in Eq. (6.10) gives

$$X_{\text{eq}}(z) = T \sum_n x_n z^{-n} \tag{6.13}$$

Similarly

$$E(z) = \sum_n c_n z^{-n} \tag{6.14}$$

and the output sequence is simply the product of Eqs. (6.13) and (6.14).

The mapping of the $s = \sigma + i\omega$ plane into the z plane is shown in Fig. 6.2(a). The entire left half of the s plane maps into points within the unit circle of the z plane. In fact, each semi-infinite strip of height $2\pi/T$ in the left-half s plane maps into points within the unit circle of the z plane. Thus, for example, the transmission zero at $z = -1/2$ maps into the periodic set of zeros in the s plane illustrated in Fig. 6.2(a). The $i\omega$ axis of real frequencies lies on the unit circle of the z plane. Therefore, in a perfectly equalized pulse the potential along this circle owing to poles and zeros of z must be constant.

In the example of the direct division for the input pulse $1 + 1/2z^{-1}$ the z transform of the equalizer response was $1 - 1/2z^{-1} + 1/4z^{-2} \ldots$. The z transform of the input pulse contains the single zero at $z = -1/2$

as shown in Fig. 6.2(b). To correct for this zero at $z = -1/2$ the equalizing network must in effect cancel the zero with a pole; however, this is impossible using the transversal-filter frequency response. The alternative means of maintaining constant potential along the circle $|z| = 1$ is for the equalizing network to insert a set of equally spaced zeros along the circle $|z| = 1/2$, except for the point $z = -1/2$ where the input pulse has its zero. Thus every point along the unit circle will, as the equalizer length becomes infinite, experience an equal depression in potential from the concentric circle of zeros at $|z| = 1/2$. The circle of zeros forming as a result of the first few terms of the equalizer response $1 - 1/2z^{-1} + 1/4z^{-2} \ldots$ is shown in Fig. 6.2(b).

When all the zeros of the input pulse lie inside the unit circle, as in the previous example, the method of direct division may be

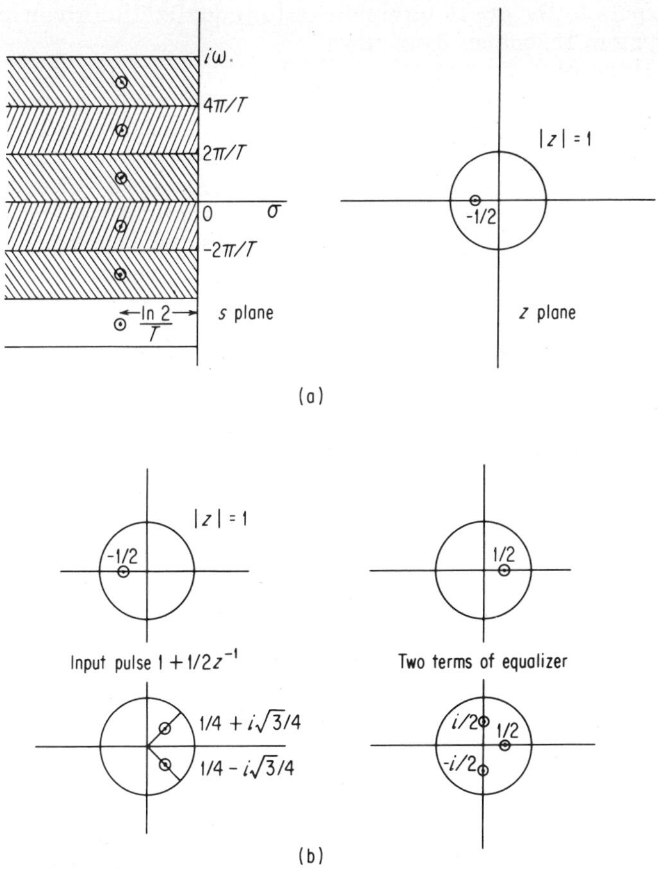

Fig. 6.2. Z-plane effects of equalization. (a) Mapping of the $s = \sigma + i\omega$ plane into the z plane where $z = e^{sT}$. (b) Circle of zeros formed as a result of pulse equalization.

used to find the required equalizer response. If all the zeros lie outside the unit circle, direct division can be accomplished by reversing the time series involved. However, in general a pulse will contain zeros both outside and inside the unit circle, and the only way of proceeding is to factor $E(z)$ into the product of two terms—one containing zeros outside and the other containing zeros inside the unit circle. Each term can then be handled separately and the two resultant equalizer characteristics combined.

The locations of the z-plane zeros of the input pulse are also useful in determining the rate of convergence of the equalizer time series. If the input pulse has a zero on the unit circle, the equalizer response cannot converge for some real frequency. If the input pulse does not have a zero on the unit circle, the pulse can be equalized, and the rate of convergence of the equalizer is determined by the distances from the unit circle of the zeros. This corresponds to the previous condition for equalization given in terms of the system frequency response.

Unfortunately, finding the zeros of high-order polynomials is a difficult task, and the tap settings of long equalizers are much more easily found by the frequency-domain approach outlined in Eqs. (6.10) through (6.12). However, both of these approaches are unrealistic in assuming that an essentially infinite transversal filter is available for pulse correction. In practice the length of the equalizer is limited not only by economic considerations but also by the accuracy with which each coefficient c_j can be set. The accumulation of errors in tap coefficients for long equalizers can more than offset the added correcting ability of the longer equalizer. Thus we need to study optimum tap settings for finite equalizers and examine the accuracy requirements for these tap settings.

As has been previously mentioned, except for special cases intersymbol interference cannot be completely eliminated with a finite transversal filter. Therefore we shall consider the minimization of the residual peak and rms distortion with finite equalizers.

6.1.1 Minimization of Peak Distortion

As previously defined the peak distortion for the pulse $h(t)$ is

$$D = \frac{1}{h_0} \sum_n{}' |h_n| \tag{6.15}$$

The peak eye closure for an L-level PAM system is $(L-1)D$, and the eye opening is $[1-(L-1)D]$. The input pulse to the equalizer will be assumed to have initial peak distortion D_0. The reference sample x_0 will also be normalized at unity. The problem is to determine the tap-gain coefficients $c_{-N}, \ldots, c_N$ for a $(2N+1)$ tap transversal equalizer which minimizes the final peak distortion, Eq. (6.15), where

$$h_n = \sum_{j=-N}^{N} c_j x_{n-j} \tag{6.16}$$

There will be an arbitrary gain factor involved in the equalizer setting; therefore the central sample pulse h_0 will be constrained to be unity.

$$h_0 = 1 = \sum_{j=-N}^{N} c_j x_{-j} \tag{6.17}$$

For mathematical convenience, the center tap gain c_0 can be used to satisfy this constraint.

$$c_0 = 1 - \sum_{j=-N}^{N}{}' c_j x_{-j} \tag{6.18}$$

We substitute Eq. (6.18) into Eq. (6.16) to obtain

$$h_n = \sum_{j=-N}^{N}{}' c_j (x_{n-j} - x_n x_{-j}) + x_n \tag{6.19}$$

and since h_0 is unity, the output pulse distortion becomes

$$D = \sum_{n=-\infty}^{\infty}{}' \left| \sum_{j=-N}^{N}{}' c_j (x_{n-j} - x_n x_{-j}) + x_n \right| \tag{6.20}$$

Now that we have eliminated c_0 to satisfy the constraint on h_0, we desire to minimize D in Eq. (6.20) over the $2N$ variables c_j, $|j| \leq N$, $j \neq 0$.

First observe that the peak distortion D is a continuous, piecewise linear function of the tap gains c_j. We can rewrite Eq. (6.20) in the form

$$D = \sum_{j=-N}^{N}{}' c_j \left[\sum_{n=-\infty}^{\infty}{}' (x_{n-j} - x_n x_{-j}) \operatorname{sgn} h_n \right] + \left[\sum_{n=-\infty}^{\infty}{}' x_n \operatorname{sgn} h_n \right] \tag{6.21}$$

where

$$\operatorname{sgn} h_n = \begin{cases} +1 & h_n \geq 0 \\ -1 & h_n < 0 \end{cases} \tag{6.22}$$

In this equation the coefficients of the c_j's are constant over certain regions of the $2N$-dimensional space of definition for $\{c_j\}$. Breakpoints where the coefficients assume new values occur whenever an output sample h_n changes sign. A minimum of D cannot occur between breakpoints where the function is linear; thus at least one value h_{k1} must be zero at the minimum. The equation

$h_{k1} = 0$ may be used to eliminate one of the variables c_i. The reduced equation is of the same piecewise-linear form, requiring at least one more output sample $h_{k2} = 0$, etc. We arrive at the conclusion that at least $2N$ samples of the output time sequence must be zero at the minimum. But $2N$ equations of the form $h_{ki} = 0$, where $i = 1, \ldots, 2N$, are sufficient to determine the values of the tap gains $\{c_j\}$. *Thus the setting of a (2N + 1)-tap equalizer which minimizes peak output pulse distortion forces 2N zeros at sampling points of the output pulse response.*

The question remains as to the values of the k_i's: Which $2N$ zeros does one force in the output response to achieve minimum peak distortion? This question may be resolved by the use of linear-programming techniques. Discussion of these techniques would lead this section somewhat too far afield; but briefly, each output sample h_n can be written as the difference between two nonnegative numbers h_n^+ and h_n^-.

$$h_n = h_n^+ - h_n^- \qquad h_n^+ \geq 0 \qquad h_n^- \geq 0 \tag{6.23}$$

If at the same time, for all n, either $h_n^+ = 0$ or $h_n^- = 0$,

$$D = \sum_n{}' (h_n^+ + h_n^-) \tag{6.24}$$

It can be shown that the process of minimizing a function of the form of Eq. (6.24) with the nonnegative constraints on h_n^+ and h_n^- will force either $h_n^+ = 0$ or $h_n^- = 0$ for all n. Therefore Eq. (6.24) is a valid objective function; by writing all the output samples h_n in terms of an arbitrary subset of $2N$ of these samples (only $2N$ of the output samples can be linearly independent), the distortion minimization can be put into the standard format for linear programming. That is, the minimization of a linear function can be made subject to a set of linear constraint equations using nonnegative variables. This background should provide the interested reader with sufficient information to obtain the necessary further details from the ample literature on linear programming.

Fortunately the necessity for linear programming in equalization is rare in practice; Theorem 6-1 shows why this is so.

Theorem 6-1: *If* $D_0 < 1$, *then the minimum distortion* D *must occur for those* $2N$ *tap-gain settings which cause* $h_n = 0$, $|n| \leq N$, $n \neq 0$.

A proof of this theorem may be found in Lucky.[6] The theorem says in essence that if the *binary eye* is not closed before equalization, use of the $2N$ free coefficients of the equalizer to force zeros in corresponding locations of the output pulse response is optimum.

The properties of the distortion as a function of tap-gain settings are further elucidated by Theorem 6-2.

Theorem 6-2: The peak distortion, Eq. (6.20), is a convex function of the $2N$ *variables* c_n, $|n| \leq N$, $n \neq 0$.

To prove this theorem we denote settings of the equalizer by the $2N$-component vectors $\boldsymbol{\alpha}$ and $\boldsymbol{\sigma}$. To prove convexity of D it is necessary to show that for any two settings $\boldsymbol{\alpha}$ and $\boldsymbol{\sigma}$ and for all λ, $0 \leq \lambda \leq 1$,

$$D[\lambda \boldsymbol{\alpha} + (1 - \lambda)\boldsymbol{\sigma}] \leq \lambda D(\boldsymbol{\alpha}) + (1 - \lambda) D(\boldsymbol{\sigma}) \tag{6.25}$$

This equation would show that the distortion always lies on or beneath a chord joining values of distortion in $2N$ space.

From Eq. (6.20)

$$D[\lambda \boldsymbol{\alpha} + (1 - \lambda)\boldsymbol{\sigma}] = \sum_{n=-\infty}^{\infty}{}' \left| \sum_{j=-N}^{N}{}' [\lambda\alpha_j + (1 - \lambda)\sigma_j](x_{n-j} - x_n x_{-j}) + x_n \right| \tag{6.26}$$

$$D[\lambda \boldsymbol{\alpha} + (1 - \lambda)\boldsymbol{\sigma}] = \sum_{n=-\infty}^{\infty}{}' \left| \lambda \left\{ \sum_{j=-N}^{N}{}' \alpha_j (x_{n-j} - x_n x_{-j}) + x_n \right\} + (1 - \lambda) \left\{ \sum_{j=-N}^{N}{}' \sigma_j (x_{n-j} - x_n x_{-j}) + x_n \right\} \right| \tag{6.27}$$

and

$$D[\lambda \boldsymbol{\alpha} + (1 - \lambda)\boldsymbol{\sigma}] \leq \lambda D(\boldsymbol{\alpha}) + (1 - \lambda) D(\boldsymbol{\sigma}) \tag{6.28}$$

QED

One of the most important properties of convex functions is that they possess no relative minima other their absolute minimum. Thus any minimum of D found by systematic search or other mathematical programming methods must be the absolute (or global) minimum of distortion.

In summary we have shown that D is a continuous, piecewise-linear, convex function of the tap gains. This function has a single minimum which must occur when $2N$ zeros appear in the output time sequence $\{h_n\}$. If the initial distortion is less than 100 percent ($D_0 < 1$), the minimum occurs when the $2N$ samples of the output time sequence which correspond in location to the $2N$ free taps on the transversal filter are simultaneously zero. This description is illustrated in a simple two-dimensional case in Fig. 6.3. In this figure the "corners" of the contour map are the points where two zeros appear in the output time sequence, while the edges are the lines along which a single zero is present. Obviously, the minimum must occur at a corner; in this case since $D_0 < 1$, the minimum is determined by that particular corner whose two output zeros are $h_1 = 0$ and $h_{-1} = 0$. Observe also that the function is piecewise-linear and convex.

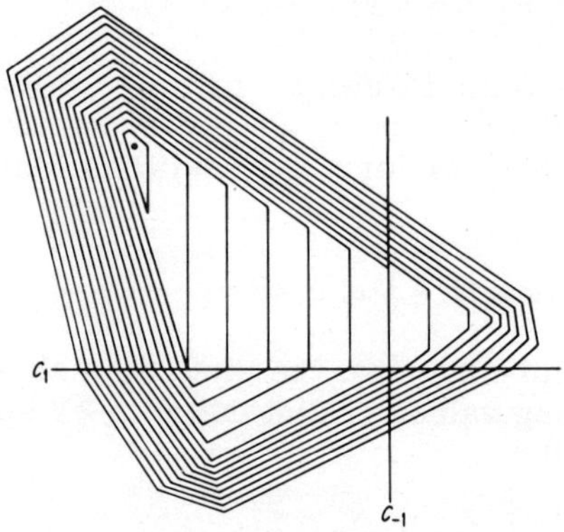

Fig. 6.3. Equal distortion contours for a two-tap example.

6.1.2 Minimization of Mean-square Distortion

The mean-square distortion of the output response is defined as

$$\mathcal{E}^2 = \frac{1}{h_0^2} \sum_n{}' h_n^2 \tag{6.29}$$

This time in minimizing $\mathcal{E}$ as a function of the tap-gain settings it is more convenient to constrain h_0 by means of a Lagrange multiplier. Thus we look for the point where $\partial V/\partial c_j = 0$ for $j = -N, \ldots, N$ where

$$V = \sum_{n=-\infty}^{\infty} h_n^2 + \lambda h_0 \tag{6.30}$$

Since

$$h_n = \sum_{j=-N}^{N} c_j x_{n-j} \tag{6.31}$$

we have

$$\frac{\partial V}{\partial c_j} = \sum_{n=-\infty}^{\infty} 2 h_n x_{n-j} + \lambda x_{-j} = 0 \tag{6.32}$$

Because the overall gain h_0 is not important in the minimization (it can be adjusted by proper scaling), it is most convenient to solve the set of equations

$$\sum_{n=-\infty}^{\infty} h_n x_{n-j} = x_{-j} \tag{6.33}$$

where $j = -N, \ldots, N$. Using Eq. (6.31) in (6.33) gives

$$\sum_{k=-N}^{N} c_k b_{j-k} = x_{-j} \tag{6.34}$$

where $j = -N, \ldots, N$ and

$$b_k = \sum_{n=-\infty}^{\infty} x_n x_{n-k} \tag{6.35}$$

The best tap gains in a mean-square sense are therefore obtained by solving simultaneously the $(2N+1)$ linear equations (6.34) and subsequently scaling the values obtained to achieve the proper pulse height.

The set of equations (6.33) has an interesting interpretation. Let ϵ_n denote the error at sample n present in the output pulse $h(t)$ as compared with an ideal pulse which has unity height and zero samples elsewhere, i.e.,

$$\left.\begin{aligned} \epsilon_n &= h_n \qquad n \neq 0 \\ \epsilon_0 &= h_0 - 1 \end{aligned}\right\} \tag{6.36}$$

Then from Eqs. (6.33) and (6.36)

$$\sum_{n=-\infty}^{\infty} \epsilon_n x_{n-j} = 0 \tag{6.37}$$

where $j = -N, \ldots, N$. Therefore the optimum setting of the equalizer will force the cross correlation between the error signal and the input pulse to be zero at each sample point within the range of the equalizer. [Any rescaling of the tap gains would, in this instance, destroy the simplicity of Eq. (6.37).]

The coefficient matrix of the simultaneous equations in Eq. (6.34) involves the infinite sum for b_k in Eq. (6.35). The b_k's are consequently more easily found for computational purposes from the equivalent frequency-domain expression, which is

$$b_k = \frac{1}{2\pi T} \int_{-\pi/T}^{\pi/T} |X_{\text{eq}}(\omega)|^2 \cos k\omega T \, d\omega \tag{6.38}$$

The x_j's on the right-hand side of the equations may also be obtained from the equivalent frequency characteristic $X_{eq}(\omega)$.

$$x_j = \frac{1}{2\pi}\int_{-\pi/T}^{\pi/T} X_{eq}(\omega)\, e^{ij\omega T}\, d\omega \tag{6.39}$$

Although the equations in this section which lead to optimum values for the coefficients c_j seem quite different from the equations which lead to minimum peak distortion ($h_n = 0, |n| \le N$, if $D_0 < 1$), in practice the two sets of results are nearly identical if the samples x_n are small for $n \neq 0$.

Minimization of total mean-square error. If the equalizer is to be set to minimize total mean-square error in the received data sequence, the effects of the equalizer on the noise must be taken into account. To arrive at an explicit solution it will be necessary for us to assume that

$$G_T(\omega) = G_R(\omega) = G^{\frac{1}{2}}(\omega) \tag{6.40}$$

where

$$\frac{1}{2\pi}\int_{-\infty}^{\infty} G(\omega)\, e^{in\omega T}\, d\omega = \delta_{n0} \tag{6.41}$$

In other words the transmitter and receiver filter are identical and together accomplish perfect Nyquist shaping. We also assume that the equalizer is to be placed at the receiver and that the transmission is L-level PAM.

The noise variance

$$\sigma_N{}^2 = \frac{N_0}{2\pi}\int_{-\infty}^{\infty} G(\omega)\,|E(\omega)|^2\, d\omega \tag{6.42}$$

can then be reduced simply to

$$\sigma_N{}^2 = N_0 \sum_{n=-N}^{N} c_n{}^2 \tag{6.43}$$

by using Eqs. (6.41) and (6.2). This expression for noise variance can then be used in Eq. (4.93) for total mean-square error and the parameter d^2 removed by using the expression for transmitted power

$$P_S = \frac{L^2 - 1}{3}\,\frac{d^2}{T} \tag{6.44}$$

When we normalize the total mean-square error by the square of the distance $x_0 d$ to the decision threshold, we arrive at

$$\sigma^2 = \frac{L^2 - 1}{3} \frac{1}{h_0^2} \left[\sum_n' h_n^2 + \frac{P_N}{P_S} \sum_n c_n^2 \right] \tag{6.45}$$

where P_N represents the noise power N_0/T in the Nyquist band.

The simultaneous equations which must be solved to find the best tap gains are similar to the previous set, Eq. (6.43). They are

$$\sum_{k=-N}^{N} c_k \left[b_{j-k} + \frac{P_N}{P_S} \delta_{kj} \right] = x_{-j} \tag{6.46}$$

where $j = -N, \ldots, N$. The coefficient matrix has been augmented by the noise-to-signal ratio times an identity matrix. To a first-order approximation for high P_S/P_N ratios this changes the tap settings about an amount P_N/P_S from the previous noiseless results. For a signal-to-noise ratio of 30 db the difference would be negligible compared to the accuracy achievable in practice in setting the gains c_j.

6.1.3 Numerical Example and Evaluation

Now we must answer the question of how well a data system with a finite equalizer can perfrom compared with an ideal system. In order to review the situation we have sketched in Fig. 6.4 block diagrams of a data system employing both the ideal linear receiver of Chap. 5 and the receiver of a conventionally equalized data system. The conventionally equalized system differs from ideal in that the portion of the matched filter at the receiver which matches the channel characteristic $C(\omega)$ is missing and that the

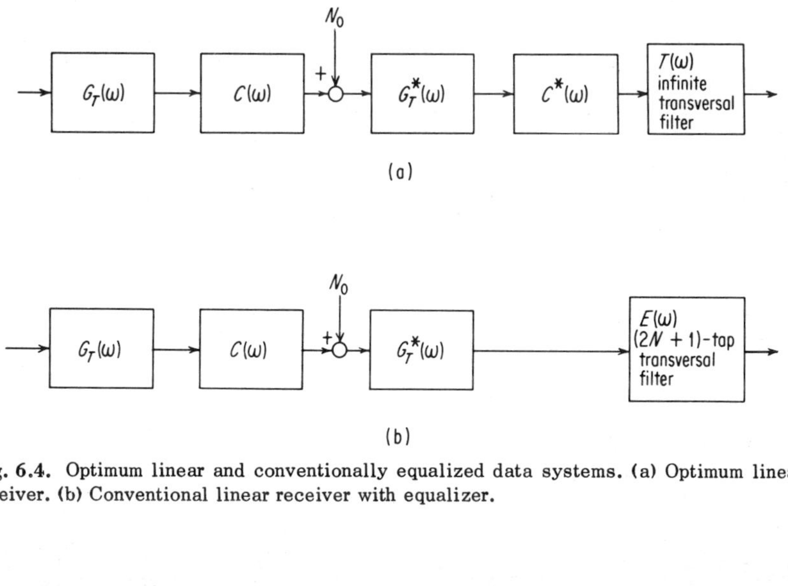

Fig. 6.4. Optimum linear and conventionally equalized data systems. (a) Optimum linear receiver. (b) Conventional linear receiver with equalizer.

transversal equalizer has only a finite number of taps available. Generally speaking we would expect the former factor to increase the noise variance in the output and the latter factor to increase the amount of intersymbol interference. Although Fig. 6.4 is drawn with the equalizer at the receiver, it should be apparent that it can be placed at the transmitter with little change in the numerical results for any specific case.

To evaluate performance for specific cases we shall employ two examples which have been used previously in Chaps. 4 and 5—parabolic delay and linear-attenuation (or slope) distortion. Naturally, nearly any kind of result one desires can be achieved by picking a proper example channel characteristic, so the ensuing results cannot be meaningfully extrapolated to other channel configurations. However, a rough estimate of the size of the equalizer required in any case can be obtained by the judicious use of echo theory (Sec. 4.2.2). The reasoning is approximately as follows: Assume that the channel response is perturbed in its amplitude characteristic by a ripple $\alpha \cos \omega\tau$. Since this response is multiplied by the equalizer frequency response

$$\sum_n c_n e^{-in\omega T}$$

the principal correction will be achieved from the tap gains c_n and c_{-n} where $n\omega T$ is closest to being equal to $\omega\tau$, or $n \approx \tau/T$. The product of the equalizer-response terms for these taps and the channel ripple will produce a cancellation term and a double-frequency residual ripple which will, of course, be of smaller amplitude. When these "echos of echos" are ignored, the relationship $n \approx \tau/T$ leads to a rule of thumb for equalizer length. The number of taps on either side of the reference tap (N) should be at least as great as the number of cycles in the Nyquist band $[-\pi/T, \pi/T]$ of the highest-frequency perturbing ripples of significance in either the channel's amplitude or phase characteristic.

Since in each of our examples the principle components of the Fourier expansion of the channel characteristic are of low frequency, a small number of taps on the equalizer can be used to effectively eliminate intersymbol interference. Unlike these mathematical channels, any real channel would exhibit a fine structure of high-frequency ripples in its frequency characteristic. This fine structure leads to far-removed echos in the channel time response which are uncorrectable with relatively short transversal equalizers. Therefore the results predicted by theory are almost always too optimistic in practice.

Residual rms distortion after equalization. Figures 6.5 and 6.6 show the residual rms distortion after equalization with a $(2N+1)$-tap equalizer for channel distortion caused by parabolic delay and slope attenuation. Each signaling shaping has been taken to be a a 60 percent raised-cosine characteristic, and the residual error $\mathcal{E}_r^2$ has been computed by solving the $(2N+1)$ simultaneous equation (6.34) with the coefficients given by Eq. (6.38). If the optimum tap gains are denoted by c_j', where $j = -N, \ldots, N$, the residual mean-square error may be easily found from the already available coefficient matrix $\{b_{ij}\}$ and

$$\mathcal{E}_r^2 = \frac{1}{h_0^2} \sum_{j=-N}^{N} \sum_{k=-N}^{N} c_j' c_k' b_{j-k} - 1 \tag{6.47}$$

The simplicity of this procedure for obtaining numerical results in comparison with the equivalent procedure for peak-distortion minimization is the reason the mean-square criterion has been used in Figs. 6.5 and 6.6. The use of the peak-distortion criterion would lead to entirely similar results.

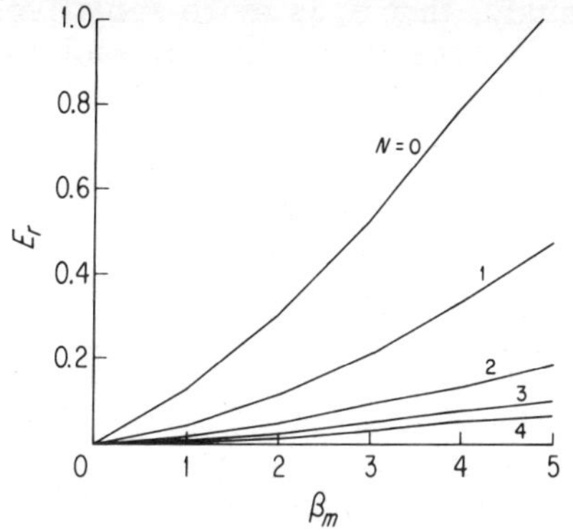

Fig. 6.5. Residual rms distortion after equalization E_r for parabolic delay β_m in pulse intervals at the Nyquist frequency.

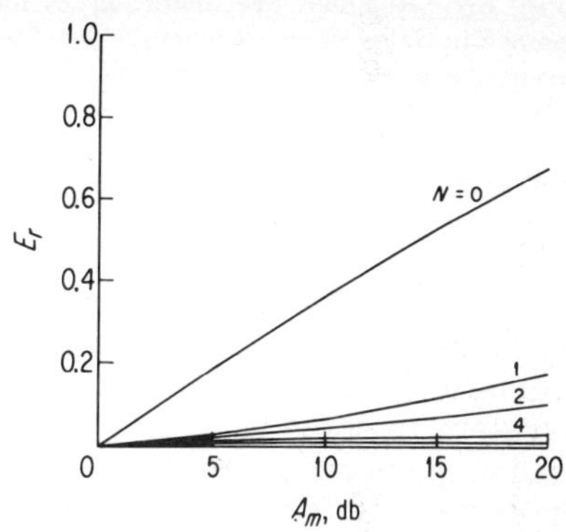

Fig. 6.6. Residual rms distortion after equalization for slope attenuation A_m at the Nyquist frequency.

One of the difficulties with residual-distortion computation thus far unmentioned is the problem of selecting an optimum reference time t_0. Both the optimum tap gains and the residual distortion depend on the initial choice of t_0. Naturally t_0 should be selected to minimize residual distortion, but the relationship is quite complicated [remember that ωt_0 must be subtracted from the phase characteristic in obtaining $X_{eq}(\omega)$], and no procedure other than trial-and-error or search techniques is known to effect this optimization. The problem of optimizing t_0 for slope-attenuation distortion is trivial since the time response must be symmetrical, and therefore $t_0 = 0$. However, considerable effort has been expended in finding the optimum timing for the parabolic-delay example of Fig. 6.5.

In most real systems the channel distortion itself will depend upon a further parameter, usually carrier phasing, which also must be optimized with respect to the final residual distortion. This means that for a practical calculation of residual distortion for, say, the VSB system treated in Chap. 7, a two-dimensional trial-and-error procedure involving the solution of a set of $(2N+1)$ simultaneous equations at each point is required. The calculation is, needless to say, quite tedious. Some idea of the factors involved may be obtained from an examination of the variation of residual distortion with t_0 for a specific parabolic-delay example in Fig. 6.7. Notice that t_0 is less critical when more taps are available for equalization. We would also find, by futher study, that $\mathcal{E}_r$ is more sensitive to t_0 both when the roll-off is sharper and when the distortion of the channel is greater. These problems of obtaining best timing and phasing are troublesome not only in analysis but also in the actual system implementation.

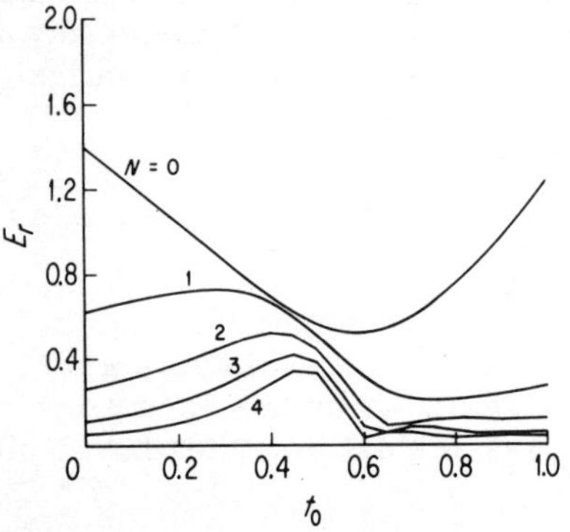

Fig. 6.7. Residual rms distortion E_r for parabolic delay β_m as a function of reference sampling time t_o. Parabolic delay $\beta_m = 3$; raised-cosine shaping $\alpha = 0.6$.

Noise performance of the conventionally equalized system. Principally because of the lack of a filter matched to the channel characteristic, the noise performance of the conventionally equalized system is somewhat poorer than for the ideal linear receiver. The noise performance is also affected by the substitution of a finite transversal filter for an infinite one. However, the effect on noise performance is usually negligible when the length of the equalizer is sufficient to ensure suppression of intersymbol interference since the missing tap gains would be very small and the noise variance varies as the square of these gains.

In order to give an idea of the amount of degradation in possible noise performance resulting from the lack of channel matching, we have plotted the noise enhancement factor $1/\beta$ (see Table 5-1) both for the ideal linear system which removes intersymbol interference and for the conventionally equalized system. These curves are shown in Figs. 6.8 and 6.9 for the examples of parabolic delay and of slope attenuation. It can be seen that very little improvement in performance can be obtained by the addition of the channel-matching

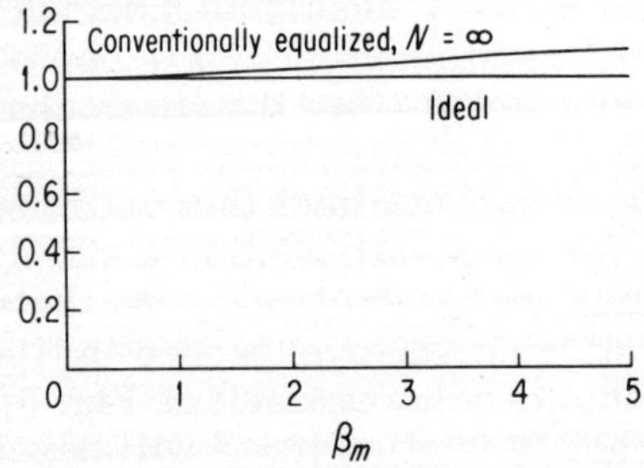

Fig. 6.8. Noise performance for parabolic delay β_m at Nyquist frequency.

filter if the channel distortion is not too severe. We have previously seen in Chap. 5 that the optimum linear receiver performance is close to the matched-filter bound. Therefore, in many cases where the channel distortion is mild (say, $D_0 < 1$) and the signal-to-noise ratio is large, the performance of a conventionally equalized system with a small number of taps is nearly equivalent to the best possible performance attainable with the given transmitted signal.

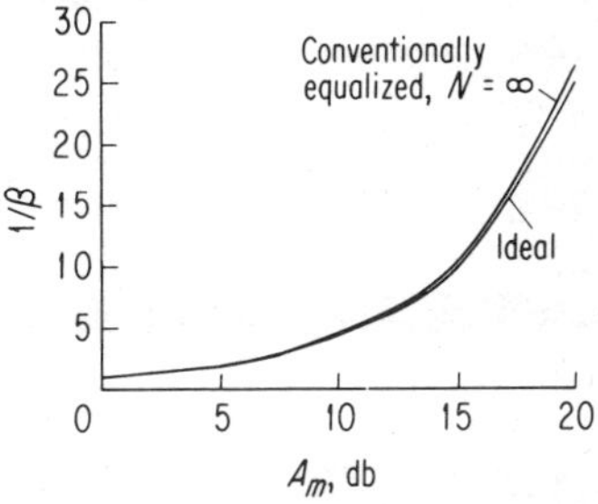

Fig. 6.9. Noise performance for slope attenuation A_m at Nyquist frequency.

Having seen that the matched filter for the channel has little effect on noise performance when channel distortion is not severe, we should consider its effect on the residual distortion $\mathcal{E}_r$. Here the result is dependent on the type of channel considered. Bear in mind that the matched filter will square amplitude characteristics and eliminate phase distortion. Therefore, for parabolic delay, the presence of a channel-matching filter would make the residual distortion identically zero and eliminate the requirement for a transversal equalizer. On the other hand, for slope-attenuation distortion, the presence of the matched filter would make matters worse for the transversal filter. More taps on the equalizer would be required to effect a given level of residual distortion. Any real channel would, of course, contain both amplitude and phase distortion: the former would be made worse and the latter eliminated. The incremental gain or loss and the added expense of an adaptive channel-matching filter should be carefully studied in any specific environment.

6.2 Equalizer-adjustment Algorithms

Having examined the problem of establishing optimum equalizer settings, we now must consider the procedure by which the equalizer is to be automated in order to arrive at these settings. There are many possibilities: the equalizer can be placed at the transmitter or at the receiver; it can be digital or analog; it can minimize peak or rms distortion; it can be set during a training period during data transmission, or by a continuous reference signal; etc. It will be impossible in this text to catalog the many varieties and hybrids of the equalizer family. Instead we shall concentrate on the basic aspects these equalizers have in common. Their most basic similarity is that they are all solvers of simultaneous linear equations.

Whether it is rms or peak distortion that is to be minimized, the solution of a set of simultaneous equations is required. In minimization of peak distortion the equalizer tap gains must be manipulated to zero the error voltages ϵ_n, where $n = -N, \ldots, N$. (Thus $h_n = 0$ for $1 \leq |n| \leq N$, $h_0 = 1$.) This assumes, of course, that the initial distortion is less than 100 percent, although in practice much higher values are encountered before this solution becomes suboptimal. This type of equalization is sometimes referred to as *zero forcing*. On the other hand, for rms equalization the tap gains must be adjusted to zero the cross correlations between the error voltages ϵ_n, and the input samples x_n at delays of $-N, \ldots, N$. In practice, the results of either type of equalization are nearly identical when $D_0 < 1$. It appears that zero forcing can be accomplished somewhat more easily than rms minimization, however rms equalization is not limited by $D_0 < 1$ and is generally less sensitive to perturbations in timing and phasing during equalizer setup.

Since in either case the equalizer acts as a solver of simultaneous equations, we can, without loss of generality, assume henceforth that a zero-forcing (peak-distortion minimizing) equalizer is to be considered. The equations that must be solved can be succinctly stated in the matrix form

$$Xc = 1 \tag{6.48}$$

where X is a $(2N+1)$th-order square matrix whose ijth element is x_{i-j}, c is the $(2N+1)$-element column vector whose jth element is c_j, and 1 is the $(2N+1)$-element column vector whose components are all zero except the zeroth, which is unity. Since $D_0 < 1$, the matrix X is diagonally dominated and consequently can be shown to possess a nonzero determinant. Furthermore, since the off-diagonal elements of X are relatively small, it is possible to conceive of various approximate expressions for the solution vector c which can be instrumented. For example, the vector with components $c_j = -x_j/x_0^2$, $j \neq 0$, and $c_0 = 1/x_0$ is a good approximate solution (for small x_j, $j \neq 0$) which has been used in an automatic equalizer by DiToro.[2]

Most highly accurate and simply instrumented methods for automatically solving the simultaneous equations involve iterative techniques. The very nature of the problem almost ensures the success of nearly any iterative technique of tap gain adjustment. Since the distortion D is a convex function of c, any sequence of adjustments may be used to reduce the output distortion. When it is no longer possible to reduce distortion by the movement of any tap gain, the optimum position has been reached. The situation is further simplified by the condition $D_0 < 1$, which ensures that the various tap gains are only "loosely" coupled in determining output distortion. In fact it can be shown that the tap gain c_j will affect output sample h_j more than the *combined* effects of all other tap gains moved an equal amount. This does not mean that there is no interaction but only that the effect of tap gain c_j on output sample h_k $(j \neq k)$ is small compared with the effect of gain c_k.

In practice the transversal equalizer may be adjusted easily, but not with great accuracy, by hand. Usually the taps c_{-1} and c_1 are adjusted first, followed by c_{-2} and c_2, etc. It is then necessary to return to c_{-1} and c_1 and readjust. The entire process may require several repetitions, but all adjustments may be conveniently made by observing the effect on the eye opening as displayed by an oscilloscope. Schreiner, Funk and Hopner[3] have designed an automatic equalizer based on a similar adjustment algorithm. In this algorithm tap c_1 is adjusted to zero-sample h_1, tap c_2 is used to zero sample h_2, etc. The process must be repeated several times because of the interactions involved, but eventual convergence is always assured.

More generally, and for faster adjustment, all $(2N+1)$ tap gains of the equalizer can be adjusted simultaneously. In this case the tap-gain vector $\mathbf{c}$ is incremented by an amount $\Delta\mathbf{c}$ which is a function of the error vector $\boldsymbol{\epsilon} = \mathbf{h} - \mathbf{1}$ in each state of the iterative procedure. The new error vector corresponding to the new tap-gain vector $\mathbf{c} + \Delta\mathbf{c}$ is used to derive a new tap gain increment, etc. Let $\mathbf{c}^k$ be the kth such tap gain vector; then we write

$$\mathbf{c}^k = \mathbf{c}^{k-1} - G(\boldsymbol{\epsilon}^{k-1}) \tag{6.49}$$

where

$$\boldsymbol{\epsilon}^{k-1} = X\mathbf{c}^{k-1} - \mathbf{1} \tag{6.50}$$

and G is an appropriate vector-valued operator.

Because of the relatively weak coupling between the tap gains, there is little point in using anything other than a "diagonal" structure for G; i.e., let

$$G\{\epsilon_{-N}, \ldots, \epsilon_N\} = \{g(\epsilon_{-N}), \ldots, g(\epsilon_N)\} \tag{6.51}$$

With this specification the $(2N+1)$-dimensional process described by Eq. (6.49) turns into $2N+1$ identical, but interacting, one-dimensional processes of the form

$$c_j^k = c_j^{k-1} - g(\epsilon_j^{k-1}) \tag{6.52}$$

There is a wide range of allowable functions g for which each of the one-dimensional iterative processes in Eq. (6.52) will converge inspite of the interaction of the other $2N$ processes. For these control functions error components $\epsilon_j^k \to 0$ as $k \to \infty$ for $|j| \leq N$.

The action of a real equalizer is considerably different from this ideal description as a result of the inevitable noise in the measurement of the error vector. Whether the error components are measured by test pulses or by cross correlation techniques (more on this later), the noisy process can be described by the iterations

$$c_j^k = c_j^{k-1} - g(\epsilon_j^{k-1} + \eta_j^{k-1}) \tag{6.53}$$

where η_j^k is the noise sample associated with the kth measurement of the jth error component. The effect of the noise components is to cause the tap gains, and consequently the error components, to exhibit random walks. When equalization begins, the error components are usually much larger, on the average, than the noise components, and the iterative process smoothly reduces the error components. However, a point is usually quickly reached where the error components are comparable in size to the noise samples. From this point on the iterations become more and more random. Generally if the process converges, the expected value of ϵ_j will approach zero, while its variance may asymptotically approach a constant value.

A typical illustration depicting the probability density of an error component at various times during equalization is illustrated in Fig. 6.10. Whether the equalizer is turned off at a random time or allowed to run continuously, it is clear that each tap gain contributes to residual distortion through its probabilistic nonoptimum setting. We define the average residual distortion per tap D_s as the expected value of the error magnitude $|\epsilon_j|$ after settling. The average final residual distortion of the output pulse is

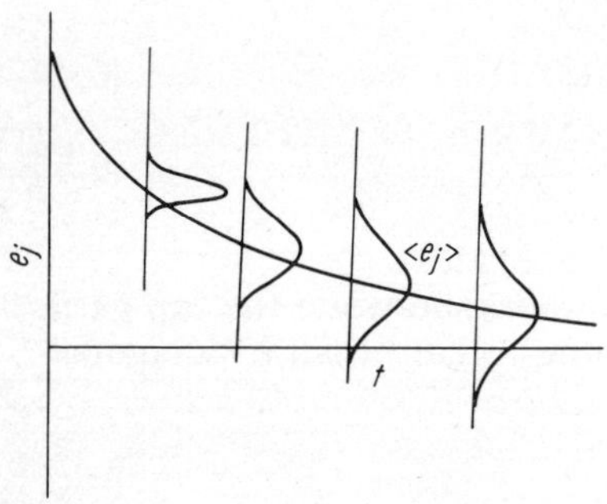

Fig. 6.10. Probability densities for an error component during equalization.

$$D_r \cong 2ND_s + D_{\min} \tag{6.54}$$

where $D_{\min}$ is the distortion when the error components are all zero, corresponding to the optimum tap setting. This relation is based on the approximation that small deviations in each tap gain will not significantly alter the distortion component associated with the output samples that were not to be zeroed.

The importance of the residual distortion can be emphasized by a specific example. Suppose that 16-level transmission is being used and that a half-open eye is desired. The output distortion D_r must then be reduced to 3 percent or less, implying that the average residual distortion per tap must be less than $0.03/2N$, even assuming that the zero-error residual distortion $D_{\min}$ may be neglected. In order to achieve a low-enough value for $D_{\min}$ at least 12 taps might be required on a voice telephone line; thus D_s must be held to 1/4 percent on each tap. At this small error value the error-to-noise ratio is 52 db below the signal-to-noise ratio in the output pulse. For a typical value of 30 db for SNR, this puts the required error-to-noise ratio at −22 db. Clearly a great deal of averaging must be done to achieve this kind of accuracy. No matter how the averaging is accomplished, the result must be a certain slowness in the settling of the equalizer to its equilibrium position. In some cases the engineer may be concerned with the amount of time required for settling, while in other examples this time may be immaterial.

In the next sections a few simple choices for the control function $g(\epsilon)$ will be examined as to their convergence, residual error, and settling time. These factors must be weighed against their simplicity of implementation and the specific requirements of the environment in which they are to be used.

6.2.1 Fixed-increment Adjustment

One of the simplest control functions to implement is given by

$$g(e) = \Delta \operatorname{sgn} \epsilon \tag{6.55}$$

Here each tap gain c_j is simply moved by an increment Δ opposing the direction of the error ϵ_j, i.e.,

$$c_j^k = c_j^{k-1} - \Delta \operatorname{sgn} \epsilon_j^{k-1} \tag{6.56}$$

Because the allowable values for the tap gain c_j are quantized with step size Δ, the physical realization of the variable-gain element can be, for example, a resistance ladder network controlled by an up-down counter. However, the quantization prohibits perfect convergence in the sense that $\epsilon_j^k \to 0$ as $k \to \infty$ in the absence of noise. It can be simply shown that if $D_0 < 1$, each tap gain *must* converge to within 2Δ of the optimum value, the average residual distortion per tap being about $\Delta/2$ (again, in the absence of noise).

When noise is present in the measurements of the error components, we have

$$c_j^k = c_j^{k-1} - \Delta \operatorname{sgn}[\epsilon_j^{k-1} + \eta_j^{k-1}] \tag{6.57}$$

where we might assume that η_k^k, the kth noise sample at tap j, is gaussian with zero mean and variance σ^2. The action of the equalizer adjustment algorithm is extremely complex. The error vector and the tap-gain vector exhibit random walks in a $(2N+1)$-dimensional space. The problem can be considerably simplified by assuming that there is no tap interaction and by treating each of the $(2N+1)$ components as an independent, one-dimensional, random walk. This type of approximation will be good for small initial distortion and will tend to underestimate the error and settling time when the initial distortion is large.

In dealing with the $(2N+1)$ identical one-dimensional random walks it will be convenient to drop the j index (denoting which tap is being considered) and to quantize both ϵ and c in integral multiples of Δ. We use the following notation:

$$\text{Prob}[\epsilon^m = k\Delta] = p_m(k) \tag{6.58}$$

$U(k)$ is the probability of "up" command when e is $k\Delta$

(the probability that $e + \eta < 0$)

$$U(k) = \frac{1}{\sqrt{2\pi}\,\sigma} \int_{-\infty}^{-k} e^{-t^2/2\sigma^2}\, dt = Q\left(\frac{k\Delta}{\sigma}\right) \tag{6.59}$$

The probability of a down command is then $[1 - U(k)]$, and the action of an error component can be described by the difference equation

$$p_{m+1}(k) = p_m(k-1)U(k-1) + p_m(k+1)[1 - U(k+1)] \tag{6.60}$$

When the equalizer begins operation, the error ϵ starts at, say, position $l\Delta$; thus Eq. (6.60) is subject to the initial condition

$$p_0(k) = \begin{cases} 1 & k = l \\ 0 & k \neq l \end{cases} \tag{6.61}$$

The difference equation (6.60) defines a probability distribution $p_m(k)$ which spreads out as time (m) progresses. The mean of the distribution approaches zero, and the variance rises asymptotically to a constant. Figure 6.11 shows the final system distortion per tap

$$D_S = \lim_{m \to \infty} \Delta \sum_k |k|\, p_m(k) \tag{6.62}$$

as a function of the increment Δ as obtained on a digital computer for various values of the signal-to-noise ratio $1/\sigma^2$. Analytical expressions for the behavior of the error ϵ can be derived approximately in the low- and high-noise regions. In the low-noise region, where equalization begins, the error decreases by Δ each step and the settling time is simply

$$T_S\Big|_{\text{low noise}} = \frac{\max_j |\epsilon_j|}{\Delta} \tag{6.63}$$

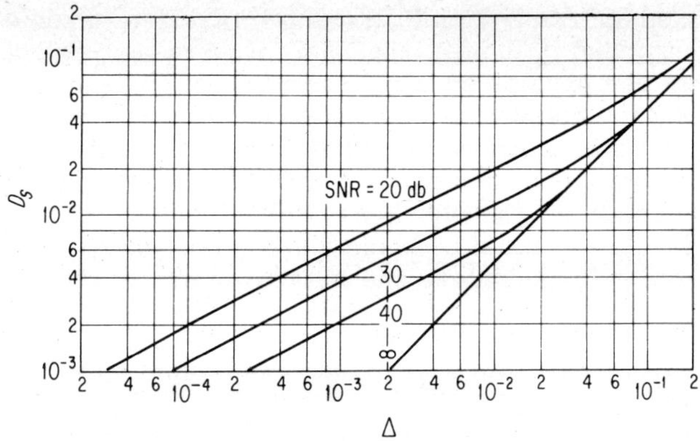

Fig. 6.11. D_S as a function of step quantization Δ for various SNR.

In the high-noise region, where the equalizer ends, a great many steps are required before the mean value of error can be reduced suitably. If $Q(k\Delta/\sigma)$ is approximated for small values of the argument, the difference equation (6.60) can be put into the form of the classical equation for the motion of an elastically bound particle. As a function of the number of iterations m, the average and mean-square value of the error are

$$\langle \epsilon \rangle = l\Delta \left(1 - \frac{2\Delta}{\sqrt{2\pi}\sigma}\right)^m \tag{6.64}$$

$$\langle \epsilon^2 \rangle = (l\Delta)^2 \left(1 - \frac{4\Delta}{\sqrt{2\pi}\,\sigma}\right)^m + \frac{\sqrt{2\pi}}{4}\Delta\sigma \left[1 - \left(1 - \frac{4\Delta}{\sqrt{2\pi}\,\sigma}\right)^m\right] \tag{6.65}$$

As $m \to \infty$ the limiting distribution is obtained where

$$\lim_{m \to \infty} \langle \epsilon \rangle = 0 \tag{6.66}$$

$$\lim_{m \to \infty} \langle \epsilon^2 \rangle = \frac{\sqrt{2\pi}}{4} \Delta\sigma \tag{6.67}$$

If we let the step-to-noise ratio Δ/σ tend to zero, the final probability distribution, now continuous, is gaussian, and the system distortion per tap becomes

$$D_S = \lim_{m \to \infty} \langle |\epsilon| \rangle = \left[\frac{\sigma\Delta}{\sqrt{2\pi}}\right]^{1/2} \tag{6.68}$$

The number of iterations required for the equalizer to settle to a given percentage of the final values is proportional to σ/Δ.

The fixed-increment method has been treated in some detail to show the profound influence that noise in the measurement of error components can have on equalizer action. The realities of equalizer design are that many taps may be required to reduce intersymbol interference and that therefore each tap gain must be set extremely accurately—perhaps an order of magnitude less than the relative noise appearing at the equalizer output. In the fixed-increment method the step Δ must be made quite small and the settling time becomes quite large unless the measurement of the error components themselves can be made accurately. If the error components are measured with sufficient accuracy, the system distortion per tap can be reduced to near the minimum possible value of $\Delta/2$.

6.2.2 Amplitude-dependent Algorithms

The fixed-increment method of adjusting the equalizer is an extremely simple technique. However, the unmodified technique can be quite inefficient from the point of view of settling time required for a given residual distortion. In order to achieve great accuracy in the final setting, the increment Δ must be quite small since an error of $\Delta/2$ is the minimum expected error attainable. However, during the initial period of equalization, when the error components are on the average larger than the noise samples, the small increments Δ are wasteful of time. One immediate and simple recourse which has been used in practice is to employ two different-sized increments Δ_1 and Δ_2. For an initial period of τ sec of operation a large increment Δ_1 is used; a "gearshift" is then employed to change to a smaller increment Δ_2 for final adjustment.

Another alternative to increase the efficiency of the iterative procedure is to employ a control function $g(e)$ that is amplitude-dependent, giving a larger correction when the error is large and a smaller correction when the error is small. For example, a two-step function would be

$$g(e) = \begin{cases} \Delta_1 \operatorname{sgn} \epsilon & |\epsilon| < \epsilon_0 \\ \Delta_2 \operatorname{sgn} \epsilon & |\epsilon| > \epsilon_0 \end{cases} \tag{6.69}$$

This function would also be quite simple to instrument, but it is not immediately apparent whether or not any benefit is achieved over the fixed-increment-with-gearshift method. In the latter method there is no possibility, as in the former, of a noise impulse causing a large step Δ_1 to take place near equalizer turn-off time resulting in an abnormally large residual error. If this possibility can be minimized, such multistep techniques as Eq. (6.69) offer improved settling-time performance over the fixed-increment method at a cost of increased equipment complexity.

A natural outgrowth of these considerations would be the use of a linear control function

$$g(e) = \alpha\epsilon \qquad (6.70)$$

as might be achieved by using motors to control tap gains. An equalizer using this principle has in fact been suggested by Davy and Saltzberg.* Gersho[7] has investigated the linear-control function and has shown that the maximum error component converges exponentially to zero in the fashion

$$\max_j |\epsilon_j{}^k| \leq P\beta^k \qquad (6.71)$$

where

$$P = \max_i |\epsilon_i{}^0| \qquad (6.72)$$

and

$$\beta = |1 - \alpha| + \alpha D_0 \qquad (6.73)$$

so long as the gain constant α is in the range

$$0 < \alpha < \frac{2}{1 + D_0} \qquad (6.74)$$

The number of iterations required for the largest error magnitude to be less than δ is bounded by

$$T_\delta \leq \frac{\log(P/\delta)}{\log(1/\beta)} \qquad (6.75)$$

All these equations are, of course, derived for the noiseless case. When gaussian noise of variance σ^2 is present in the measurement of the error components, Gersho[7] has shown the residual distortion per tap to be bounded by

$$D_S < 2\sigma\sqrt{\frac{\alpha}{1 - D_0}} \qquad (6.76)$$

Equation (6.76) shows that the residual distortion increases with the square root of the gain constant α. Naturally the settling time, Eq. (6.75), has exactly the opposite behavior. Fastest convergence is obtained with $\alpha = 1$, but to increase the accuracy it is necessary to use smaller values for α. Again an improved performance can be obtained by using two or more values of α and switching between them either as a function of time or of error amplitude.

---

* U. S. patent 3,289,108; November 29, 1966.

The nonlinear control function $g(\epsilon)$ which results in some suitably defined "best" performance is not known. However, Gersho[7] has shown that the iterative procedure will always converge if the control function $g(\epsilon)$ satisfies the following conditions:

1. $g(\epsilon)$ is monotonic nondecreasing,
2. $g(\epsilon)$ is an odd function of ϵ,
3. $g(\epsilon)$ lies within the bounds, $\phi e \leq g(\epsilon) \leq \epsilon$ for $\epsilon > 0$ and for some $\phi > 0$.

The settling time for this control function can be shown to satisfy the bounds of Eq. (6.75) with $\alpha = \phi$.

6.3 Equalization Systems

We shall weakly distinguish two types of automatic equalization. In *preset equalization* the equalizer is adjusted prior to, or during breaks in, data transmission by the use of a prearranged test signal. In *adaptive equalization* the equalizer continues to adjust itself during the transmission of data using either a low-level reference signal or by operating directly on the data signal itself.

Since the previous section discussed iterative techniques for equalizer adjustment, the only remaining consideration in the design of the equalization system is the method by which the error voltages ϵ_n are measured. (Note that $\epsilon_n = h_n$, except for $\epsilon_0 = h_0 - 1$.) There are clearly many techniques and implementations for automating the equalization system. We shall give two examples in the following sections on preset and adaptive equalization. These examples are only intended as such—two ways that have been successfully used in practice. The rapid change in device technology will determine the most efficient and economic system of the future.

6.3.1 Preset Equalization

In the simplest preset equalizer the error components are measured by transmitting test pulses through the system and inspecting the output of the transversal filter at the sampling times. Such a preset equalizer is shown in Fig. 6.12. In this system the fixed-increment-iteration method is used while a sequence of widely separated test pulses is transmitted. At the output of the transversal filter the pulses are sampled at times selected by a timing circuit triggered by a peak detector. The center sample is sliced at +1 (or whatever reference voltage is desired) to produce ϵ_0 while the other samples are sliced at zero level. The polarities of the error voltages are then stored in a shift register. At the end of each test pulse a gate is opened, and the contents of the shift register are used to command the tap gains along the transversal filter to "count up" (increase gain by Δ) or "count down" (decrease gain by Δ).

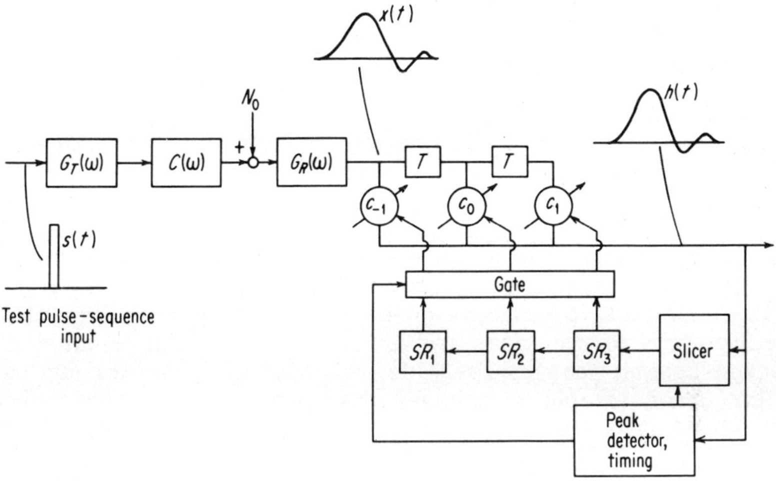

Fig. 6.12. Block diagram of preset equalizer with fixed-increment adjustment.

The residual distortion per tap for this type of iteration was discussed in the previous section. In a typical high-efficiency application we might require that $D_S = 1/4$ percent while the signal-to-noise ratio in the output pulse might be 30 db. We see from Fig. 6.11 that an increment of $\Delta = 0.0004$ is required. The expected error of 1/4 percent represents about six such steps. Should this increment be used throughout the equalization interval, about 1,000 test pulses would be needed before the largest error component was brought down to the noise level. A gearshift in Δ could be used to decrease this figure considerably.

The step size of 0.0004 in this example would be difficult to realize in practice. Instead, it might be preferable to average the effect of many test pulses before making any changes in the equalizer setting. If sufficient averaging of test pulses were accomplished, the maximum value of $2D_S = 0.005$ could be used for Δ. Sufficient averaging would mean that the probability of a correct adjustment when the error voltage was Δ would have to be on the order of 0.9 or so. Let $p_c(\epsilon)$ be the probability of a count in the proper direction when the error voltage is ϵ. In the equalizer of Fig. 6.12 we have

$$p_c(\epsilon) = 1 - P[\eta > \epsilon] = 1 - Q\left(\frac{\epsilon}{\sigma}\right) \tag{6.77}$$

When the error is small relative to σ, this is approximately

$$p_c(\epsilon) \cong \frac{1}{2} + \frac{\epsilon}{\sigma\sqrt{2\pi}} \tag{6.78}$$

which, for the numbers we have given, would be very little greater than the random count of $p_c(\epsilon) = 1/2$.

The error voltages could be averaged, say K at a time, to effect a reduction of σ by $\sqrt{K}$. One of the difficulties with this idea is that the operation is slowed by a factor of K throughout the equalization interval, whether or not the error is small relative to the noise. A method used in practice to achieve the benefits of averaging without suffering the attendant inefficient settling process is *sequential testing*. In a sequential test the sample size itself is a random variable determined by the input data. As applied to the equalizer, thresholds of $+v$ and $-v$ volts would be set and running sums of the measured error voltages would be constantly checked against these thresholds. Whenever the sum for any particular error sample ϵ_j crossed the upper or lower threshold, a corresponding decision would be made to increase or decrease tap gain c_j by Δ. This type of test has the advantage of requiring less time, on the average, than a test of a similar power using a fixed time interval. The big advantage in equalization is that the sequential test is able to terminate itself quickly when the error sample is large relative to the noise and to effect the slower test when so required because of the submersion of the error sample into the noise.

As usually implemented, the sequential test is digitalized so that instead of storing the actual error samples $(\epsilon_j^k + \eta_j^k)$, the polarity sgn $(\epsilon_j^k + \eta_j^k)$ is used to increment by one unit an up-down counter of capacity $2C$ counts. Whenever the counter overflows, a decision is made that $\epsilon_j^k > 0$, and the counter is reset to its middle position. Similarly, on underflow the decision $\epsilon_j^k < 0$ is made, and the counter again resets to its middle position. The minimum number of counts needed to effect a decision is C. This number would be close to the actual figure required during the early stages of equalization. On the other hand, it can be shown that when the input counts are entirely random $(\epsilon = 0)$ an average of C^2 counts is required before over or underflow. The equalizer would, on correct adjustment, end by requiring about this number of samples per movement of tho tap gain. An illustration of the action of the overflow counters is shown in Fg. 6.13.

The problem of over and underflows is identical to the classical problem of the gambler's ruin. If p is the probability of an up count and q is the probability of a down count, the probability of an overflow occurring before an underflow is given by Feller[8] as

$$p_c' = 1 - \frac{(q/p)^{2C} - (q/p)^C}{(q/p)^{2C} - 1} \tag{6.79}$$

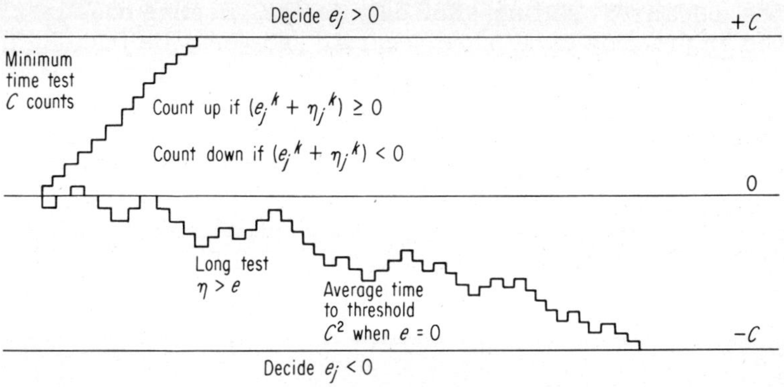

Fig. 6.13. Action of overflow counters.

In the case in which we are most interested, p is the probability $p_c(\epsilon)$ in Eq. (6.78) and is only slightly larger than 1/2. Under this circumstance an excellent approximation to Eq. (6.79) is

$$p_c'(\epsilon) \cong 1 - \exp\left[\frac{-4C\epsilon}{\sigma\sqrt{2\pi}}\right] \tag{6.80}$$

An appropriate counter size C can be selected in order to make $p_c'(\Delta)$ about 0.9. The average number of counts required before a decision is also given by Feller as

$$\overline{K} = \frac{C}{q-p} - \frac{2C}{q-p}\left[\frac{1-(q/p)^C}{1-(q/p)^{2C}}\right] \tag{6.81}$$

for which a satisfactory approximation when p is given by Eq. (6.78) is

$$\overline{K} \cong \frac{C\sigma}{\epsilon}\sqrt{\frac{\pi}{2}} \tag{6.82}$$

When $\epsilon = 0$, $\overline{K}$ may be shown to be equal to C^2.

6.3.2 Adaptive Equalization

In adaptive equalization the error voltages ϵ_n are continually estimated during the course of normal data transmission, and corrections to the equalizer are effected as required. Such a closed-loop system has an inherent accuracy advantage over the preset type of equalizer. Besides having the ability to adapt to any changes in channel transmission characteristics during a data call, the adaptive equalizer obviates the necessity for a long training period and reduces the effects of channel nonlinearities and slicing bias on the equalizer setting. Not only is the adaptive equalizer observed in practice to be more accurate and versatile than the preset variety but it also actually appears that adaptive equalizers can be built more cheaply than similar preset equalizers. This latter fact comes about because the existence of a continuous error signal in the equalizer loop makes it possible to use volatile memory in the tap-gain settings. It is then no longer necessary, as in preset operation, to hold a gain setting over an indefinite period of time.

The most common means of identifying the impulse response sample $h(\tau)$ of a linear system is by cross correlating the output voltage with the known input signal delayed by τ sec. Similarly, in adaptive equalization to obtain a maximum likelihood estimate of the error sample ϵ_j the input data sequence $\{a_{n-j}\}$ is cross correlated with the difference voltage between the actual received sample y_n and the desired sample a_n over a sequence of length K. Specifically we consider the estimates

$$\hat{\epsilon}_j = \frac{1}{\overline{a^2} K} \sum_{n=1}^{K} a_{n-j}(y_n - a_n) \tag{6.83}$$

Substituting the expression

$$y_n = \sum_i a_i h_{n-i} + \eta_n \tag{6.84}$$

we have for the expected value of $\hat{\epsilon}_j$

$$\hat{\epsilon}_j = \frac{1}{\overline{a^2} K} \left\{ \sum_{n=1}^{K} \sum_i a_i a_{n-j} h_{n-i} + \sum_{n=1}^{K} \left[\langle a_{n-j} \eta_n \rangle - \langle a_{n-j} a_n \rangle \right] \right\} \tag{6.85}$$

Provided $\langle a_n a_m \rangle = \overline{a^2} \delta_{nm}$, it can be seen that

$$\langle \hat{\epsilon}_j \rangle = h_j - \delta_{j0} = \epsilon_j \tag{6.86}$$

This result suggests an efficient means of adjusting a preset equalizer. Instead of transmitting a sequence of isolated test pulses, a quasi-random pattern is sent. At the receiver the same quasi-random sequence is available through, say, a maximal length sequence generator. This sequence is subtracted from the sequence of received samples $\{y_n\}$, and the resulting voltages are correlated with the sequence $\{a_{n-j}\}$ to derive an estimate of the error sample ϵ_j. By this means the preset equalizer can be set to a given accuracy in less time than by means of test-pulse transmission. This is because the test pulses must be spaced so far apart in time to prevent any overlap that the average power transmitted must be very low (since the pulses themselves must be peak-power-limited to avoid any nonlinear effects in the transmission channel).

In normal, random data transmission the transmitted sequence $\{a_n\}$ is not available. However, a very good facsimile of this sequence is presumably being decided at the output of the threshold detector. For the purposes of adaptive equalization the delivered data at the receiver can be used as if it were the correct transmitted sequence $\{a_n\}$ to obtain the estimate of the error components in Eq. (6.83). Such a learning procedure has been termed *decision-directed* since the receiver attempts to learn by employing its own decisions. If these decisions are relatively good, say $P_e < 0.1$, the equalizer can generally obtain a good enough indication of the error components through the correlation procedure to improve its tap-gain settings. The improved tap-gain settings result in a lower error rate and correspondingly better estimates of the error components, etc. However, the reverse effect can also occur, and the equalizer settings can completely lose acquisition of the channel.

An adaptive equalizer based on this principle is shown in Fig. 6.14. In an adaptive equalizer which may have to track a time-varying channel we are concerned with settling times and accuracies even more than with the preset equalizers. For the equalizer of Fig. 6.14, the settling times and accuracies can be estimated from the variance of $\hat{\epsilon}_j$. This variance can be shown to be

$$\sigma_K^2(j) = \langle (\hat{\epsilon}_j - \epsilon_j)^2 \rangle = \frac{1}{K} \sum_{\substack{n=-\infty \\ n \neq j}}^{\infty} \epsilon_n^2 + \frac{2}{K^2} \sum_{n=1}^{K-1} (K-n)\epsilon_{j+n}\epsilon_{j-n} + \frac{\sigma^2}{K\overline{a^2}} \tag{6.87}$$

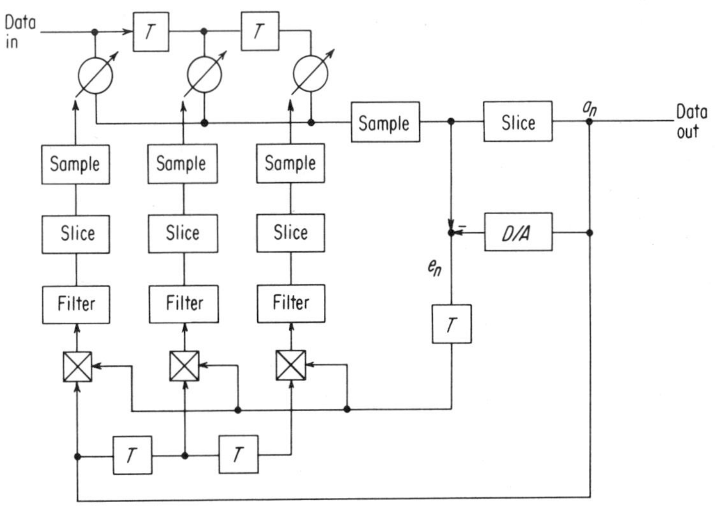

Fig. 6.14. Periodically sampled adaptive equalizer.

For a large averaging time K we might assume that the estimate $\hat{\epsilon}_j$ is normally distributed with mean ϵ_j and variance $\sigma_K^2(j)$. The probability $p_c(\epsilon,K)$ of a correct adjustment after K is then

$$p_c(\epsilon,K) \cong 1 - Q\left(\frac{\epsilon}{\sigma_K}\right) \tag{6.88}$$

As an example assume that the equalizer is very near its optimum settings so that $\epsilon_j = \pm 0.5\Delta$ with equal probability when $|j| \leq N$, and is zero otherwise. Then we have

$$\langle \sigma_K^2(j) \rangle = \frac{1}{K}\left(\frac{N\Delta^2}{2} + \frac{\sigma^2}{a^2}\right) \tag{6.89}$$

The first term $N\Delta^2/2$, represents a sort of "intererror" interference in the estimate of ϵ_j, but when the errors are small, this term is usually negligible relative to the second term which represents noise. In a typical voice telephone line application with a signal-to-noise ratio of 30 db an averaging time K of several thousand symbols would be required to make $p_c(0.005) = 0.9$, which would ensure roughly 1/4 percent accuracy per tap.

Since it would be inefficient to use a long averaging time K when the error components are large compared with the noise, the use of overflow counters or some other sort of sequential testing is recommended in place of the fixed-sample-size test just discussed. The necessity for analog-to-digital conversion and analog multiplication in the adaptive equalizer can also be removed by polarity coincidence correlation as illustrated in the digital adaptive equalizer of Fig. 6.15. The equalizer shown is for a 16-level system.

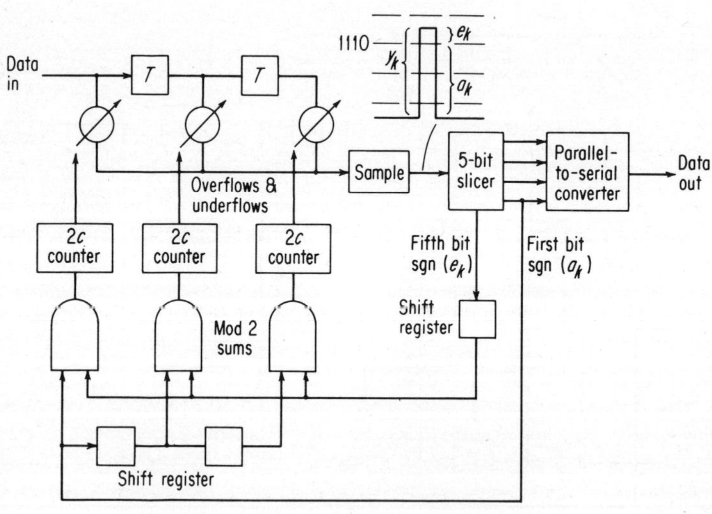

Fig. 6.15. Three-tap digital adaptive equalizer.

The digital correlation involves forming the sum

$$v_j = \sum_{n=1}^{K} \text{sgn}\,(a_{n-j})\ \text{sgn}\,(y_n - a_n) \tag{6.90}$$

whose expected value may be shown to be proportional to ϵ_j. However, in the fixed-increment method illustrated it is only necessary that sgn (v_j) = sgn (ϵ_j) for a correct increment of the overflow counters. In this system the polarities sgn (a_n) and sgn $(y_n - a_n)$ are easily obtained by the addition of one extra stage of slicing over the conventional four stages required for 16-level transmission. The first bit out of the slicer in the Gray code representation for 16-level transmission is the symbol polarity sgn (a_n), while the addition of a fifth stage produces the polarity of the difference voltage sgn $(y_n - a_n)$. (Because of the folding involved in the conventional slicer, it is necessary to add to the fifth bit a parity bit calculated from the first four bits.)

By making a number of suitable approximations, it has been shown[9] that the probability $p_c(\epsilon)$ of a correct increment to the storage counters for the equalizer of Fig. 6.15 when the error voltage is ϵ is

$$p_c(e) \cong \frac{1}{2} + \frac{\epsilon L d}{2\sigma\sqrt{2\pi}} \tag{6.91}$$

where there are L levels, spaced $2d$ apart, and the noise has variance σ^2. By using overflow counters of total capacity $2C$ counts the final probability of correctly adjusting the equalizer tap is approximately

$$p_c'(e) \cong 1 - \exp\left[-\frac{2C\epsilon dL}{\sigma\sqrt{2\pi}}\right] \tag{6.92}$$

from which the required storage capacity C may be calculated. The time required for an adjustment in the equalizer is a random variable whose mean as a function of ϵ varies between C and C^2 symbols. The mean number of counts per adjustment is approximately

$$\overline{K} = \frac{C\sigma\sqrt{2\pi}}{dL\epsilon} \tag{6.93}$$

In the adaptive equalizer the linear-control-function iterative technique is simply instrumented by closing the loop on the estimated error signal with a variable gain device for c_j such as a field-effect transistor. Such an equalizer is quite simple, but it does have a difficult time establishing initial equalization. This is a failing common to most decision-directed methods. Once correct equalization is acquired, the error-component estimates are accurate, and the equalization loops track easily unless the channel changes faster than the equalizer reaction time permits. One way around this difficulty is to use a hybrid system which permits a brief preset period during which a quasi-random sequence is transmitted and regenerated at the receiver. When equalization is sufficiently good (usually an open eye is required), the equalizer is changed over to the adaptive mode and data transmission may be begun.

One further problem with decision-directed adaptive equalization is the dependence of the equalizer loop on the transmission of "random" data. Unfortunately many sources of data are characterized by long periods of such repetitive patterns as all 1s or the dotting sequence 101010, etc. During these periods the frequency spectrum of the line signal consists solely of a few discrete spectral components. Clearly it is impossible for the equalizer to be able to identify more than a few isolated points of the channel's frequency characteristic. Since the equalizers discussed in this section are forced to continue some sort of movement, the suspicion arises that a sort of runaway in equalizer settings might be involved. Fortunately it can be shown that this is not so. The equalizer will tend to oscillate about its former position, acquired by random data, until the repetitive sequence ends and meaningful data are once again transmitted. In practice a runaway of equalizer settings resulting from poor data statistics is very rarely observed.

6.4 Summary

In this chapter we discussed transversal-filter equalization. The questions raised were the following: What are the best settings for the equalizer? How well does the equalized system perform relative to the optimum? How should the equalizer be automated so as to achieve these best settings? It was shown that in both rms- and peak-distortion minimization the transversal-filter tap gains satisfy a set of linear simultaneous equations. For peak-distortion minimization the equalizer must force zeros at sampling times of the output pulse response, while in rms-distortion minimization the equalizer must force to zero the cross correlation between input and error voltages at the sampling times.

It was shown in several examples that if channel distortion is not severe, a finite transversal filter can provide near optimum performance. The automation of the equalizer was then discussed. Since the equalizer must solve with great accuracy a set of simultaneous linear equations, iterative techniques are most appropriate. Several examples were shown involving preset and adaptive equalization systems utilizing iterative methods for distortion minimization. The emphasis was placed on the generally severe accuracy requirements for setting of tap gains and the attendant interaction with equalizer response time.

The field of automatic equalization is a new and rapidly expanding area which is being critically affected by changing device technology. As the economics of devices change, so do the best engineering solutions of equalization problems. The great wave of the present is the press toward digitalization and integrated circuits. The automatic equalizer is a system which is particularly amenable to this philosophy. The entire equalizer can be easily digitalized and certain operations time-shared to result in the most economic package of today. Tomorrow this philosophy may change, but the basic questions of accuracy, settling time, system identification, and optimum equalizer settings must still confront the engineer much in the fashion discussed in this chapter.

Notes

Only implementation for minimization of peak distortion has been described in the text. For an example of mean-square distortion minimization see Coll and George.[10] Although we have also restricted this chapter to baseband synchronous-data equalization, various automatic equalizers have been designed for other data systems. Equalization for four-phase data transmission is the subject of Kettel,[11] while O'Neill and Saltzberg[12] describes equalization for quadrature carrier systems. A more general type of passband channel equalization is given in Lucky and Rudin.[13] This latter type of equalizer can be used for any data modem but is not as efficient for synchronous, baseband transmission as the varieties described in this chapter.

7

Linear Modulation Systems

The previous three chapters have discussed pulse-amplitude-modulation systems at baseband. Most real communication channels have very poor transmission in the neighborhood of zero frequency and hence are regarded as passband channels. Therefore, it is necessary to consider techniques of shifting the baseband data signal to more suitable frequency ranges for transmission over typical communications media.

In this chapter we consider linear-frequency translation methods. Emphasis will be placed on basic understanding rather than on particular details of instrumentation. Since the performance of linear-modulation and -demodulation systems in the presence of various impairments (such as noise and delay distortion) can be deduced from an equivalent baseband system, no detailed analysis of performance will be undertaken in this chapter. An equivalent baseband structure will be derived which will enable the reader to correspond performance results derived in the chapter on baseband systems with results pertaining to linear-modulation systems. Only impairments unique to the modulation process, such as phase and frequency errors in the recovered carrier, will be discussed in detail.

7.1 Methods of Modulation

Two types of modulation are recognized, linear and nonlinear. Linear-modulation techniques are well suited for high-speed data transmission because of their efficient use of available bandwidth. In order to utilize the bandwidth fully, coherent detection techniques must be employed. This implies that carrier frequency and phase must be precisely known at the receiver. These requirements tend to increase the modem complexity.

Nonlinear techniques such as frequency and phase modulation, on the other hand, can effectively be operated in an incoherent mode, thus eliminating the carrier recovery problems associated with efficient linear-modulation methods. However, as we shall see, these techniques are not efficient from a bandwidth point of view and therefore are not well suited for high-speed applications.

In this chapter we examine the properties of linear amplitude modulation, while in Chaps. 8 and 9 we consider nonlinear angle modulation.

7.1.1 Signal Generation and Representation

Linearly modulated waves may be produced by multiplying the information-carrying signal by a sinusoidal carrier signal and processing the product by a linear filter. Figure 7.1 shows a block diagram of a general linear modulator. The response of this system to a baseband signal $g(t)$ is

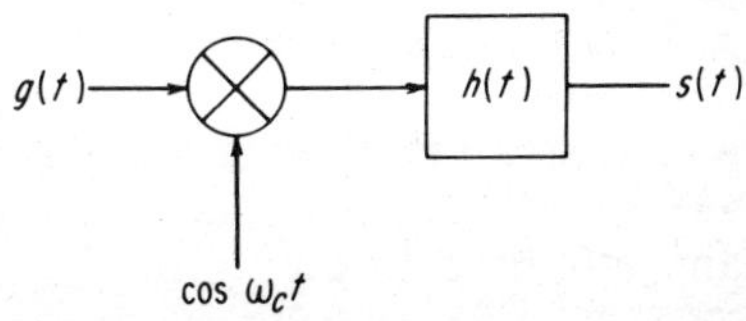

Fig. 7.1. Block diagram showing a general linear modulator.

$$s(t) = \int_{-\infty}^{\infty} h(\tau) g(t - \tau) \cos(\omega_c t - \omega_c \tau)\, d\tau$$

$$= \cos \omega_c t \int_{-\infty}^{\infty} h(\tau) g(t - \tau) \cos \omega_c \tau \, d\tau$$

$$+ \sin \omega_c t \int_{-\infty}^{\infty} h(\tau) g(t - \tau) \sin \omega_c \tau \, d\tau \qquad (7.1)$$

Introducing two related low-pass filters whose impulse responses are

$$h_1(t) = h(t) \cos \omega_c t \quad \text{and} \quad h_2(t) = h(t) \sin \omega_c t \qquad (7.2)$$

allows Eq. (7.1) to be written as

$$s(t) = [g(t) * h_1(t)] \cos \omega_c t + [g(t) * h_2(t)] \sin \omega_c t \qquad (7.3)$$

where $$g(t) * h(t) = \int_{-\infty}^{\infty} g(\tau) h(t - \tau)\, d\tau$$

It is customary to refer to $g(t) * h_1(t)$ as the *in-phase component* of the modulated signal and $g(t) * h_2(t)$ as the *quadrature component.*

The system response could just as well have been written in the frequency domain where the frequency shifting process is exhibited explicitly; that is,

$$S(\omega) = \frac{1}{2} H(\omega)[G(\omega + \omega_c) + G(\omega - \omega_c)] \qquad (7.4)$$

where $H(\omega)$ is the transfer characteristic of the filter $h(t)$, and $G(\omega)$ is the Fourier transform of the input signal $g(t)$. The time- and frequency-domain representations will be interchanged freely; convenience will dictate which will be used in any particular situation.

The notion of linearity is self-evident from the above representation since if the sum $g_1(t) + g_2(t)$ is admitted to the modulator, the response is a linear combination of the individual responses. In other words, there are no product terms involving both $g_1(t)$ and $g_2(t)$ in the output.

An alternate means of generating linearly modulated signals suggested by Eq. (7.3) is shown in Fig. 7.2. In this representation the information-bearing signal is processed by two related filters whose outputs are multiplied by cos $\omega_c t$ and sin $\omega_c t$, respectively, and summed to give $s(t)$.

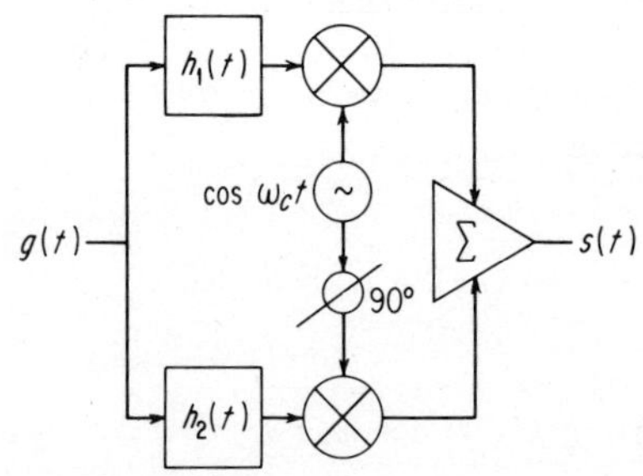

Fig. 7.2. A block diagram showing an alternate means of generating a linearly modulated signal.

In practice, of course, the transmitting bandpass filter $h(t)$ may also be used to shape the baseband signal waveform to some desired characteristic as discussed in previous chapters. Also, since the channel is assumed to be linear, its characteristics may be combined with the transmitting filter to form an overall linear-transmission system. The inclusion of the channel in the arbitrary filter characteristic $H(\omega)$ is convenient and often simplifies the analysis. As discussed in previous chapters the exact channel characteristics are usually not known, and therefore equalization may be required. For purposes of this discussion it will be assumed that the channel is equalized and the baseband signal is appropriately shaped so that one has complete liberty to design the transmitting filter $h(t)$, to achieve the type of linearly modulated signal desired.

By specializing $h(t)$, it is possible to generate the various familiar types of linearly modulated signals such as double-sideband amplitude-modulation (DSBAM), double-sideband suppressed-carrier (DSBSC), vestigial-sideband (VSB), and single-sideband (SSB). These various modulation methods will now be investigated.

7.1.2 Double-sideband Modulation

Suppose $h(t)$ is chosen to be an ideal filter whose impulse response is a delta function $\delta(t)$. It is then apparent from Eq. (7.1) that the output of the modulator is simply

$$s(t) = [g(t) \cos \omega_c t] * \delta(t) = g(t) \cos \omega_c t \tag{7.5}$$

The signal $s(t)$ has frequency components symmetrically disposed on either side of the carrier frequency ω_c; that is, it is a *double-sideband* (DSB) *signal*. The frequency spectra of a representative $g(t)$ and $s(t)$ are shown in Fig. 7.3. The signal $s(t)$ is said to be un-

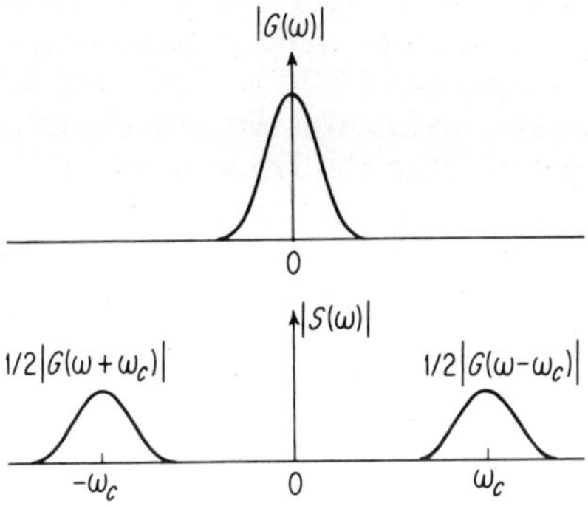

Fig. 7.3. Frequency spectra of $g(t)$ and $s(t)$.

supressed- or suppressed-carrier depending on whether the signal $g(t)$ has or does not have an average, or dc, component. This component, denoted by g_d, is given by

$$g_d = \lim_{T \to \infty} \frac{1}{2T} \int_{-T}^{T} g(t)\, dt \tag{7.6}$$

It is possible to define a new signal $y(t)$ such that $g(t) = g_d + y(t)$, where $y(t)$ has no dc component. Using this change in notation, we write Eq. (7.5) as

$$s(t) = g_d \cos \omega_c t + y(t) \cos \omega_c t \tag{7.7}$$

and consequently the frequency spectrum becomes

$$S(\omega) = \frac{g_d}{2} \delta(\omega - \omega_c) + \frac{g_d}{2} \delta(\omega + \omega) + \frac{1}{2} Y(\omega - \omega_c) + \frac{1}{2} Y(\omega + \omega_c) \tag{7.8}$$

When g_d is not equal to 0 it can be seen that in addition to the information-bearing components in the spectrum, steady-state sinusoidal components are also present.

From the spectral representation it should be clear that DSB expands the bandwidth of the information-carrying signal by a factor of 2. For every frequency component above the carrier frequency there is one below. This seems redundant, and suggests the possibility of halving the bandwidth of the signal $s(t)$ by eliminating the upper or the lower sideband. This suggestion is the basis of the more efficient linear-modulation techniques discussed in the following sections.

7.1.3 Single-sideband Modulation

A passband signal which results from a linear-modulation process and has frequency components only on one side of the carrier is said to be *single-sideband* (SSB) *signal*. The bandwidth of such a signal equals the base bandwidth. We now examine the structure of these signals by specializing the filter $h(t)$ in Fig. (7.1).

It is useful at this stage to introduce the notion of the Hilbert transform[1] of a signal. Much of the work dealing with single- and vestigial-sideband modulation becomes very simple when this concept is employed. However, a rigorous mathematical treatment of Hilbert transforms will not be attempted. Rather, the concept will merely be defined, and some of its important properties will be stated without proof.

Let $g(t)$ be a well-behaved function on the real t-axis. Its Hilbert transform is defined as

$$H[g(t)] = \hat{g}(t) = \frac{1}{\pi}\int_{-\infty}^{\infty} \frac{g(\tau)}{t-\tau}\, d\tau \tag{7.9}$$

where the integral is interpreted in the Cauchy principal-value sense. One can very easily ascertain the following properties:

1. $H[\cos(\omega_c t + \phi)] = \sin(\omega_c t + \phi)$
2. $H[\sin(\omega_c t + \phi)] = -\cos(\omega_c t + \phi)$
3. If $g(t)$ is band-limited to $|\omega| < \omega_c$,

$$H[g(t)\cos\omega_c t] = g(t)\sin\omega_c t$$

and

$$H[g(t)\sin\omega_c t] = -g(t)\cos\omega_c t$$

4. If $G(\omega)$ is the Fourier transform of $g(t)$, the Fourier transform of $\hat{g}(t)$ is

$$G_h(\omega) = -iG(\omega)\,\text{sgn}\,\omega = \begin{cases} -iG(\omega) & \omega > 0 \\ 0 & \omega = 0 \\ iG(\omega) & \omega < 0 \end{cases}$$

A filter having transfer function $-i$ sgn ω is often referred to as a *Hilbert filter*. These are all the properties that are needed in the sequel.

One approach which will guarantee that the signal emerging from the linear modulator is SSB is to choose the filter $h(t)$ so that it completely attenuates frequencies either above or below the carrier. The transfer function of one such filter is shown in Fig. 7.4.

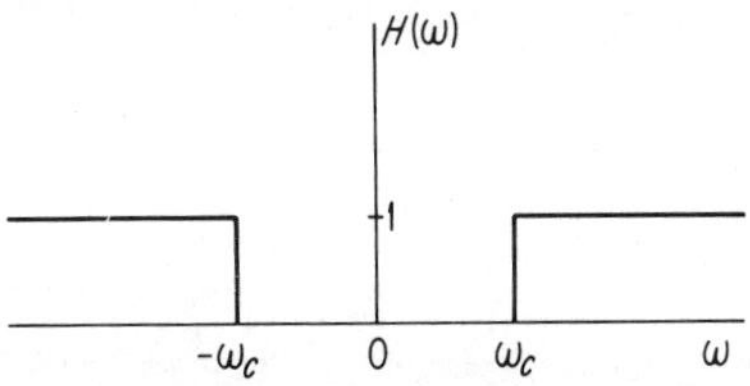

Fig. 7.4. A filter characteristic that generates an SSB signal.

Mathematically,

$$H(\omega) = \begin{cases} 0 & |\omega| \leq \omega_c \\ 1 & |\omega| > \omega_c \end{cases} \tag{7.10}$$

or

$$H((\omega) = 1 - R(\omega) \tag{7.11}$$

where

$$R(\omega) = \begin{cases} 1 & |\omega| \leq \omega_c \\ 0 & |\omega| > \omega_c \end{cases}$$

It will now be demonstrated that the resulting time function at the output of this filter is the original DSB signal plus the Hilbert transform of $g(t)$ in quadrature.

The impulse response of $H(\omega)$ is simply

$$\begin{aligned} h(t) &= \frac{1}{2\pi} \int_{-\infty}^{\infty} e^{i\omega t}[1 - R(\omega)]\, d\omega \\ &= \delta(t) - \frac{1}{2\pi} \int_{-\omega_c}^{\omega_c} e^{i\omega t}\, d\omega \\ &= \delta(t) - \frac{1}{\pi} \frac{\sin \omega_c t}{t} \end{aligned} \tag{7.12}$$

The signal $s_u(t)$ at the output of $h(t)$ is therefore

$$\begin{aligned} s_u(t) &= h(t) * g(t) \cos \omega_c t \\ &= g(t) \cos \omega_c t - \frac{1}{\pi} \int_{-\infty}^{\infty} \frac{g(\tau) \cos \omega_c \tau \sin (\omega_c t - \omega_c \tau)}{t - \tau}\, d\tau \end{aligned} \tag{7.13}$$

Denoting the second term in Eq. (7.13) by $a(t)$ and using elementary trigonometric identities yields

$$a(t) = \frac{1}{\pi} \sin \omega_c t \int_{-\infty}^{\infty} \frac{g(\tau) \cos \omega_c \tau \cos \omega_c \tau}{t - \tau} d\tau - \frac{1}{\pi} \cos \omega_c t \int_{-\infty}^{\infty} \frac{g(\tau) \cos \omega_c \tau \sin \omega_c \tau}{t - \tau} d\tau$$

$$= \frac{1}{2} \sin \omega_c t \left[\frac{1}{\pi} \int_{-\infty}^{\infty} \frac{g(t)}{t - \tau} d\tau + \frac{1}{\pi} \int_{-\infty}^{\infty} \frac{g(\tau) \cos 2\omega_c \tau}{t - \tau} d\tau \right] - \frac{1}{2\pi} \cos \omega_c t \int_{-\infty}^{\infty} \frac{g(\tau) \sin 2\omega_c \tau}{t - \tau} d\tau \tag{7.14}$$

Employing the definition of the Hilbert transform and property 3 above allows Eq. (7.14) to be written as

$$\begin{aligned} a(t) &= \frac{1}{2} \sin \omega_c t \left[\hat{g}(t) + g(t) \sin 2\omega_c t \right] + \frac{1}{2} \cos \omega_c t [g(t) \cos 2\omega_c t] \\ &= \frac{1}{2} \hat{g}(t) \sin \omega_c t + \frac{1}{2} g(t) [\sin \omega_c t \sin 2\omega_c t + \cos \omega_c t \cos 2\omega_c t] \\ &= \frac{1}{2} \hat{g}(t) \sin \omega_c t + \frac{1}{2} g(t) \cos \omega_c t \end{aligned} \tag{7.15}$$

When this result is substituted in Eq. (7.13), the output signal finally becomes

$$s_u(t) = \frac{1}{2} g(t) \cos \omega_c t - \frac{1}{2} \hat{g}(t) \sin \omega_c t \tag{7.16}$$

which proves the original assertion.

Conceptually the signal $\hat{g}(t)$ may be thought of as having been obtained by passing $g(t)$ through a filter having impulse response $1/t\pi$. The action of this filter is to shift the phases of all frequencies in the signal by 90°. Spectra of a typical $g(t)$ and $s_u(t)$ are shown in Fig. 7.5(a) and (b), respectively.

To gain further insight into SSB modulation, the representation of the signal at the output of a filter which eliminates the upper, rather than the lower, sideband will now be derived using the frequency-domain approach. In order to eliminate the upper sideband, the transfer function of the filter is taken as $R(\omega)$, which equals unity when $|\omega| \leq \omega_c$ and zero otherwise. The Fourier transfer of the output now becomes

$$S_l(\omega) = R(\omega) \left[\frac{1}{2} G(\omega + \omega_c) + \frac{1}{2} G(\omega - \omega_c) \right] \tag{7.17}$$

and the output time function is simply the inverse Fourier transform of $S_l(\omega)$ given by

$$s_l(t) = \frac{1}{4\pi}\int_{-\infty}^{\infty} R(\omega)[G(\omega + \omega_c) + G(\omega - \omega_c)]\, e^{i\omega t}\, d\omega$$

$$= \frac{1}{4\pi}\int_{-\omega_c}^{\omega_c} G(\omega + \omega_c)\, e^{i\omega t}\, d\omega + \frac{1}{4\pi}\int_{-\omega_c}^{\omega_c} G(\omega - \omega_c)\, e^{i\omega t}\, d\omega \tag{7.18}$$

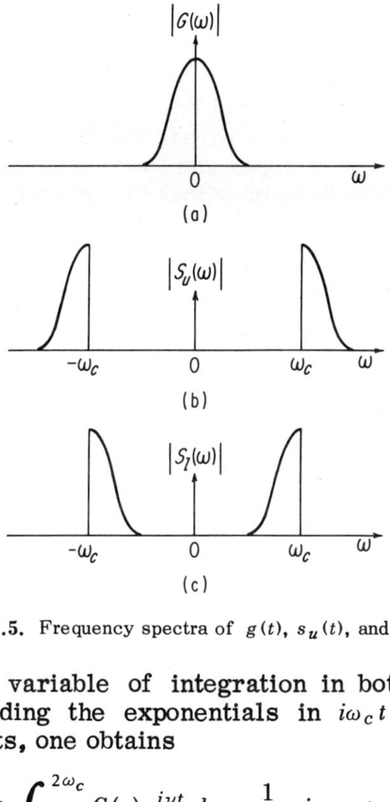

Fig. 7.5. Frequency spectra of $g(t)$, $s_u(t)$, and $s_l(t)$.

By changing the variable of integration in both of the integrals above and expanding the exponentials in $i\omega_c t$ into $\cos\omega_c t$ and $\sin\omega_c t$ components, one obtains

$$s_l(t) = \frac{1}{4\pi}\cos\omega_c t\int_{-2\omega_c}^{2\omega_c} G(\nu)\, e^{i\nu t}\, d\nu - \frac{1}{4\pi}\sin\omega_c t\int_{-2\omega_c}^{2\omega_c} G_h(\nu)\, e^{i\nu t}\, d\nu \tag{7.19}$$

The first integral is just $g(t)$ while the second integral represents the output of a Hilbert filter when the input is $g(t)$. Thus the lower-sideband signal is now written as

$$s_l(t) = \frac{1}{2}\, g(t)\cos\omega_c t + \frac{1}{2}\,\hat{g}(t)\sin\omega_c t \tag{7.20}$$

Its spectrum is shown in Fig. 7.5(c).

It should not be surprising that the lower-sideband signal has the same representation as the upper sideband except for a minus sign. Clearly the sum of the two sidebands must equal the original double-sideband signal.

Equations (7.16) and (7.20) suggest an alternate means of generating an SSB signal. One generates the Hilbert transform of the baseband signal and adds it to, or subtracts it from, the original signal in quadrature.

Thus far very little has been said about the physical realization of the filters necessary to achieve SSB modulation. For example, a filter having impulse response $1/\pi t$ is not physically realizable and yet is required to generate the Hilbert transform of a signal. Also it is impossible to build filters which have a step function in their frequency response. Thus, if the baseband modulating signals contain significant energy around zero frequency, it becomes extremely difficult to separate the upper and lower sidebands. Consequently SSB modulation becomes impractical for this most important class of signals. However, we shall cite one exceptional application of SSB when partial-response operation is employed. This technique is described by Becker, Kretzmer, and Sheehan.[2] Even if one could approximate very nearly a sharp cutoff filter, the corresponding time function would decay very slowly at infinity, resulting in large sensitivity to phase error in the demodulation process. These concepts will be more fully explained in subsequent sections when vestigial-sideband modulation is discussed.

We remark at this point that the original baseband signal can be recovered from the SSB signal if the in-phase component can be separated from the quadrature component. This, of course, requires precise knowledge of the phase and frequency of the carrier.

7.1.4 Vestigial-sideband Modulation

The practical limitations of SSB modulation and the inherent bandwidth wastefulness of DSB modulation impel one to seek a compromise. *Vestigial-sideband* (VSB) *modulation* has the implementation advantages of DSB and yet requires only slightly more bandwidth than SSB. In this method a residual portion of the unwanted sideband is transmitted along with the complete other sideband. This has the effect of relaxing the stringent requirements on sharp cutoff filters, and, as we shall see later, increases the margin on phase jitter.

We proceed to develop the mathematical structure of an arbitrary VSB signal by again specializing the filter at the output of the linear modulator. It will be shown that the VSB signal is, like the SSB signal, a linear combination of an in-phase signal and a quadrature signal. Toward this end, consider a symmetrical filter $H_\beta(\omega)$ as shown in Fig. 7.6. In terms of transfer characteristics

H_l and H_r, $H_\beta(\omega)$ can be written as

$$H_\beta(\omega) = \begin{cases} H_l(\omega) & -\omega_c - \beta \leq \omega \leq -\omega_c + \beta \\ 1 & -\omega_c + \beta \leq \omega \leq \omega_c - \beta \\ H_r(\omega) & \omega_c - \beta \leq \omega \leq \omega_c + \beta \end{cases} \tag{7.21}$$

This filter has a smooth roll-off the amount of which depends on the parameter β. In the limit as β approaches zero, H_l and H_r vanish, and $H_\beta(\omega)$ approaches the SSB filter. If $\beta \neq 0$, this particular filter will transmit a portion of the upper sideband, while the characteristic, $1 - H_\beta$, transmits a vestige of the lower sideband.

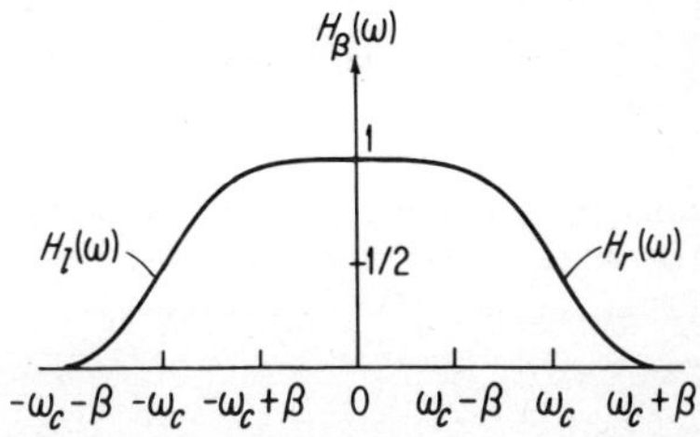

Fig. 7.6. Filter characteristic required to generate a VSB signal.

We proceed as before. The output Fourier transform is

$$S_\beta(\omega) = \frac{1}{2} H_\beta(\omega)[G(\omega + \omega_c) + G(\omega - \omega_c)] \tag{7.22}$$

and the corresponding time function

$$s_\beta(t) = \frac{1}{4\pi} \int_{-\infty}^{\infty} H_\beta(\omega)[G(\omega + \omega_c) + G(\omega - \omega_c)] e^{i\omega t} \, d\omega \tag{7.23}$$

Under the assumption that $g(t)$ is band-limited to $|\omega| < \omega_c$, Eq. (7.23) can be written as

$$s_\beta(t) = \frac{1}{4\pi} \int_{-\omega_c - \beta}^{0} H_\beta(\omega) G(\omega + \omega_c) e^{i\omega t} \, d\omega + \frac{1}{4\pi} \int_{0}^{\omega_c + \beta} H_\beta(\omega) G(\omega - \omega_c) e^{i\omega t} \, d\omega \tag{7.24}$$

By changing the variable of integration from ω to $\nu - \omega_c$ in the first integral and from ω to $\nu + \omega_c$ in the second, one obtains

$$s_\beta(t) = \frac{1}{4\pi} e^{-i\omega_c t} \int_{-\beta}^{\omega_c} H_\beta(\nu - \omega_c) G(\nu) e^{i\nu t} \, d\nu + \frac{1}{4\pi} e^{i\omega_c t} \int_{-\omega_c}^{\beta} H_\beta(\nu + \omega_c) G(\nu) e^{i\nu t} \, d\nu \tag{7.25}$$

Substituting the definition of H_β in terms of H_l and H_r in Eq. (7.25) gives

$$s_\beta(t) = \frac{1}{4\pi} e^{-i\omega_c t} \int_{-\beta}^{\beta} H_l(\nu - \omega_c)\, G(\nu)\, e^{i\nu t}\, d\nu$$
$$+ \frac{1}{4\pi} e^{-i\omega_c t} \int_{\beta}^{\omega_c} G(\nu)\, e^{i\nu t}\, d\nu$$
$$+ \frac{1}{4\pi} e^{i\omega_c t} \int_{-\omega_c}^{-\beta} G(\nu)\, e^{i\nu t}\, d\nu$$
$$+ \frac{1}{4\pi} e^{i\omega_c t} \int_{-\beta}^{\beta} H_r(\nu + \omega_c)\, G(\nu)\, e^{i\nu t}\, d\nu \tag{7.26}$$

Writing $e^{i\omega_c t}$ as $\cos \omega_c t + i \sin \omega_c t$ and grouping terms yields

$$s_\beta(t) = \frac{1}{4\pi} \cos \omega_c t \Bigg[\int_{-\omega_c}^{-\beta} G(\nu)\, e^{i\nu t}\, d\nu + \int_{\beta}^{\omega_c} G(\nu)\, e^{i\nu t}\, d\nu$$
$$+ \int_{-\beta}^{\beta} G(\nu) [H_l(\nu - \omega_c) + H_r(\nu + \omega_c)]\, e^{i\nu t}\, d\nu \Bigg]$$
$$- \frac{i}{4\pi} \sin \omega_c t \Bigg[\int_{-\beta}^{\beta} G(\nu)\, e^{i\nu t} [H_l(\nu - \omega_c) - H_r(\nu + \omega_c)]\, d\nu$$
$$- \int_{-\omega_c}^{-\beta} G(\nu)\, e^{i\nu t}\, d\nu + \int_{\beta}^{\omega_c} G(\nu)\, e^{i\nu t}\, d\nu \Bigg] \tag{7.27}$$

It can be seen from the above representation that in order to obtain $g(t)$ without distortion, the following relation is sufficient:

$$H_l(\nu - \omega_c) + H_r(\nu + \omega_c) = 1 \qquad |\nu| < \beta \tag{7.28}$$

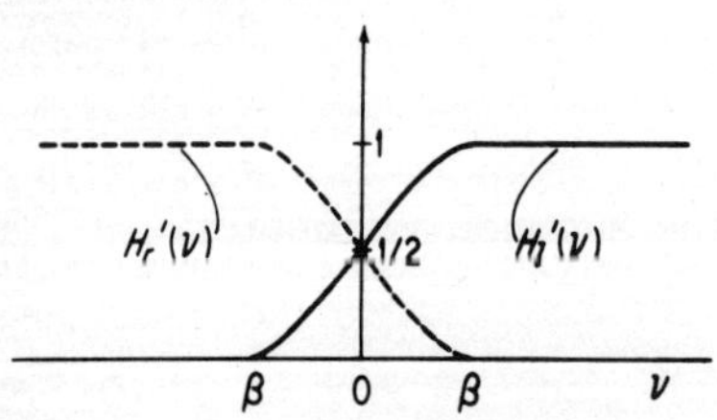

Fig. 7.7. Frequency characteristics of $H_l'(\nu)$ and $H_r'(\nu)$ as a function of ν.

For convenience, set $H_l(\nu - \omega_c) = H_l'(\nu)$ and $H_r(\nu + \omega_c) = H_r'(\nu)$. Fig. 7.7 shows the transfer characteristics $H_r'(\nu)$ and $H_l'(\nu)$. It is clear from the figure that $H_l'(-\nu) \equiv H_r'(\nu)$ for $H_\beta(\omega)$ real. For further clarity set

$$H_l'(\nu) = \frac{1}{2} + f_0(\nu) \qquad |\nu| < \beta \tag{7.29}$$

where $f_0(\nu)$ is an arbitrary odd function of frequency ν satisfying $f_0(\beta) = 1/2$.

Using this representation yields the final result

$$s_\beta(t) = \frac{1}{2} g(t) \cos \omega_c t + \frac{1}{2} g_\beta(t) \sin \omega_c t \tag{7.30}$$

where
$$g_\beta(t) = \frac{1}{2\pi} \int_{-\omega_c}^{\omega_c} W_\beta(\omega) G(\omega) e^{i\omega t} \, d\omega$$

and
$$W_\beta(\omega) = \begin{cases} -2if_0(\omega) & |\omega| < \beta \\ -i \operatorname{sgn}(\omega) & |\omega| > \beta \end{cases}$$

This representation in Eq. (7.30) looks very similar to a SSB signal except that here $g_\beta(t)$ is not the Hilbert transform of $g(t)$, although it is again obtained by passing $g(t)$ through a linear filter. It is clear that in the limit as $\beta \to 0$, $W_\beta(\omega) G(\omega) \to G_h(\omega)$, the output of the Hilbert filter. The spectrum of $S_\beta(t)$ is shown in Fig. 7.8. It is important to realize that the in-phase component is not independent of the quadrature component. Therein lies the reason these signals have favorable spectral distribution. Had these components been completely independent, the linear combination would have been a DSB signal. A complete data terminal using VSB modulation is described in Becker.[3]

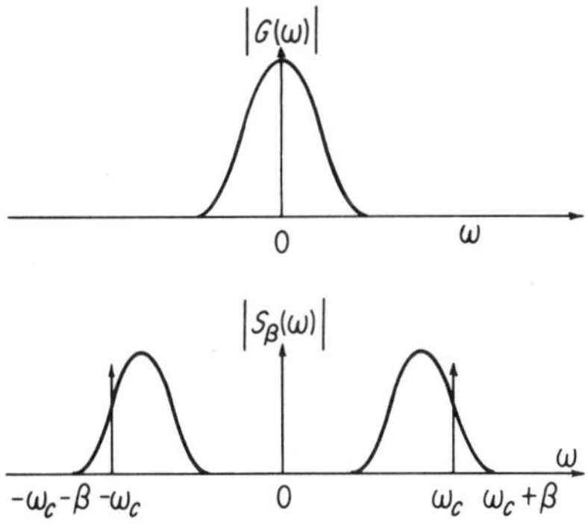

Fig. 7.8. The frequency spectrum of a typical VSB signal.

7.1.5 Quadrature-amplitude Modulation

A modulation system closely resembling VSB but not conforming exactly to the definition of linear modulation is sometimes used in data transmission. This type of modulation system is referred to as *quadrature-amplitude modulation* (QAM) because the structure of the modulated signal is of the same form as that of two independent DSB suppressed-carrier signals in quadrature. In this type of modulation two independent baseband data sources are transmitted in the same bandwidth. This is possible because the one signal linearly modulates the envelope of a cosine wave, while the other independently modulates the envelope of a sine wave. More specifically, denote the two baseband signals by $x_1(t)$ and $x_2(t)$, respectively. The modulated signal $q(t)$ has the representation

$$q(t) = x_1(t) \cos \omega_c t + x_2(t) \sin \omega_c t \tag{7.31}$$

Each component in $q(t)$ may be generated by the general configuration shown in Fig. 7.1. The representation above differs in one very important respect from the VSB signal of Eq. (7.30). Namely, the in-phase and quadrature components are independent.

To complete the general discussion of linear modulation systems, the various demodulation procedures will be discussed next.

7.2 Methods of Demodulation

It is customary to distinguish between two general types of demodulators, coherent and incoherent. Coherent demodulators are the most efficient and are widely used in high-speed data transmission. We therefore focus attention in this chapter on these methods. In Chaps. 8 and 9 we examine incoherent types of demodulators used in angle-modulation systems.

7.2.1 Coherent Demodulators

The representation of linearly modulated signals as developed in the previous sections suggests a straightforward means of deriving the original baseband signal. It is merely necessary to multiply the modulated signal by $\cos \omega_c t$ and pass the product through a low-pass filter. This filter extracts the baseband signal from the product by eliminating components centered at twice the carrier frequency. Of course, this can only be accomplished when the carrier frequency is larger than the base bandwidth. To facilitate faithful extraction of the baseband signal, the demodulator must be in phase synchronism with the received signal. This means that carrier phase and frequency must be precisely known to the receiver. This form of demodulation is often referred to as *homodyne demodulation*. In most practical systems precise phase information is unavailable at the receiver and must therefore be obtained by either applying additional nonlinear processing on the received signal (in DSB systems) or by deriving this information from low-level transmitted pilot tones (as is often done in SSB and VSB systems).

For purposes of the present discussion it will be assumed that carrier phase and frequency are known to the receiver. The means of recovery of these parameters will be discussed in a subsequent section. Demodulators which use carrier-phase and frequency information in recovering information-bearing signals are sometimes termed *coherent demodulators*. A block diagram of such a demodulator is shown in Fig. 7.9.

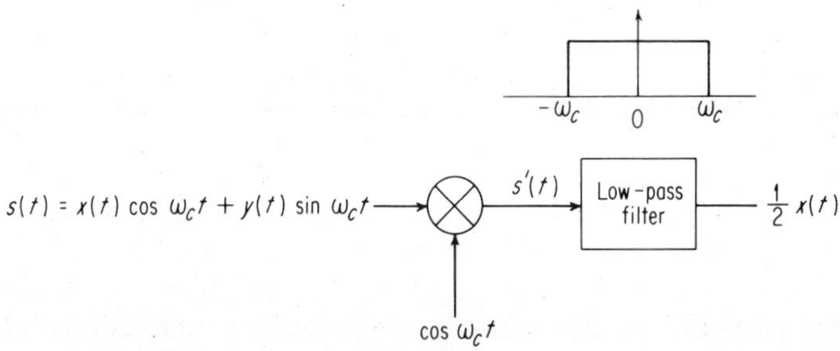

Fig. 7.9. Block diagram showing a coherent demodulator.

In practice the low-pass filter in Fig. 7.9 may also provide some additional baseband signal shaping. The possibility of allocating shaping functions to these various filters will become clearer when an equivalent baseband system is derived. For the discussion in this section this filter only serves to eliminate the double-frequency components at the output of the multiplier. To understand the operation of the demodulator, write the output of the demodulator when the input is an arbitrary linearly modulated signal as

$$\begin{aligned} s'(t) &= x(t) \cos^2 \omega_c t + y(t) \cos \omega_c t \sin \omega_c t \\ &= \frac{1}{2} x(t) + \frac{1}{2} x(t) \cos 2\omega_c t + \frac{1}{2} y(t) \sin 2\omega_c t \end{aligned} \tag{7.32}$$

If $x(t)$ and $y(t)$ are band-limited to $|\omega| < \omega' < \omega_c$ and if the transfer function of the low-pass filter is unity up to $\omega = \omega'$ and zero when $|\omega| > \omega'$, $s'(t)$ will equal $x(t)/2$ as shown in Fig. 7.9.

The important item to be observed is the role played by the local oscillator generating the sinusoidal signal $\cos \omega_c t$. For suppose there is a phase error θ between the received signal and the local oscillator. As a result of this error the received signal will be demodulated with $\cos [\omega_c t + \theta]$, and the signal after low-pass filtering will be of the form

$$s'(t,\theta) = x(t) \frac{\cos \theta}{2} + y(t) \frac{\sin \theta}{2} \tag{7.33}$$

which depends on the phase error. The second term in Eq. (7.33) represents distortion which ultimately results in intersymbol interference. In a subsequent section curves will be presented showing the closure of the eye as a function of this phase error for various linear-modulation systems. For efficient coherent demodulation one strives to make the phase error as close to zero as possible.

When the structure of the received signal is DSB and an additional pilot tone is transmitted at the carrier frequency, the baseband signal may be extracted by an envelope detector.* This form of nonlinear demodulation does not require phase information and is therefore incoherent. The pilot tone must be of intensity greater than, or equal to, the maximum intensity of the baseband signal for perfect demodulation. If this condition is violated, the demodulated signal will be distorted. The behavior of this demodulation process, analyzed in detail in "Data Transmission" by Bennett and Davey,[4] is shown to have performance inferior to coherent demodulators. This demodulation technique is rarely used in data-communications systems because of its inherent inefficiency and relatively poor performance in the presence of additive noise. For these reasons no further discussion of this system will be presented.

7.2.2 An Equivalent Baseband System

Before taking up methods of carrier phase and frequency recovery an equivalent baseband system will be derived. From the discussion of linear-modulation techniques, it may be inferred that it is possible to reflect the action of linear modulation and demodulation back to baseband. The objective of this section is to indicate how this may be done, thus justifying previous remarks concerning equivalence of performance of linearly modulated systems employing coherent demodulators with that of baseband systems.

Consider the block diagram in Fig. 7.10 representing an overall data-communication system employing linear modulation and coherent demodulation. In this block diagram $C(\omega)$ is the channel transfer function, and $H(\omega)$ and $R(\omega)$ are the transmitting- and receiving-filter transfer characteristics. These three filters may be combined into a single, composite characteristic $Q(\omega)$. The transfer functions $T(\omega)$ and $L(\omega)$ represent baseband and post-detection filters, while $D_1(\omega)$ and $D_0(\omega)$ are the transfer functions of input and output data signals, respectively. Symbollically

$$d_1(t) = \sum_n a_n g_1(t - nT) \quad \text{and} \quad D_1(\omega) = G_1(\omega) \sum_n a_n e^{-i\omega nT} \tag{7.34}$$

---

*Voelcker[5] describes a method of demodulating VSB signals, which applies to a certain class of modulating signals, using only envelope information.

are the representations of the input signal where, as before, $\{a_n\}$ are the input symbols. The output data signal can likewise be represented as

$$d_0(t) = \sum_n a_n g_2(t - nT) \quad \text{and} \quad D_0(\omega) = G_2(\omega) \sum_n a_n e^{-i\omega nT} \tag{7.35}$$

where $G_2(\omega)$ is an equivalent baseband characteristic related to the filters shown in Fig. 7.10. The following string of multiplications can be used to obtain an expression for $G_2(\omega)$.

$$D_1(\omega)T_1(\omega) \xrightarrow{\cos \omega_c t} \frac{Q(\omega)}{2}[D_1(\omega + \omega_c)T_1(\omega + \omega_c) + D_1(\omega - \omega_c)T_1(\omega - \omega_c)]$$

$$\xrightarrow{\cos \omega_c t} \frac{1}{4} L(\omega)\Big\{Q(\omega - \omega_c)[D_1(\omega)T(\omega) + D_1(\omega - 2\omega_c)T_1(\omega - 2\omega_c)]$$

$$+ Q(\omega + \omega_c)[D_1(\omega + 2\omega_c)T_1(\omega + 2\omega_c) + D_1(\omega)T_1(\omega)]\Big\} \tag{7.36}$$

Since $L(\omega)$ and $T_1(\omega)$ are low-pass filters, products like $L(\omega) \cdot D_1(\omega \pm 2\omega_c)\, T_1(\omega \pm 2\omega_c)$ are zero because of the band-limited nature of $D_1(\omega)\, T_1(\omega)$. Therefore

$$D_0(\omega) = \frac{1}{4} L(\omega)\, T(\omega)\, D_1(\omega)[Q(\omega - \omega_c) + Q(\omega + \omega_c)] \tag{7.37}$$

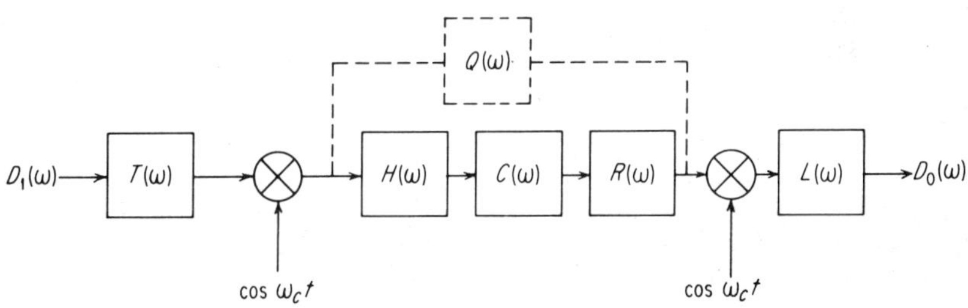

Fig. 7.10. A block diagram showing an overall digital communications system employing linear modulation and demodulating.

For bandpass channels, $Q(\omega - \omega_c) + Q(\omega + \omega_c) = A(\omega)$ is a low-pass characteristic, and therefore

$$D_0(\omega) = \frac{1}{4} L(\omega)\, T(\omega)\, A(\omega)\, G_1(\omega) \sum_n a_n e^{-i\omega nT} \tag{7.38}$$

where $G_2(\omega) = L(\omega)\, T(\omega) G_1(\omega) A(\omega)/4$ is the equivalent low-pass filter.

Figure 7.11 is a block diagram of the equivalent baseband system.

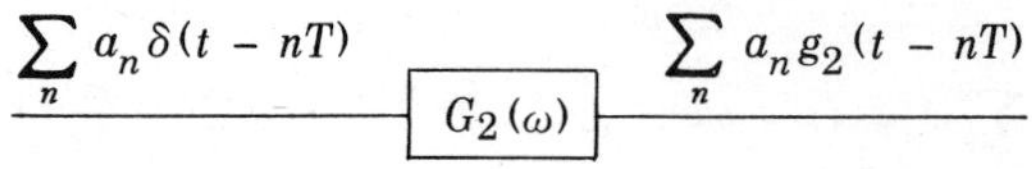

Fig. 7.11. A block diagram showing a carrier recovery system for DSB transmission.

Because of this equivalence, all the analysis used in the chapters on baseband systems is directly applicable. It should also be clear that, because of the linearity of the system, if noise is added anywhere in the system, its effect can be reflected back to baseband.

In the next section methods of deriving carrier phase and frequency will be examined.

7.3 Recovery of Carrier Phase and Frequency

We have seen that coherent demodulators require knowledge of carrier phase and frequency. This information must often be separately communicated to the receiver. In this section we consider various methods of sending and extracting these parameters. We conclude the section by examining the degradations suffered when phase information is not exactly known.

7.3.1 Direct Methods

In DSB modulation systems operating in the suppressed-carrier (SC) mode the problem of deriving a synchronous carrier from the modulated signal is relatively straightforward. Two methods of deriving this information in DSB modulation will be discussed.

A very simple circuit which may be used to derive a carrier wave of the right frequency and phase is shown in Fig. 7.12. Recall that the DSBSC signal has no quadrature component. Therefore in the circuit shown the information-carrying signal must be eliminated in order to derive a carrier tone. To see how this is accomplished

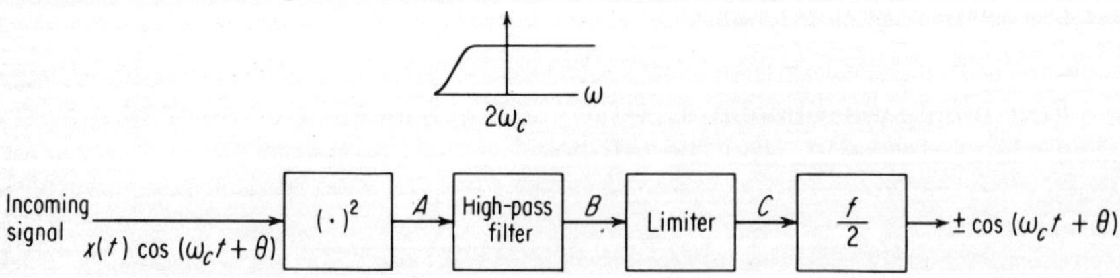

Fig. 7.12. A block diagram showing a carrier recovery system for DSB transmission.

observe that at point A the incoming signal is first squared to obtain

$$a(t) = \frac{x^2(t)}{2} + \frac{x^2(t)}{2} \cos[2\omega_c t + 2\theta] \tag{7.39}$$

The high-pass filter eliminates the baseband signal, which is now twice the bandwidth of the original signal $x(t)$. This elimination can be achieved only if the carrier frequency is larger than twice the bandwidth of $x(t)$. At point B, the signal (centered at $2\omega_c$) is ideally limited to obtain

$$b(t) = A\cos[2\omega_c t + 2\theta] \tag{7.40}$$

where A is a constant. This sinusoidal signal of fixed amplitude and twice the original carrier frequency is fed into a frequency divider to obtain a carrier used to demodulate the original signal.

A slight difficulty is the ambiguous polarity of the recovered carrier. It is equally correct to write the carrier wave as $\cos[\omega_c t + \phi(t) + 2\pi k]$, where k is an arbitrary integer. The frequency divider will divide the $2\pi k$ factor by 2 to yield πk. This inserts an ambiguity of π in the resulting phase whenever k is an odd integer. This is an important consideration since, when the data is recovered, symbols having positive signs are given negative signs and vice versa. To remedy this awkwardness, various precautionary techniques are possible. For one the initial data symbols can be so encoded as to specify only "change information." For example, in the binary case a change in signs of two successive symbols is encoded as a 1, no change in signs as a 0. It can be seen that irrespective of the polarity of carrier signal, symbols will be decoded correctly. This is called *differential coding* and will be discussed more fully in the chapter dealing with phase modulation. Another method is to establish the correct polarity of the carrier initially and then monitor for π radian changes.

The above method of carrier recovery will not work with VSB or the other modulation methods discussed in previous sections, because the quadrature component inhibits the exact determination of the carrier phase. To see this, consider the DSB signal $x(t)\cos(\omega_c t + \theta)$ plus a quadrature component, say, $y(t)\sin(\omega_c t + \theta)$. The composite signal can be represented in terms of a positive envelope $r(t)$ as

$$p(t) = r(t)\cos[\omega_c t + \theta + \phi(t)] \tag{7.41}$$

where $\quad r(t) = [x^2(t) + y^2(t)]^{\frac{1}{2}} \quad$ and $\quad \phi(t) = \arctan\dfrac{y(t)}{x(t)}$

In passing this signal through the circuit shown in Fig. 7.12, one obtains at point B

$$b(t) = A\cos[2\omega_c t + 2\theta + 2\phi(t)] \tag{7.42}$$

Clearly $\phi(t)$ depends on $x(t)$ and $y(t)$ and is only zero when $y(t)$ is zero, in which case the original signal [Eq. (7.41)] is DSB.

An alternative method[6] of deriving phase information from a DSB signal again operating in an SC mode is shown in Fig. 7.13.

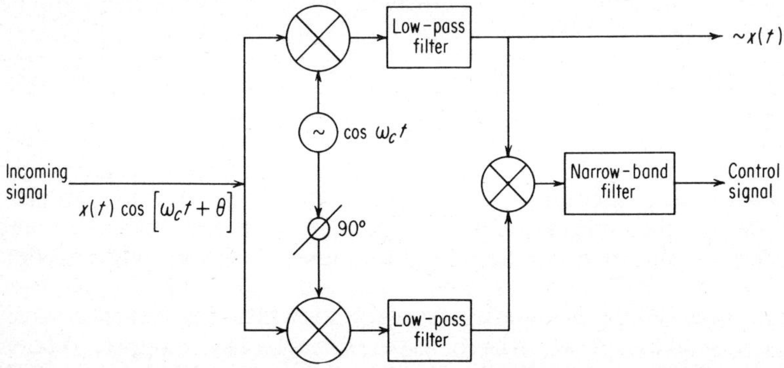

Fig. 7.13. A block diagram showing Costas'[6] method of deriving carrier phase.

In this circuit it is assumed that the carrier frequency is known a priori. The incoming signal is multiplied by cos $(\omega_c t + \theta)$ in one arm and by sin $(\omega_c t + \theta)$ in the other arm, where θ is the phase error. The products in each arm are low-pass-filtered, and the outputs are multiplied to produce an output signal whose dc term is proportional to the local oscillator phase error. The final product is of the form

$$w(t) = x^2(t)\sin 2\theta \tag{7.43}$$

When $w(t)$ is time-averaged, the output is proportional to sin 2θ; when θ is small, it is directly proportional to θ. If the local oscillator is in exact synchronism with the incoming signal, the output of the upper arm is the desired demodulated signal. Since the lower arm of the circuit is in quadrature with the incoming signal, its output is identically zero. The voltage or current from the averaging filter may be used to adjust the phase of the local oscillator to minimize the phase error.

7.3.2 Insertion and Extraction of Pilot Tones

In VSB or SSB transmission auxiliary pilot tones are often transmitted to maintain continuous control of the phase and frequency error. Modulators and channels are seldom stable enough; therefore continuous monitoring of carrier phase and frequency is required to maintain coherent demodulation. Various means are available for accomplishing this task. Some particular circuits are discussed in detail in Bennett and Davey.[4]

A problem unique to VSB transmission arises when a pilot tone is transmitted at the carrier frequency. To extract the pilot tone, say, with a notch filter, the region around the carrier must be free of data signal energy. If this condition is not fulfilled, the pilot tone may at times be canceled by this energy, and the system will fail to demodulate the data signal coherently. To ensure that this does not happen, the baseband data signal is passed through a high-pass filter which guarantees that no data-signal energy will exist around the carrier frequency. The loss of low-frequency energy has the effect of adding a great deal of intersymbol interference. This condition can be corrected at the receiver by employing *digitized feedback*[7] or, as it is also known, *dc restoration*. This technique and its effect on performance will be discussed in Sec. 7.3.3.

In some applications it is possible to transmit low-level pilot tones at the band edge, thus avoiding the low-frequency restoration problem. The pilot tones are recovered from two narrowband filters tuned to these frequencies. Ideally the bandwidth of these filters can be made as narrow as desired, resulting in tones free of data signal contamination and additive noise. However, because of phase jitter, which is unavoidable on practical communications media, the filter bandwidths cannot be made arbitrarily narrow. In a typical situation the received VSB signal plus the pilot tones may be represented as

$$\begin{aligned} s(t) = {} & x(t)\cos[\omega_c t + \Delta\omega t + \theta(t) + \phi] \\ & + y(t)\sin[\omega_c t + \Delta\omega t + \theta(t) + \phi] \\ & + a\cos[\omega_1 t + \Delta\omega t + \theta(t) + \phi_1] \\ & + a\cos[\omega_2 t + \Delta\omega t + \theta(t) + \phi_2] \end{aligned} \tag{7.44}$$

where $\Delta\omega$ is a small frequency shift commonly encountered on carrier facilities, $\theta(t)$ is phase jitter which is typically a slowly varying signal relative to the signaling rate, and ϕ, ϕ_1, and ϕ_2 are constant phase shifts. The object of a carrier recovery system is to derive $\cos[\omega_c t + \Delta\omega t + \theta(t) + \phi]$ by processing the pilot tones. The individual frequencies ω_1 and ω_2 are related to the signaling rate R and the parameter β as shown in Fig. 7.14. Vestigial and baseband roll-off are assumed to be equal.

In order to indicate how these tones are processed, assume that they can be separated from the information-carrying signal essentially undistorted. Once they are separately available, they are multiplied and filtered to obtain a cosine wave of the difference

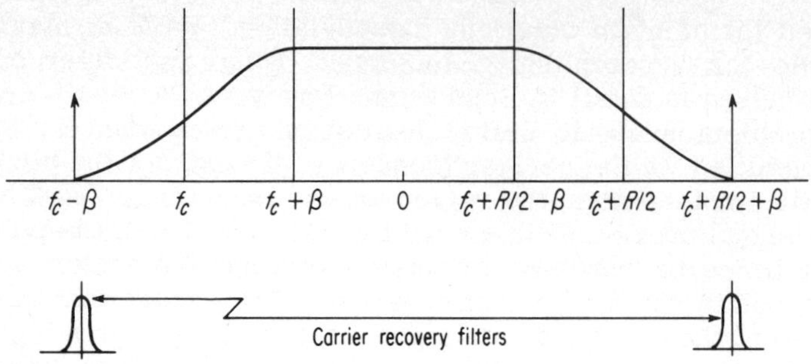

Fig. 7.14. Positive spectral distribution of a VSB signal with two pilot tones at the bandedges. R is the Nyquist rate.

frequency. As a result of these operations, the receiver has available a tone devoid of frequency shift and phase jitter of the form

$$c(t) \sim \cos\left[2\pi\left(\frac{R}{2} + 2\beta\right)t + \phi_2 - \phi_1\right] \sim \cos[2\pi\beta qt + \phi_2 - \phi_1] \tag{7.45}$$

where

$$q = \frac{R}{2} + 2\beta$$

The next operation in the processing is to divide the frequency of $c(t)$ by q to obtain $\cos\,[2\pi\beta t + (\phi_2 - \phi_1)/q]$. (Note that ω_c is known a priori.) The resulting tone is now used to multiply the tone at the upper bandedge. This time the resulting tone is

$$c_1(t) \sim \cos\left[2\pi\left(f_c + \frac{R}{2}\right)t + \Delta\omega t + \theta(t) + \gamma\right] \tag{7.46}$$

where

$$\gamma = \phi_2 - \frac{\phi_2 - \phi_1}{q}$$

This tone has the correct frequency shift and phase jitter to be used for demodulating the VSB signal and differs only from the received tone by a constant phase shift γ. This phase difference can be readily tuned out by initially transmitting only a tone at the carrier frequency and then adjusting the phase of $c_1(t)$ to the received tone.

While these methods by no means exhaust the area of carrier recovery, they should serve to familiarize the reader with some of the problems involved.

7.3.3 DC Restoration

As previously mentioned, to facilitate carrier recovery in VSB transmission, the dc and low-frequency components of the passband signal are sometimes removed before modulation. Satisfactory carrier recovery can then be accomplished with relatively low carrier-signal power. The demodulated baseband signal is processed by a feedback circuit which essentially reinserts the low-frequency components removed at the transmitter. This circuit, shown in Fig. 7.15, is referred to as the *dc restorer*. It consists of a sample-and-hold circuit and a slicer in the forward path. The response of a low-pass filter to the slicer output is added to the input of the sampling circuit to form the positive feedback loop.

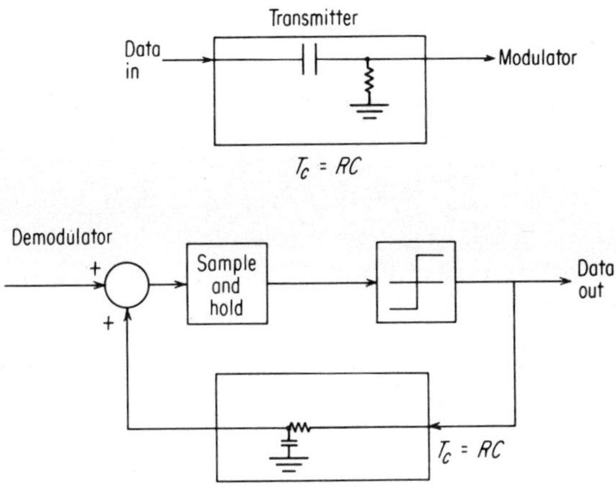

Fig. 7.15. Block diagram showing a dc restoration system.

Assume that the input to the dc restorer is the convolution of a high-pass-filter impulse response $g(t)$ with a random sequence of (± 1) rectangular pulses. If $\{a_j\}$ is the binary data sequence and $p(t)$ is the rectangular pulse, the input to the restorer may be written as

$$s(t) = \sum_j a_j \int_0^\infty g(\tau)\, p(t - jT - \tau)\, d\tau \tag{7.47}$$

where $1/T$ is the data rate.

With reference to Fig. 7.15 the output of the low-pass filter in the feedback loop is

$$c(t) = \sum_j b_j \int_0^\infty h(\tau)\, p\left(t - jT + \frac{T}{2} - \tau\right) d\tau \tag{7.48}$$

where $h(t)$ is the low-pass-filter impulse response and the sequence $\{b_j\}$ is the output binary data. In this equation the output data is delayed relative to the input data by a 1/2-bit interval. This results from the slicer transitions made at the sampling time in the center of the input data pulses. Since synchronous operation is assumed, the waveforms in Eqs. (7.47) and (7.48) need be considered only at the sampling times. Let $s_i = s(t - iT)$; the feedback voltage $c(t) + s(t)$ at the ith sampling time is then

$$s_i + c_i = \sum_{j=0}^{j=i} (a_j g_{i-j} + b_j h_{i-j}) \tag{7.49}$$

where
$$h_i = \int_0^\infty h(\tau)\, p\left(iT + \frac{T}{2} - \tau\right) d\tau$$

and
$$g_i = \int_0^\infty g(\tau)\, p(iT - \tau)\, d\tau$$

In practice $h_0 = 0$; thus Eq. (7.49) can be written in the form

$$a_i g_0 + \sum_{j=0}^{j=i-1} (a_j g_{i-j} + b_j h_{i-j}) = a_i g_0 + e_i \tag{7.50}$$

The feedback voltage or current at the ith sampling time therefore consists of the ith-input bit a_i times g_0 plus a contribution e_i resulting from the last $i-1$ bits. If no errors have occurred in bits 0 through $i-1$ and if g_0 is approximately 1, e_i equals zero, the input to the slicer is a_i, and the low-frequency components of the data pulse have been restored.

Complete restoration requires that $g_0 = 1$ and $g_j + h_j = 0$, where $j = 1, \ldots, i-1$. In the dc restorer shown in Fig. 7.15 RC filters are used; therefore

$$h(t) = \frac{1}{T_c} e^{-t/T_c} \quad \text{and} \quad g(t) = \delta(t) - \frac{1}{T_c} e^{-t/T_c} \tag{7.51}$$

where $T_c = RC$ is the time constant of each filter.

In terms of these specific filters Eq. (7.49) may be expressed as

$$h_i = \begin{cases} 1 & i = 0 \\ \dfrac{1-r}{r} r^i & i \neq 0 \end{cases} \quad \text{and} \quad g_i = \begin{cases} r^{\frac{1}{2}} & i = 0 \\ r^{\frac{1}{2}} \dfrac{r-1}{r} r^i & i \neq 0 \end{cases} \tag{7.52}$$

where
$$r = e^{-T/T_c}$$

It is seen from Eq. (7.52) that except for the factor $r^{\frac{1}{2}}$ multiplying g_i, $g_0 = 1$ and $g_i + h_i = 0$ when $i \neq 0$. The factor $r^{\frac{1}{2}}$ arises from the one-half-bit delay present in the restoration circuit. For T_c considerably greater than T (the usual range of interest), r is approximately 1, and the effect is small. For smaller T_c, the circuit could be simply modified by scaling h_i such that $g_i + h_i = 0$ if $i \neq 0$. To simplify the discussion further, it will be assumed that

$g_i + h_i = 0$ when $i \neq 0$; thus if no errors have occurred in the past, the error voltage e_i equals zero.

To determine the effect of errors, assume an error has occurred in the mth bit. The feedback voltage at bit i is then equal to $g_0 a_i + \{a_m g_{i-m} + b_m h_{i-m}\}$. Depending on the polarity of a_i, the noise margin of the system may be increased or decreased by the term $\{a_m g_{i-m} + b_m h_{i-m}\}$. Thus the probability of error at the ith bit depends on the presence of all errors occurring in the past. Furthermore, errors in the immediate past greatly increase the probability of error at the present bit; the result is that errors tend to occur in small bursts.

It is seen from Eq. (7.49) that if

$$\sum_{j=0}^{k} |h_j| > \sum_{j=0}^{k} |g_j|$$

for some k, the feedback term may override the input signal, a_i and the system will produce a steady train of 1s regardless of the input. For the filters considered here this is impossible. In practice, diodes are inserted in the forward path which saturate at the ± 1 levels. This prevents the circuit from persisting in one state during startup or test sequences.

A system equation for the dc restorer can be written by observing that the ith output b_i is ± 1, depending on whether the feedback voltage immediately preceding the slicer is greater or less than zero; thus

$$b_i = \operatorname{sgn}(g_0 a_i + n_i + e_i) \tag{7.53}$$

where $i = 0, 1, \ldots,$ and

$$e_i = \sum_{j=0}^{i-1} \left(a_j g_{i-j} + b_j h_{i-j}\right) \tag{7.54}$$

and n_i is the ith sample of the added-noise process.

The exact evaluation of the error rate of this circuit depends on the distribution of noise samples n_i and is in general very complicated because of the nonlinear feedback circuit. When the noise samples are independent, gaussian random variables, iterative procedures for calculating the probability of error are given in Zador[8] and Aaron and Simon.[9] Experimental results with gaussian noise indicate that it is possible to attain nearly optimum performance with this circuit. The degradations involved are on the order of fractions of 1 db. It is to be noted, however, that because of the feedback circuit, errors tend to propagate and are not independent. This type of circuit is successfully used in a practical wideband data set reported on by Fracassi and Froehlich.[10] Experimental results with an earlier version of such a circuit are described by Becker, Davey, and Saltzberg.[11]

7.3.4 Degradations Resulting from Phase Errors

In this section we examine the effect of phase errors on the performance of linear demodulators. In Eq. (7.33) we saw that the demodulated baseband wave depends on the phase error. Its effect on DSB (for which $y(t)$ is zero) is to multiply the desired signal $x(t)$ by a factor $1/2 \cos \theta$. This is just a gain factor which causes a reduction in noise immunity but does not introduce intersymbol interference. However, in the other types of modulation the effect of phase jitter is more severe since, in addition to reducing noise immunity, it carries along the quadrature term $y(t)$ and thereby causes intersymbol interference.

Here we shall examine this effect on the eye opening of the baseband data signal and assume that the channel is otherwise perfect. We therefore represent $x(t)$ in Eq. (7.33) as

$$x(t) = \sum_n a_n g(t - nT) \tag{7.55}$$

where $g(t)$ satisfies $g[T(n - m)] = \delta_{nm}$.

In VSB and SSB modulation, the quadrature component $y(t)$ is represented as

$$y(t) = \sum_n a_n g(t - nT) * h_\alpha(t) \tag{7.56}$$

where $h_\alpha(t)$ denotes a linear operation on $g(t)$. The precise form of this linear filter depends on the roll-off factor α as well as on the type of modulation employed. In quadrature modulation, $h_\alpha(t) = \delta(t)$ and

$$y(t) = \sum_n b_n g(t - nT) \tag{7.57}$$

where $\{b_n\}$ is a sequence of input symbols independent of $\{a_n\}$. Here the effect of phase error is to cause cross-channel interference rather than intersymbol interference, i.e., interference between the x and y channels. Observe that as long as $g(t)$ is a perfect pulse, the cross-channel interference is independent of roll-off.

We use Eqs. (7.55) and (7.56) to write Eq. (7.33) in the form

$$s'(t, \theta) = \frac{\cos\theta}{2} \sum_n a_n g(t - nT) + \frac{\sin\theta}{2} \sum_n a_n g(t - nT) * h_\alpha(t) \tag{7.58}$$

For example, to obtain the zeroth symbol a_0, sample Eq. (7.58) at $t = 0$ to obtain

$$s'(0,\theta) = \frac{\cos\theta}{2}\left[a_0 + \tan\theta \sum_n a_n g(t - nT) * h_\alpha(t)\Big|_{t=0}\right] \tag{7.59}$$

The second term represents intersymbol or cross-channel interference and is zero only when $\theta = 0$. By assuming that a_0 equals ±1, the bracketed term in Eq. (7.59) is the normalized eye opening as discussed in previous chapters. The quantity

$$D(\theta,\alpha) = \tan\theta \sum_n |[g(t - nT) * h_\alpha(t)]_{t=0}| \tag{7.60}$$

is the maximum value of distortion over all possible sequences $\{a_n\}$ and $\{b_n\}$.

A family of curves showing D as a function of θ for a raised-cosine pulse with different values of α (refer to Sec. 4.1 for the definition of α) is shown* in Fig. 7-16. These curves were calculated for the various modulation systems examined. As can be observed, even an infinitesimal phase error will close the eye in SSB transmission. The degradation turns out to be graceful for

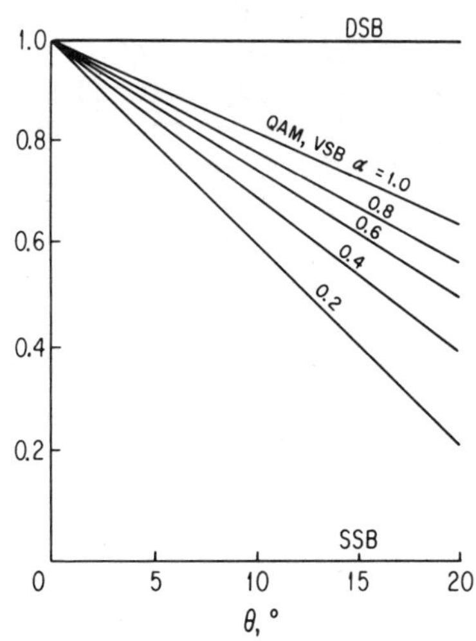

Fig. 7.16. Relative eye opening D as a function of error θ for various linear modulation systems when the eye is sampled at time origin.

* These curves were supplied to the authors by R. D. Howson, of the Bell Telephone Laboratories, Holmdel, N. J.

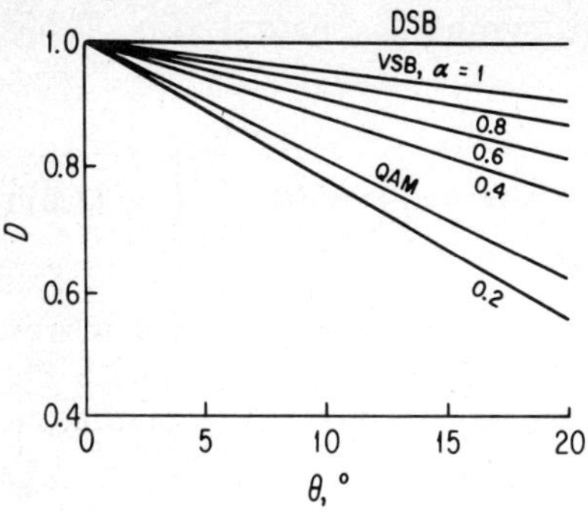

Fig. 7.17. Relative eye opening D as a function of error θ for various linear modulation systems when the eye is sampled at its optimum time.

any value of α other than zero. The more bandwidth allowed, the less the degradation per increment of phase error.

As was pointed out in Chap. 6 the best sampling instant and the phase error are intimately related. If for a given phase error the sampling time is optimized in Eq. (7.58) to obtain the maximum eye opening, an improvement in performance may be attained. This improvement is in evidence in Fig. 7.17 where the same quantities as in Fig. 7.10 are plotted when the eye is sampled at its maximum opening. In the QAM system the optimum sampling time is zero.

7.4 Illustrative Examples

At this point it might be instructive to illustrate some of the basic ideas discussed in the previous sections by a few concrete examples. For instance, suppose an ideal bandpass channel of 1,800 Hz is to be used for the transmission of two independent binary data sources emitting 1,200 bps. The baseband data waves are presented in the form of sequences of rectangular pulses having amplitude ± 1 occurring at random. Mathematically, the two baseband data waveforms are prepresented as

$$d_1(t) = \sum_n a_n g\left(t - \frac{n}{1{,}200}\right) \quad \text{and} \quad d_2(t) = \sum_n b_n g\left(t - \frac{n}{1{,}200}\right) \tag{7.61}$$

where $\{a_n\}$ and $\{b_n\}$ are two independent sequences of ± 1s and

$$g(t) = \begin{cases} 1 & 0 \le t \le \dfrac{1}{1{,}200} \\ 0 & \text{otherwise} \end{cases}$$

In the frequency domain

$$D_1(f) = \frac{e^{-i\pi(f/1{,}200)}}{1{,}200} \frac{\sin \pi (f/1{,}200)}{\pi (f/1{,}200)} \sum_n a_n e^{-i2\pi n(f/1{,}200)}$$

and $$D_2(f) = \frac{e^{-i\pi(f/1,200)}}{1,200} \frac{\sin \pi\,(f/1,200)}{\pi(f/1,200)} \sum_n b_n e^{-i2\pi n(f/1,200)} \tag{7.62}$$

The channel transfer function for positive frequencies is

$$H(f) = \begin{cases} 1 & f_c - 900 \le f \le f_c + 900 \\ 0 & \text{otherwise} \end{cases} \tag{7.63}$$

where $f_c > 900$ Hz.

Having a description of the baseband signal and channel, we must now select a modulation technique. The choices include, for example, the use of QAM or VSB in quadrature. QAM will be illustrated first, and then VSB.

7.4.1 Example of Quadrature-amplitude Modulation

It is convenient to shape the baseband signal before modulation in such a way that the magnitude of its Fourier transform has a flat characteristic across the band. To accomplish this shaping admit $d_1(t)$ and $d_2(t)$ to filters having identical transfer functions:

$$F(f) = \begin{cases} 1200\, e^{i\pi(f/1,200)} \dfrac{\pi f/1,200}{\sin \pi\,(f/1,200)} & |f| \le 900 \\ \text{arbitrary function of } f & |f| > 900 \end{cases} \tag{7.64}$$

This filter may be realized approximately since no sharp cutoffs are involved. Observe that the first zero of $\sin \pi f/1,200$ occurs when $f = 1,200$. When $d_1(t)$ and $d_2(t)$ are shaped by the filters specified, Eq. (7.62) becomes

$$D_1'(f) = \begin{cases} R(f) \sum_n a_n e^{-i2\pi n(f/1,200)} & |f| \le 900 \\ \text{Arbitrary} & |f| > 900 \end{cases}$$

$$D_2'(f) = \begin{cases} R(f) \sum_n b_n e^{-i2\pi n(f/1,200)} & |f| \le 900 \\ \text{Arbitrary} & |f| > 900 \end{cases} \tag{7.65}$$

where $$R(f) = \begin{cases} 1 & |f| \le 900 \\ 0 & \text{otherwise} \end{cases}$$

The baseband signal as represented in Eq. (7.65) is still not ready for the modulator. Now it must be shaped to achieve the desired Nyquist symmetry condition. Since the signaling speed is 1,200 bps, the signal must be rolled off symmetrically around 600 Hz. For this purpose a raised-cosine roll-off characteristic like that in Eq. (7.66) is used:

$$P(f) = \begin{cases} 1 & |f| \leq 300 \\ \cos^2\left[\dfrac{\pi}{1{,}200}(f - 300)\right] & 300 \leq |f| \leq 900 \end{cases} \tag{7.66}$$

Combining the filter characteristics of Eqs. (7.64) and (7.66) yields a composite characteristic; namely,

$$G(f) = F(f)P(f) = \begin{cases} 1{,}200\, e^{i\pi \frac{f}{1{,}200}} \dfrac{\dfrac{\pi f}{1{,}200}}{\sin \dfrac{\pi f}{1{,}200}} & |f| < 300 \\ 1{,}200\, e^{i\pi \frac{f}{1{,}200}} \dfrac{\dfrac{\pi f}{1{,}200}}{\sin \dfrac{\pi f}{1{,}200}} \cos^2\left[\dfrac{\pi(f - 300)}{1{,}200}\right] & 300 \leq |f| \leq 900 \end{cases} \tag{7.67}$$

The data signals emerging from $G(f)$ are properly shaped so that the sequence of bits $\{a_n\}$ and $\{b_n\}$ can be extracted without intersymbol interference. The next step is to apply the two data signals to a quadrature modulator and place the spectrum around the carrier frequency f_c. This amounts to multiplying the respective data waves by $\cos 2\pi f_c t$ and $\sin 2\pi f_c t$ and then eliminating double-frequency terms. The received signal at the output of the channel is

$$2S(f) = D_1''(f + f_c) + D_1''(f - f_c) + iD_2''(f + f_c) - iD_2''(f - f_c) \tag{7.68}$$

where $$D_1''(f) = H(f)G(f) \sum_n a_n e^{-i2\pi n(f/1{,}200)}$$

and $$D_2''(f) = H(f)G(f) \sum_n b_n e^{-i2\pi n(f/1{,}200)}$$

A synchronous demodulator will extract $D_1'(f)$ and $D_2'(f)$ from $S(f)$ by multiplying the received wave by $\cos 2\pi f_c t$ and $\sin 2\pi f_c t$, respectively, and then eliminating double-frequency terms. To derive the data, the baseband signals are synchronously sampled. This particular transmission system is shown in Fig. 7.18.

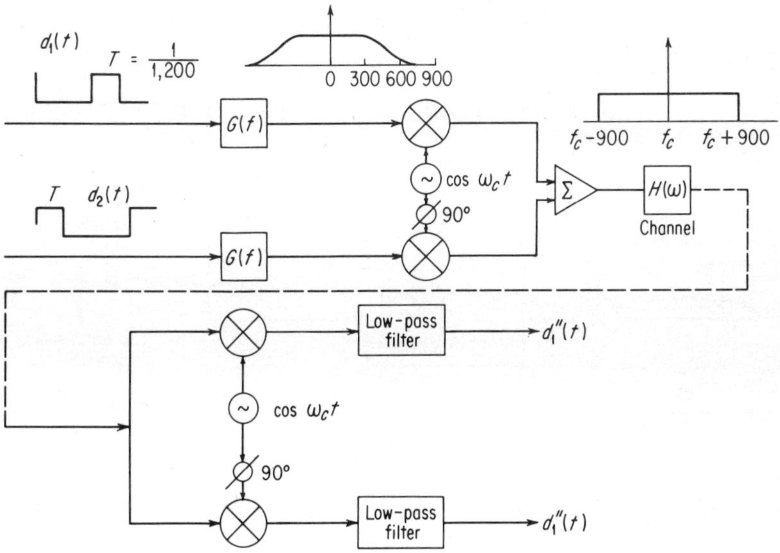

Fig. 7.18. Block diagram showing a QAM data-transmission system.

7.4.2 Example of Vestigial-sideband Modulation

Another modulation method which accomplishes the same task as the QAM system of the preceding section employs VSB modulation. However, observe that the channel does not provide enough bandwidth to accommodate both data signals, and therefore some overlapping of the spectra will have to occur. Becker of Bell Telephone Laboratories demonstrated the possibility of designing the two VSB data channels such that, even though the spectra overlap, separation at the receiver is possible. This technique, called *VSB in quadrature*, is illustrated in what follows. A block diagram of the system is shown in Fig. 7.19.

Consider the filter characteristic

$$F(f) = \begin{cases} \cos\left(\dfrac{\pi}{1,200} f\right) & |f| \le 600 \\ 0 & |f| > 600 \end{cases} \tag{7.69}$$

This characteristic is again within practical attainment since $F(f)$ is continuous at ± 600 Hz. Suppose two such filters are placed in

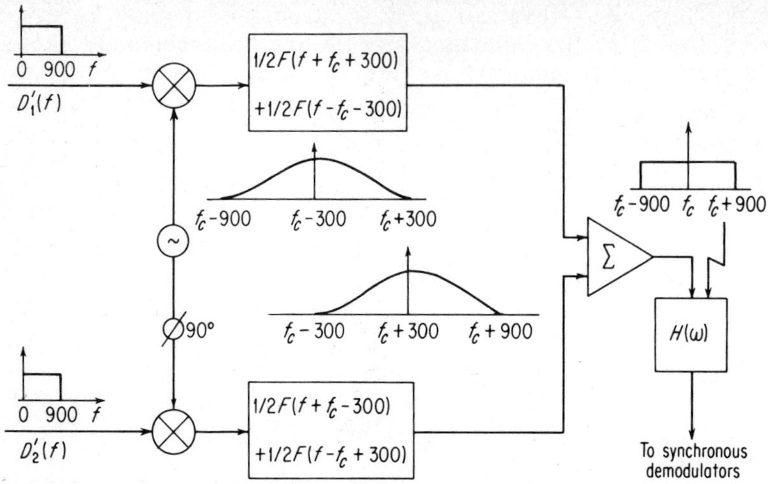

Fig. 7.19. Block diagram showing a Quadrature VSB data-transmission system.

the passband, one located at $f = f_c + 300$ and the other at $f = f_c - 300$. The transfer functions of these filters in the passband are given, respectively, by

$$F_1(f) = \tfrac{1}{2}F(f - f_c - 300) + \tfrac{1}{2}F(f + f_c + 300)$$
$$F_2(f) = \tfrac{1}{2}F(f - f_c + 300) + \tfrac{1}{2}F(f + f_c - 300) \tag{7.70}$$

The modulation procedure is as follows. Multiply the first data source, whose spectrum is given in Eq. (7.65) by $\cos 2\pi f_c t$, and admit the result to $F_1(f)$ to obtain

$$S_1(f) = \tfrac{1}{4}D_1'(f + f_c)F(f + f_c + 300) + \tfrac{1}{4}D_1'(f - f_c)F(f - f_c - 300) \tag{7.71}$$

It can be seen that the range of frequencies in $S_1(f)$ is confined to an interval $f_c - 900 \le f \le f_c + 300$. The other data signal, whose spectrum is also given in Eq. (7.65) is multiplied by $\sin 2\pi f_c$, which is in quadrature with the first data channel. When the product is processed by $F_2(f)$, the following spectrum results:

$$S_2(f) = \frac{i}{4}D_2'(f + f_c)F(f + f_c - 300) - \frac{i}{4}D_2'(f - f_c)F(f - f_c + 300) \tag{7.72}$$

The range of frequencies in $S_2(f)$ is now confined to an interval $f_c - 300 \le f \le f_c + 900$. Now, since the medium has unity transmission over the range of frequencies in $S_1(f) + S_2(f)$, the received signal is the sum of Eqs. (7.71) and (7.72).

Two synchronous demodulators are required to extract the information from $S_1(f) + S_2(f)$. Preceding the demodulators, two identical receiving filters, as specified in Eq. (7.70), are provided. Passing $S_1(f) + S_2(f)$ through $F_1(f)$ yields

$$
\begin{aligned}
8S_{01}(f) = {} & D_1'(f + f_c)F^2(f + f_c + 300) + D_1'(f - f_c)F^2(f - f_c - 300) \\
& + iD_2(f + f_c)F(f + f_c - 300)F(f + f_c + 300) \\
& - iD_2(f - f_c)F(f - f_c + 300)F(f - f_c - 300) \qquad (7.73)
\end{aligned}
$$

The other data channel yields $8S_{02}(f)$ when $S_1(f) + S_2(f)$ is passed through $F_2(f)$. The signal $S_{02}(f)$ is obtained from $S_{01}(f)$ by interchanging the subscripts in D_1' and D_2' since a perfectly symmetrical situation prevails.

When $S_{01}(f)$ is processed by a coherent demodulator, the output is proportional to

$$S_{01}'(f) = D_1'(f)[F^2(f + 300) + F^2(f - 300)] \qquad (7.74)$$

The terms involving the other data channel in Eq. (7.73) cancel. Observe that the filter characteristic

$$F_0(f) = F^2(f + 300) + F^2(f - 300) \qquad (7.75)$$

is a 50 percent roll-off raised-cosine filter satisfying the Nyquist criterion for transmission of 1,200 signals per second without intersymbol interference.

7.4.3 Example of Single-sideband Modulation

As another example, the application of SSB will be considered. The use of SSB is only feasible when the baseband data signal does not have a dc component. It is possible to design pulses which lack low-frequency energy and yet are suitable for data transmission, provided partial response coding and decoding (as described in Chap. 4) is employed. To illustrate the application, consider the transmission of 2,400 bps over an ideal 1,200 Hz bandpass channel. For this application let us examine the following baseband pulse response.

$$p(t) = \frac{\sin 2\pi F[t - (1/2F)]}{2\pi F[t - (1/2F)]} - \frac{\sin 2\pi F[t + (1/2F)]}{2\pi F[t + (1/2F)]} \qquad (7.76)$$

where $F = 1{,}200$. The average value of this pulse is 0 and therefore is suitable for SSB application. The Fourier transform of $p(t)$ is

$$P(f) = \begin{cases} 2Te^{i\pi/2} \sin 2\pi fT & |f| < 1{,}200 \\ Te^{i\pi/2} \sin 2\pi fT & |f| = 1{,}200 \\ 0 & \text{elsewhere} \end{cases} \tag{7.77}$$

where $T = 1/2400$. In order to synthesize a SSB signal the Hilbert transform of $p(t)$ is required. A straightforward procedure for calculating this transform is to multiply $p(f)$ by a Hilbert filter characteristic and then calculate the inverse Fourier transform. This calculation yields

$$\hat{p}(t) = -\frac{T}{\pi} \int_{-2\pi\cdot 1{,}200}^{2\pi\cdot 1{,}200} [i \operatorname{sgn} \omega][i \sin \omega T] e^{i\omega t}\, d\omega \tag{7.78}$$

Since a sgn (·) is an odd function, Eq. (7.78) may be reduced to

$$\begin{aligned} \hat{p}(t) &= \frac{2T}{\pi} \int_0^{2\pi\cdot 1{,}200} \sin \omega T \cos \omega T\, d\omega \\ &= \frac{T}{\pi} \int_0^{2\pi\cdot 1{,}200} \sin \omega (t + T)\, d\omega + \frac{T}{\pi} \int_0^{2\pi\cdot 1{,}200} \sin \omega (t - T)\, d\omega \\ &= -\frac{\cos 2\pi F[t + (1/2F)]}{2\pi F[t + (1/2F)]} - \frac{\cos 2\pi F[t - (1/2F)]}{2\pi F[t - (1/2F)]} \end{aligned} \tag{7.79}$$

With the pulse $p(t)$ and its Hilbert transform, two baseband data signals are formed:

$$d(t) = \sum_n a_n p(t - nT) \quad \text{and} \quad \hat{d}(t) = \sum_n a_n \hat{p}(t - nT) \tag{7.80}$$

The SSB signal is now formed by multiplying $d(t)$ by $1/2(\cos \omega_c t)$ and $\hat{d}(t)$ by $1/2(\sin \omega_c t)$ and taking the difference to obtain

$$s(t) = \frac{1}{2} d(t) \cos \omega_c t - \frac{1}{2} \hat{d}(t) \sin \omega_c t \tag{7.81}$$

The above example merely serves to illustrate the principles developed earlier. This is not to imply that in actual practice the signal $s(t)$ will necessarily be formed in this manner. A more expedient way to achieve the same thing might be to perform the desired shaping in the passband. This should be done by first forming the DSB signal which might have constant amplitude characteristics across the band and then modifying this signal by an appropriate filter to form the SSB signal. Because the pulse $p(t)$ has no dc component, such a filter can be synthesized.

In this particular system a low-level carrier signal could be added so that coherent demodulation at the receiver can be carried out. The task of carrier recovery in this system would be relatively easy since there is very little data energy in the vicinity of the carrier. A narrowband filter centered at this frequency would extract the pilot tone, which would then be multiplied by the received signal. Finally the demodulated signal $d(t)$ would then be further processed to derive the data sequence $\{a_n\}$.

7.5 Summary

This chapter presented a discussion of various linear-modulation techniques. These techniques are used to translate the baseband spectrum to a suitable passband region. We noted in particular how this frequency translation could be accomplished efficiently by the use of single- or vestigial-sideband modulation. We discussed coherent detection procedures and methods of deriving carrier frequency and phase. In order to connect the material presented in the previous two chapters with the material in this chapter, we derived an equivalent baseband system. This served to unify the performance analysis of baseband- and linear-modulation systems. We examined degradations caused by recovered carrier-phase errors and presented curves that showed eye closure as a function of phase error for the various modulation techniques. The chapter was concluded with a few examples to clarify the ideas presented.

Notes

In "Data Transmission" by Bennett and Davey,[4] the reader can find more detailed analyses of certain topics touched on lightly in this chapter. A more extensive treatment of modulation theory and systems is given in "Modulation Theory" by Black.[12] An interesting way of looking at the modulation process and an extensive bibliography is given in Voelker.[13] Oswald[14] gives an historical account of single-sideband transmission. Early work on pulse transmission by AM is given by Sunde.[15]

8

Frequency Modulation

Frequency-modulation (FM) *techniques* find a wide range of application in digital data-transmission systems. Their appeal derives primarily from instrumentation advantages which result principally from the incoherent nature of the demodulation process and the relative ease of signal generation. The various demodulators operate directly on the received wave without the need to recover carrier frequency and phase as in some linear-modulation systems. However, FM is generally not used in high-speed data transmission since, from a bandwidth utilization point of view, it is not as efficient a technique as some linear-modulation methods. When bandwidth is not the major consideration, FM can provide excellent performance with minimum equipment complexity. It also is to some degree immune to certain channel impairments such as frequency shift and certain nonlinearities. Since only the frequency of a sine wave is modulated, the amplitude, and consequently, the power level of the modulated signal, is constant. These attributes make FM well suited for applications where the peak power must be limited and independent of the modulating signal.

In this chapter various FM techniques and properties will be investigated. Unlike linear-modulation systems, no equivalent baseband system has been found for frequency modulation. Realizations of frequency modulation and detection usually involve nonlinear processing at the transmitter and receiver; therefore the principle of superposition does not hold. These facts make the analysis of FM systems more complicated than the linear-modulation systems. Whereas in linear-modulation systems the modulation process can be viewed as simply transporting frequency bands up and down the frequency scale of the medium, in FM systems the modulation and demodulation processes are intimately related to the signal selection and processing problem.

8.1 The Digital FM Signal

We saw in the previous chapter how AM waves are produced and mathematically represented. The linear nature of the modulation process permitted a very simple mathematical representation which made the analysis of spectral properties straightforward. In this section we first discuss the generation and representation of FM waves and conclude with a detailed examination of their spectral properties.

8.1.1 Constant-envelope Continuous-phase Signals

Frequency-modulated waves are readily generated by varying the frequency of an oscillator. Ideally, the oscillator frequency follows the baseband signal $x(t)$ which implies an output signal representable as

$$v(t) = A \cos \left[\omega_c t + \omega_d \int_0^t x(t')\, dt' + \theta \right] \tag{8.1}$$

In this representation, A is the carrier amplitude, ω_c is the angular frequency of the oscillator in the absence of modulation, θ is the phase of the carrier at $t = 0$, and ω_d is a conversion factor relating frequency displacement to baseband-signal voltage. When $x(t)$ is normalized to unity, ω_d is called the *peak frequency deviation*. The argument of the sinusoidal wave is termed the *instantaneous phase* while the time derivative of the phase is called *frequency*. Notice that the phase in Eq. (8.1) is a continuous function of time. Clearly the frequency need not be continuous. Since the phase is differentiable, the existence of the first derivative of $v(t)$ is guaranteed. This implies that the power spectrum of $v(t)$ must fall off at least as the inverse fourth power of frequency as the latter becomes large. For example, consider an $x(t)$ composed of a sequence of $\pm$ rectangular pulses. The first derivative of $x(t)$ does not exist; therefore for large frequencies, as is well known, the power spectrum of $x(t)$ goes down as $1/\omega^2$. However, if this wave is used to modulate the frequency of a carrier, the power spectrum of the resulting wave goes down as $1/\omega^4$ as ω becomes large. Spectral properties of digital FM waves will be treated more extensively in Sec. 8.1.2.

The baseband data signal $x(t)$ is represented as in previous chapters by

$$x(t) = \sum_n a_n g(t - nT) \tag{8.2}$$

where $\{a_n\} = (\ldots a_{-1}, a_0, a_1 \ldots)$ is the data sequence and the signal $g(t)$ is a suitable pulse of unit height. For analytical purposes we shall model $\{a_n\}$ as a collection of discrete-valued, independent, random variables possessing identical but otherwise arbitrary probability distributions. For example, when $\{a_n\}$ is a binary sequence, and $g(t)$ is a rectangular pulse, the wave in Eq. (8.1) represents conventional *frequency shift keying* (FSK). In order to maintain phase continuity, the phase at every transition is usually dependent on the past data sequence. As an aid in visualizing a multilevel FSK wave, consider the following wave-construction procedure. The sequence $\{a_n\}$ is first associated with a sequence of segmented cosine waves, $A \cos [\omega_c t + \omega_d a_n t + \theta_n]$, of T-sec durations. The FM wave, Eq. (8.1), is then constructed by specifying a sequence of phases $\{\theta_n\}$ as follows. The first member in the sequence has an arbitrary phase θ. The phase of the second segment must then be $\theta_2 = \theta + (\omega_1 + \omega_c)T$ and the nth phase $\theta_n = \theta + (\omega_1 + \omega_c)T + \cdots (\omega_c + \omega_{n-1})T$, where $\omega_n = \omega_d a_n$. By shifting the phase of these different segments sequentially one obtains a continuous-phase, constant-envelope FM wave.

It is also possible to generate a digital FM wave without regard to phase continuity. This may be accomplished by having many oscillators, each of which is tuned to the various desired frequencies whose outputs are controlled by a digital baseband wave. This method gives rise to undesirable transients in addition to complicating the transmitter. For these reasons it is rarely used in practice, particularly if the number of frequencies is large.

The fact that the modulating signal is not linearly related to the constant-envelope, continuous-phase FM signal poses great analytical difficulties. These difficulties are particularly apparent when studying the response of linear filters. To cope with these problems, various approximate techniques are sometimes employed.[1] These approximation techniques are well suited for investigating the performance of analog systems but are not very satisfactory for digital applications. Therefore, they will not be treated here.

8.1.2 Spectral Properties

In linear-modulation systems the spectrum of the modulated wave is simply the frequency translation of the baseband spectrum. However, since FM is a nonlinear-modulation technique, the spectral properties of the modulated signal cannot, in general, be deduced from the baseband spectrum. In this section we shall be concerned with spectral properties of digital FM waves. As will be seen, FM techniques generally alter the baseband spectral shaping and can result in a bandwidth considerably greater than the base bandwidth for some choices of modulation parameters. This property provides a capability of generating large time-bandwidth product signals. On the other hand, for some other choices of modulation parameters, bandwidths of the same order as DSB systems can be obtained.

Before this discussion is continued, the notion of bandwidth must be made clearer in this application since the spectrum of a constant-envelope FM signal cannot be strictly band-limited. A practically useful definition of bandwidth is usually employed. The frequency band outside of which the intensity of the spectrum is below some arbitrarily small fraction of its peak is usually regarded as sufficient measure of bandwidth.

In addition to furnishing bandwidth requirements, the spectral density is critically important in optimizing procedures. Also, evaluation of mutual interference between channels requires knowledge of spectral distribution. An important item in some applications is the extent to which sharp spectral peaks occur near the discrete signaling frequencies. These peaks can, under certain conditions, become delta functions indicating the presence of steady sine-wave components. Such components cause severe crosstalk, for instance, in telephone facilities, and must be treated. Thus it is of paramount importance to know where in the spectrum and under what choices of parameters discrete lines occur.

The derivation of the spectral-density function of a digital FM wave is rather involved. Therefore, the details are derived in the Appendix at the end of this chapter, and only the results will be discussed here. To make use of these results, it is only necessary to express the real FM wave in Eq. (8.1) in terms of its complex and complex conjugate components and assume that the initial phase θ is uniformly distributed on the interval $[0, 2\pi]$. The results in the Appendix are general and do not assume any particular pulse shaping. In this section attention will be restricted to rectangular pulses of T-sec duration. Although in some applications shapes other than rectangular are used, the detailed analysis with these shapes becomes unwieldy and does not provide additional physical insight.

To illustrate the behavior of the spectrum graphically consider a baseband data signal with L possible equally likely levels which are taken as

$$a_n = 2k - (L + 1) \tag{8.3}$$

where $k = 1, 2, \ldots, L$ for all n. (L is assumed to be even.) When we evaluate the appropriate functions in the Appendix at the end of this chapter for this particular case, we obtain for the normalized spectrum of the wave $v(t)$

$$\frac{V(\beta)}{A^2 T} = \begin{cases} \dfrac{I_1}{2} + \operatorname{Re}\left[\dfrac{I_2^{\,2}}{1 - C_a(\omega_d T)\, e^{-i2\pi\beta}}\right] & |C_a(\omega_d T)| < 1 \\[2ex] \dfrac{I_1}{2} + \dfrac{1}{2\pi}\left|\left\langle\dfrac{\sin\gamma}{\gamma}\right\rangle\right|^2 \displaystyle\sum_n [\delta(\lambda - 2\pi n) - 1] & |C_a(\omega_d T)| = 1 \end{cases} \tag{8.4}$$

where

$$I_1 = \left\langle \left(\frac{\sin \gamma}{\gamma} \right)^2 \right\rangle \qquad I_2 = \left\langle \frac{\sin \gamma}{\gamma} e^{-i\gamma} \right\rangle$$

$$\gamma = (\omega - \omega_c - a_n \omega_d) \frac{T}{2} \qquad \beta = (\omega - \omega_c) \frac{T}{2\pi} \qquad \lambda = (\omega_c - \omega) T + \pi$$

and $C_a(\omega_d T)$ is the characteristic function of the random variable a_n evaluated at $\omega_d T$. It is given explicitly by

$$C_a(\omega_d T) = \frac{2}{L} \sum_{k=1}^{k=L/2} \cos \left[\omega_d T (2k - 1) \right] \tag{8.5}$$

It can be observed from Eq. (8.4) that when the magnitude of the characteristic function is unity, the spectrum contains discrete lines occurring at $\beta = n - 1/2$, where $n = \pm 1, \pm 2, \ldots$ with minimum spacings of $1/T$.

In Figs. 8.1 through 8.4 several values of the two parameters (L, the number of levels, and $\Delta = \omega_d T/\pi$) are used to plot the normalized spectral density as a function of normalized frequency $\beta = (\omega - \omega_c)T/2\pi$. Because of symmetry, we only show values for

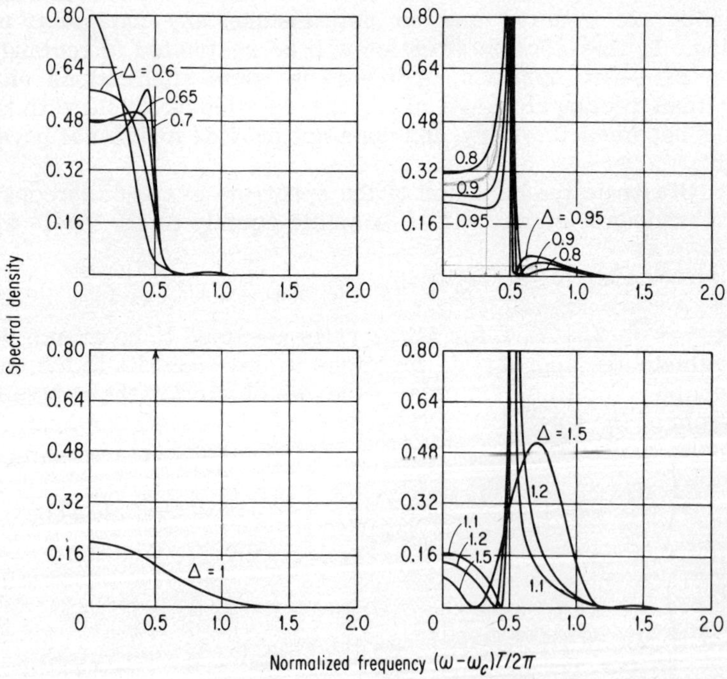

Fig. 8.1. Binary digital FM spectra.

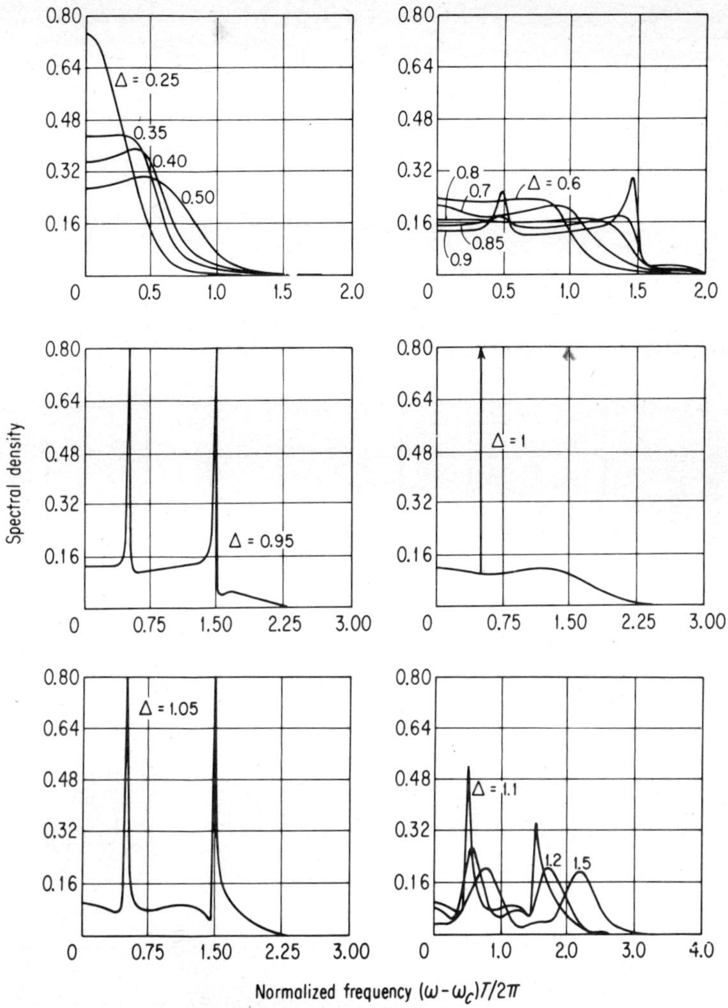

Fig. 8.2. Four-level digital FM spectra.

$\beta \geq 0$. The various curves are exhibited to indicate the way the spectral shapes change as the parameter Δ varies. Figure 8.1 shows the binary FM case. When $\Delta = 1/2$, a very smooth roll-off occurs with a center peak at the normalized carrier frequency, or $\beta = 0$. A zero occurs at $\beta = 0.75$, and another minor peak is seen to occur around $\beta = 0.85$. As Δ increases toward unity, the curves flatten and are almost rectangular in shape when $\Delta \sim 0.65$. As Δ approaches unity, the major peaks occur at $\beta = 1/2$, and the minor peaks become larger. When $\Delta = 1$, the curve is smooth again but of greater spread. Also for this value of Δ the power spectrum includes delta functions occurring at $\beta = \pm 1/2$.

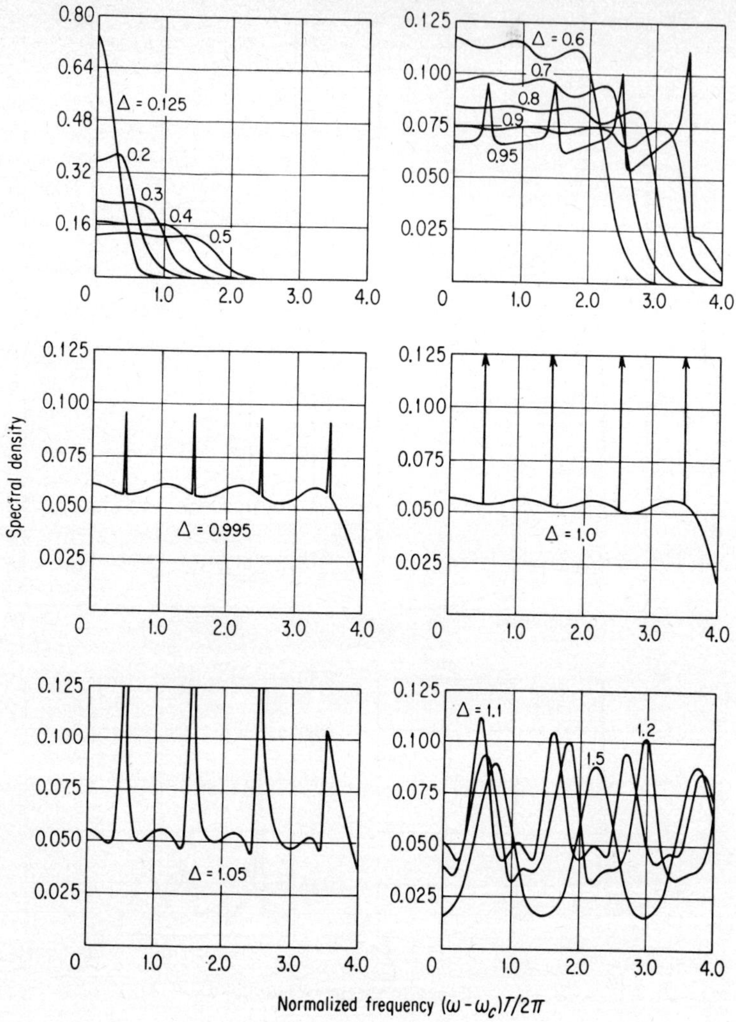

Fig. 8.3. Eight-level digital FM spectra.

The spectra for four- and eight-level FM are shown in Figs. 8.2 and 8.3, respectively. The multilevel spectra show considerable similarity to the binary ones. For small values of Δ, the spectra are narrow and decrease smoothly toward zero. These cases of small Δ will be of considerable interest later when the performance of narrowband FM is examined. For later reference we show in Fig. 8.4 the spectra for $\Delta = 1/L$ for two-, four-, and eight-level systems. It can be seen that these curves are nearly identical.

When Δ is the reciprocal of the number of levels, the principal portion of the spectrum is confined to a relatively narrow band.

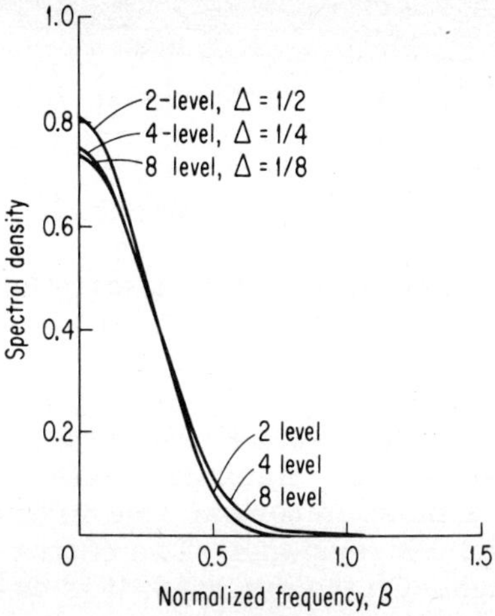

Fig. 8.4. Digital FM spectra for $\Delta = 1/L$.

Table 8-1 gives the computed value of the spectral density at various values of β as a function of L and Δ. As Δ increases toward unity, the spectrum widens in all cases and there tends to be a concentration of power about the transmitted frequencies. This concentration is especially marked in the range $1 - 1/2L \leq \Delta \leq 1 + 1/2L$. The total average power is, of course, independent of the modulation waveform and always equals $A^2/2$.

It can be seen that a variety of spectral shapes may be obtained by changing the parameter Δ and the number of levels L. In a later section we shall use this knowledge to design efficient, FM data transmission systems.

Table 8-1. Computed Spectral-density Values.

No. of levels	Spectral density at β =				
	Δ	0	0.25	0.50	0.75
2	0.500	0.810	0.500	0.090	0.000
4	0.250	0.750	0.470	0.117	0.011
8	0.125	0.735	0.430	0.124	0.013

8.2 Demodulation of FM Signals

In this section several of the more widely used FM demodulators or detectors are discussed. Emphasis is placed on examining the physical behavior of the various systems, and, where possible, simple mathematical models are derived in order to facilitate performance analysis in the presence of various kinds of channel impairments.

The choice of a particular detection method depends on the application. Various factors, such as accuracy requirements, size, and cost, influence the choice. It is not intended to treat exhaustively all methods of FM detection. Frequency demodulators such as phase- and frequency-locked loops have been omitted from the discussion primarily because they are used in wideband analog applications rather than in digital data transmission. The various detectors treated are phase-incoherent; i.e., no phase information is required at the receiver. Therein lies part of the appeal of FM.

8.2.1 Discrimination Detection

A classical method of detecting FM signals is called *discrimination detection*. This method employs a circuit referred to as a *limiter-discriminator*, whose ideal response to an FM wave is the derivative of the phase. The circuit is inherently nonlinear since no linear filter can produce a response to a bandpass signal proportional to its phase function. We shall give a brief description of the circuit, explain its operation, and then examine its performance.

The basic structure of the demodulator is shown in Fig. 8.5. The front-end filter serves to exclude out-of-band noise and can also provide compensation for linear in-band distortion suffered in transmission. This filter plays a most important role in determining the error rates in digital FM systems. Two contradictory attributes are generally sought. On the one hand, it is desired to make the filter narrowband for noise-limiting purposes and, on the other hand, wideband to minimize signal distortion. An optimum design is possible and will be examined later.

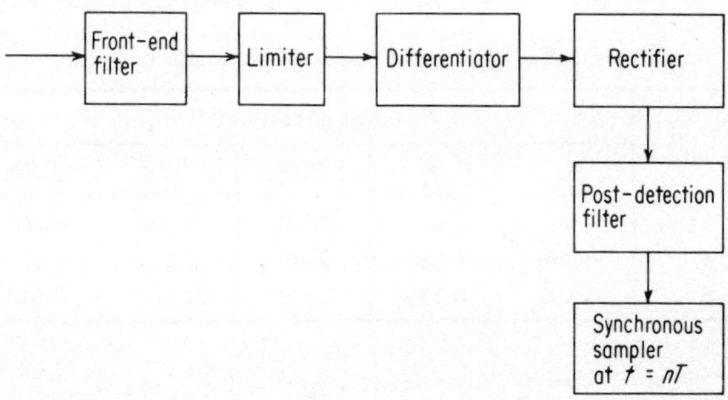

Fig. 8.5. Block diagram of a discrimination detection system.

Following the front-end filter is a limiter which renders at its output a constant-envelope wave. To see how this is achieved, consider an ideal baseband limiter whose output is a positive constant e_0 whenever the input is positive and $-e_0$ whenever the input is negative. If a wave having the representation

$$e(t) = r(t) \cos [\omega_c t + \psi(t)] \quad r(t) \geq 0 \tag{8.6}$$

is passed through this device its output is

$$e_0(t) = \begin{cases} e_0 & \cos z > 0 \\ -e_0 & \cos z \leq 0 \end{cases} \tag{8.7}$$

where $$z = \omega_c t + \psi(t)$$

Since the output is a function of cos z, which is periodic in the variable z with period 2π, it can be expanded in a Fourier series as

$$e_0(t) = \frac{4e_0}{\pi} \sum_{n=0}^{\infty} \frac{(-1)^n \cos(2n+1)z}{2n+1}$$

$$= \frac{4e_0}{\pi} \sum_{n=0}^{\infty} \frac{(-1)^n \cos\{(2n+1)[\omega_c t + \psi(t)]\}}{2n+1} \tag{8.8}$$

Notice that all the odd harmonics of the input wave appear at the output. This device, when followed by a filter whose function is to eliminate all harmonics above the third, is called a *bandpass limiter*. From Eq. (8.8) its output is seen to be

$$e_1(t) = \frac{4e_0}{\pi} \cos[\omega_c t + \psi(t)] \tag{8.9}$$

The separation of the harmonics is only possible when there is no spectral overlap. Thus if the spectrum of Eq. (8.1) is relatively narrow compared with the center frequency ω_c, separation is possible. To exhibit the signal processing further, the wave in Eq. (8.8) is differentiated to obtain

$$e_1'(t) = \frac{4e_0}{\pi}\left[\omega_c + \frac{d\psi(t)}{dt}\right]\sin[\omega_c t + \psi(t)] \tag{8.10}$$

Observe that the information-bearing signal appears in the envelope of the wave of Eq. (8.10). It can be recovered by submitting $e_1'(t)$ to an envelope detector. This is usually accomplished by first rectifying the wave and then passing the result through a low-pass filter. A unique envelope will exist only if $|d\psi/dt| \leq \omega_c$. In most practical circuits the ideal differentiator is approximated by a bandpass filter having linear amplitude and phase characteristics about the carrier frequency ω_c over the signal bandwidth.

8.2.2 Performance of the Discriminator in the Presence of Noise

The performance evaluation of FM systems is complicated by the nonlinear nature of the detection process. Even though an exact evaluation of the attainable error rates is, in general, not mathematically tractable, a great deal of insight can be gained by analyzing simple models based on suitable assumptions. When the detector operates on the received signal plus noise, the error rate may be calculated from the knowledge of the statistical structure of the output sample. The observed random variable is, in general, not gaussian—even when the input noise is gaussian. It is often tempting to approximate the output-sample probability density by the gaussian density to simplify error-rate calculations. Unfortunately, for several reasons the approximations thus obtained are not always useful. For one, the predicted error rates based on the gaussian approximation do not always agree with experiments. This by itself would be sufficient reason to doubt the validity of the results. Another cause for concern is that the approximate error-rate formulas thus obtained do not reflect the correct dependence on the important parameters of the system. For this reason a more fundamental analysis of the discriminator is necessary.

In this section we shall discuss a simple theory, based on a phenomenological model, which correctly predicts the performance of this system. Here we shall confine the analysis to the additive gaussian channel; in the next section we shall discuss FM systems operating over the dispersive gaussian channel. While the gaussian assumption serves to provide quantitative answers, the qualitative effects of the noise are not restricted to any particular statistics.

Consider the digital FM wave appearing in Eq. (8.1), and assume that the pulse shaping $g(t)$ is rectangular. Consequently, the frequency of the wave $v(t)$ is a piecewise constant function of time with allowable changes occurring only at points $t = kT$, where $k = 0, 1, \ldots$. A convenient representation for this wave, valid only over an interval, say, $kT \leq t \leq (k+1)T$, is a cosine wave of fixed frequency determined by the transmitted symbol a_k and a constant phase angle $\phi(\mathbf{a})$ whose value may depend on the past symbols. Mathematically then

$$v(t) = A \cos [\omega_c t + a_k \omega_d t + \phi(\mathbf{a}) + \theta] \qquad kT \leq t < (k+1)T \tag{8.11}$$

The discriminator operates on $v(t)$ plus a zero-mean Gaussian noise process $\eta(t)$. This process can be represented in terms of its in-phase and quadrature components about the frequency $\omega_c + a_k \omega_d$ as

$$\eta(t) = z(t) \cos (\nu_k t + \alpha) + y(t) \sin (\nu_k t + \alpha) \tag{8.12}$$

where $z(t)$ and $y(t)$ are baseband noise components, $\nu_k = \omega_c + a_k \omega_d$, and $\alpha = \phi(\mathbf{a}) + \theta$. We also assume that the noise spectrum is symmetrical about ω_c. The sum of Eqs. (8.11) and (8.12) is now expressed as

$$\begin{aligned} s(t) &= v(t) + \eta(t) \\ &= [z(t) + A] \cos (\nu_k t + \alpha) + y(t) \sin (\nu_k t + \alpha) \\ &= r(t) \cos [\nu_k t + \Psi(t) + \alpha] \qquad kT \leq t \leq (k+1)T \end{aligned} \tag{8.13}$$

where
$$r(t) = \sqrt{[z(t) + A]^2 + y^2(t)}$$

and
$$\Psi(t) = \tan^{-1} \frac{-y(t)}{z(t) + A}$$

When the noise components $z(t)$ and $y(t)$ are referred to the output of the front-end filter,* the limiter-discriminator response is ideally [see Eq. (8.10)]

$$d_0(t) = \nu_k + \frac{d\Psi(t)}{dt} = \nu_k + \frac{d}{dt}\left\{\tan^{-1}\frac{y(t)}{z(t)+A}\right\}$$

$$= \nu_k + \frac{[z(t)+A]\dot{y}(t) - y(t)\dot{z}(t)}{[z(t)+A]^2 + y^2(t)} \tag{8.14}$$

where the dots over y and z denote differentiation with respect to time. We approximate the low-pass postdetection filter and sampler by an integrate-and-dump circuit so that the symbol a_k is detected without intersymbol interference. Integrating Eq. (8.14) and dumping at $t = T(k+1)$ yields the observable quantity q' given by

$$q = q' - \nu_k T = \int_{kT}^{(k+1)T} \frac{[z(t)+A]\dot{y}(t) - y(t)\dot{z}(t)}{[z(t)+A]^2 + y^2(t)}\, dt \tag{8.15}$$

By factoring $[z(t) + A]^2$ from numerator and denominator in the above integrand and making use of the fact that the differential of $y(t)/[z(t) + A]$ equals $[(z + A)dy - ydz]/[z + A]^2$, we can express the above integral as a line integral:

$$q = \oint_{\xi(kT)}^{\xi[(k+1)T]} \frac{d\xi(t)}{1 + \xi^2(t)} \tag{8.16}$$

where

$$\xi(t) = \frac{y(t)}{z(t) + A}$$

The value of q depends not only on the end points $\xi[(k+1)T]$ and $\xi(kT)$ but also on the path. Since $\xi(t)$ is a random process, the paths taken are random; consequently q is a random variable whose statistics depend on the statistics of $\xi(t)$. Unfortunately the probability density of q, which we need in order to calculate error rates, does not depend only on the elementary statistics of $\xi(t)$. As we shall show, in addition to the elementary statistics of $\xi(t)$, the distribution of its singularities enters the calculation. The singularities of $\xi(t)$ are determined by the zero crossings of $z(t) + A$. This might discourage the analyst from pursuing the problem further since the statistics of zero crossings of most elementary

*In this analysis we assume that the front-end filter serves only to exclude out-of-band noise and does not distort the received signal.

random processes are as yet unknown. Nevertheless, we shall proceed with the analysis, keeping in mind that we are seeking not exact answers but reasonable approximations which will render physical insight to the mechanism of error production.

To see how the problem may be handled, let us take the following example. Suppose for $0 < t \leq T$ [we set $k = 0$ in Eq. (8.16) without loss of generality] we have (1) that $y(t) > 0$ and (2) that $z(t) + A$ is positive for awhile, then decreases once through zero at $t = t_0$, and thereafter remains negative. A possible plot of $\xi(t)$ as a function of t over this time interval is shown in Fig. 8.6. For this particular path we may evaluate the integral in Eq. (8.16) exactly. We first observe that it can be decomposed into the sum of two ordinary integrals; namely,

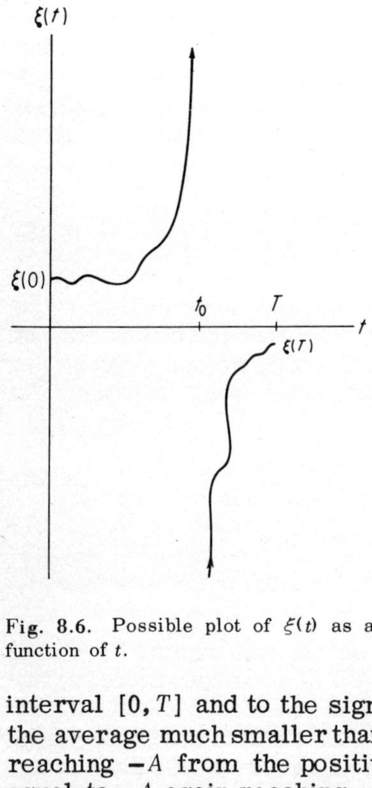

Fig. 8.6. Possible plot of $\xi(t)$ as a function of t.

$$q = \int_{-\infty}^{\infty} \frac{d\xi}{1+\xi^2} + \int_{\xi(0)}^{\xi(T)} \frac{d\xi}{1+\xi^2} \tag{8.17}$$

These are elementary integrals and can be evaluated directly. The first is just equal to π and the second to $\tan^{-1}\xi(T) - \tan^{-1}\xi(0)$, where $\tan^{-1}(\cdot)$ means the principal value of the inverse tangent function such that $|\tan^{-1}(\cdot)| \leq \pi/2$. For a more general path for which $\xi(t)$ has more than one singularity, we obtain

$$q = \tan^{-1}\xi(T) - \tan^{-1}\xi(0) + n(T)\pi \tag{8.18}$$

where $n(T)$ is an integer related to the number of times $z(t) = -A$ in the interval $[0, T]$ and to the sign of $y(t)$ when $z(t) = -A$. When $z(t)$ is on the average much smaller than A and in the rare event that $z(t) = -A$, reaching $-A$ from the positive side, it will almost immediately be equal to $-A$ again reaching $-A$ from the other direction. If $y(t)$ has not changed sign, the contribution of the "return trip" to $n(T)$ will cancel the contribution from the previous crossing of the y axis. On the other hand, it $y(t)$ does change sign, the contribution to $n(T)$ will be the same as that of the previous crossing. A clear picture of the behavior of the line integral may be obtained by visualizing the random paths in the $(z+A)-y$ plane. As long as the paths are confined to the right-hand plane, $\xi(t)$ has no singularities; therefore the value of the path integral is just $\tan^{-1}\xi(T) - \tan^{-1}\xi(0)$. However, if the paths cross the y axis but do not encircle the origin, the quantity $\pm\pi$ is added to $\tan^{-1}\xi(T) - \tan^{-1}\xi(0)$ each time a crossing occurs. The value of π is taken for crossing to the left-hand (right-hand) plane above (below) the horizontal axis and $-\pi$ for the other possible crossings. In other words, if the crossing tends to increase the angle, π is taken. If it tends to decrease the angle, $-\pi$ is taken. If the path encircles the origin, the quantity $\pm 2\pi$ is added, the sign depending on the direction of the encirclement. The net contribution to $n(T)$ of a number of paths is shown in Fig. 8.7. The paths which have $\delta n = \pm 2$ are identified as "clicks," a notion introduced by Cohn[3] and Rice,[4] who made extensive use of these ideas to study the threshold effect in analog FM receivers.

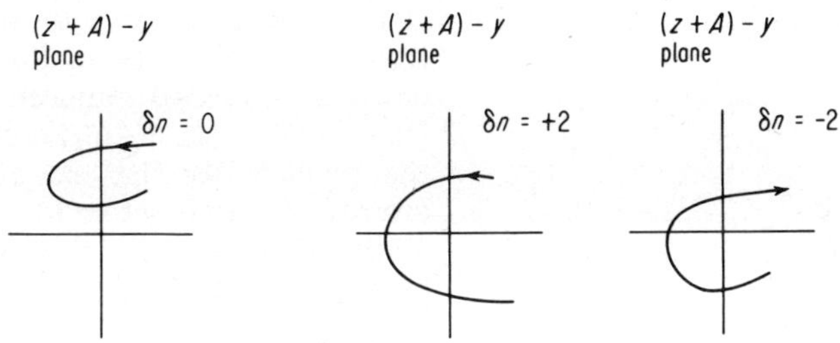

Fig. 8.7. Net changes δn in $n(T)$ caused by some typical paths.

The clicks are not the only contribution to $n(T)$, however. There is also a contribution because at $t = 0$, we may be in the middle of a large noise fluctuation and be in the left-hand plane. Immediately afterward, at $t = 0+$, we experience a contribution of ± 1 to $n(T)$. A similar situation will prevail at $t = T$, when a possibility exists of stopping the process immediately after we have crossed to the left-half plane. In the analysis to follow we shall ignore these end effects since they occur with a negligibly small probability compared with the probability of a click. In other words, encirclements of 2π, or clicks, are dominant.

We remark that q is decomposed, as seen in Eq. (8.18), into the sum of three random variables. The first two are continuous and bounded, and their probability densities are related to the elementary statistics of $z(t)$ and $y(t)$. The third random variable is discrete, and its probabilities are determined by the probabilities of threshold crossings of $z(t)$ and $y(t)$.

To permit further analysis of the model several assumptions must be made. The assumptions involve statements of independence of the three random variables in Eq. (8.18). The justification stems from two different physical arguments, one of which has to do with bandwidth and the other with signal-to-noise ratio. The first assumption states that $\tan^{-1}\xi(T)$ and $\tan^{-1}\xi(0)$ are independent. For a gaussian input process with bandwidth $W \geq 1/T$ the motivation for the assumption is clear. The second assumption states that $n(T)$ is independent of the other two random variables. Furthermore, it is assumed that the clicks are independent of one another, which may not be entirely correct. (We shall later cite an example for which this assumption does not hold.) However, if the front-end filter characteristic is flat and the signal-to-noise ratio is large, it is reasonable to assume that clicks occur rarely and, consequently, independently from one another.

In principle, the density of q can be determined by convolving the continuous density of $[\tan^{-1}\xi(T) - \tan^{-1}\xi(0)]$ with the discrete density of $\pi n(T)$. In Fig. 8.8, where a qualitative illustration of the result is shown, the density is seen to consist of a central lobe about zero extending to $\pm\pi$ on each side; this is the density of the continuous and bounded random variable $[\tan^{-1}\xi(T) - \tan^{-1}\xi(0)]$ multipled by the probability that no clicks occurred in the interval $[0, T]$. In addition, identical lobes displaced by integral multiples of 2π cover the entire q axis. These identical lobes are weighted by the probability of getting the appropriate number of clicks to effect the displacement. We remark that the density need not necessarily be symmetric about zero; i.e., the rate of occurrence of clicks which increase the phase by 2π need not be the same as the rate of occurrence of those clicks which decrease the phase by 2π.

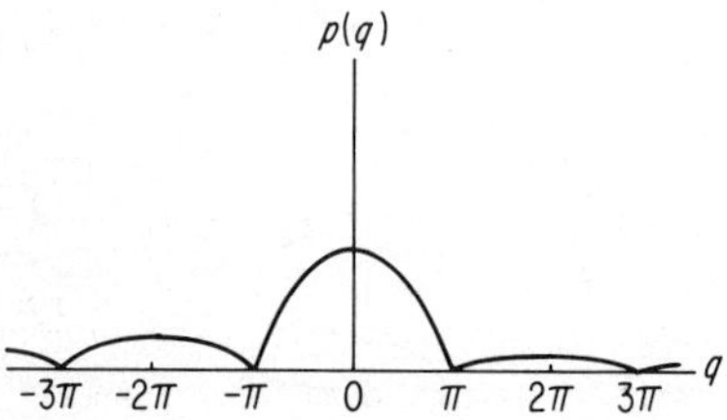

Fig. 8.8. A sketch of the probability density function of q.

The probability density of the angle $\phi = \tan^{-1}\xi(T)$, $|\phi| \leq \pi/2$ can be shown[5] to be

$$p(\phi) = \frac{e^{-\rho}}{\pi} + \sqrt{\frac{\rho}{\pi}}\cos\phi\, e^{-\rho\sin^2\phi}\operatorname{erf}(\sqrt{\rho}\cos\phi) \tag{8.19}$$

where

$$\rho = \frac{A^2}{2\sigma^2} \qquad \sigma^2 = \left\langle z^2(T)\right\rangle = \left\langle y^2(T)\right\rangle$$

and erf(·) is the error function. The probability density of $\xi(0)$ is of the identical form as Eq. (8.19).

Suppose ϕ_1 and ϕ_2 are two independent angles which obey the density of Eq. (8.19) and define an angle $\Phi = \phi_1 - \phi_2$, $|\Phi| \leq \pi$. When analyzing particular FM systems, the probability that Φ exceeds some angle $\alpha > 0$ is usually of interest. We therefore seek an expression for the probability

$$P(\alpha) = \int_{-\pi/2}^{\pi/2-\alpha} d\phi_2 \int_{\phi_2+\alpha}^{\pi/2} d\phi_1\, p(\phi_1)\, p(\phi_2) \qquad \alpha > 0 \tag{8.20}$$

In general, although it is not possible to perform these integrations exactly, a good approximation may be given which is valid when $P(\alpha) \ll 1$. Standard asymptotic techniques can be used to show[5] that

$$P(\alpha) \sim \begin{cases} \dfrac{1}{\sqrt{8\pi}} \dfrac{\cot \alpha/2}{\sqrt{\cos \alpha}} e^{-2\rho \sin^2(\alpha/2)} & 0 < \alpha < \dfrac{\pi}{2} \\ \dfrac{1}{4} e^{-\rho} & \alpha = \dfrac{\pi}{2} \\ \dfrac{e^{-\rho(1+\cos^2\alpha)}}{2\pi\sqrt{\pi}\,\rho\sqrt{\rho}\,\sin\alpha\cos^2\alpha} & \alpha > \dfrac{\pi}{2} \end{cases} \tag{8.21}$$

The final item needed to completely specify the probability distribution of q is the density of $\pi n(T)$, associated with the clicks. Rice[4] has shown that the positive and negative clicks form approximately independent Poisson processes with rates of occurrence N_+ and N_-. In general, when ν_k does not equal ω_c, the carrier frequency, N_+ is not equal to N_-. This can be seen physically since the quadrature and in-phase components of the added noise, when represented about a frequency other than the carrier† ω_c, are not statistically independent. It is also shown[5] that the probability of getting exactly $n(T)$ clicks in an interval T is, for large ρ,

$$P[n(T)] \sim \frac{e^{-N_- T}(N_- T)^{n(T)}}{[n(T)]!} \tag{8.22}$$

where

$$N_- \sim \frac{|a_k|\,\omega_d}{2\pi} e^{-\rho}$$

Furthermore, the probability of getting m or more clicks is, for large ρ, approximately the probability of getting exactly m clicks. Physically, then, the effect of the clicks is to pull the transmitted frequencies closer toward the carrier rather than away from it. Consequently, the probability density of q has most of its area on the negative axis when $a_k\omega_d$ is larger than zero.

This essentially concludes the analysis. With the aid of Eqs. (8.21) and (8.22) asymptotic error rates for any specific system can be determined. In a later section we shall refer to these formulas when examining particular FM systems.

8.2.3 Analysis for the Gaussian Dispersive Channel

The previous section established a phenomenological model for the conventional FM receiver that explains the mechanism of error

---

† The noise spectrum about ω_c is assumed to be symmetrical.

production in the presence of noise. To render the problem mathematically tractable, several assumptions had to be made. In particular, it was assumed that the FM signal was essentially undistorted prior to processing. When a dispersive channel is interposed between the transmitter and receiver, this assumption may not be valid. Although the mechanism of error production can still be explained in terms of the clicks, very few quantitative results can be obtained. The difficulty arises because the signal plus noise is subjected to a mixture of linear and nonlinear operations. The dispersive channel introduces intersymbol interference in a very complicated manner that affects the rate at which clicks are produced.

A major source of difficulty in the analyses of FM detection is the presence of the postdetection filter. If one assumes that this filter is ideal, in the sense that it does not perform significant baseband processing, the mathematical problems become tractable. In many applications this assumption is valid, particularly so when the modulated signal bandwidth is of the same order as that of the baseband signal.

Assuming therefore an ideal postdetection filter, we can start the analysis by expressing the impulse response of the dispersive channel in terms of its in-phase and quadrature components about the center frequency ω_c as

$$h(t) = h_1(t) \cos \omega_c t - h_2(t) \sin \omega_c t \tag{8.23}$$

The digital FM signal, Eq. (8.1), is likewise expanded as

$$v(t) = v_1(t) \cos \omega_c t - v_2(t) \sin \omega_c t \tag{8.24}$$

where $$v_1(t) = A \cos \left[\omega_d \int_0^t x(t')\, dt' + \theta \right]$$

and $$v_2(t) = A \sin \left[\omega_d \int_0^t x(t')\, dt' + \theta \right]$$

The noise-free input to the detector $v'(t)$, which is the convolution of $h(t)$ and $v(t)$, can be expressed as

$$v'(t) = p(t) \cos \omega_c t - q(t) \sin \omega_c t \tag{8.25}$$

Dropping the double-frequency terms, it can be ascertained that

$$p(t) = \frac{1}{2} \int_{-\infty}^{\infty} [v_1(t - \tau) h_1(\tau) - v_2(t - \tau) h_2(\tau)]\, d\tau$$

and $$q(t) = \frac{1}{2} \int_{-\infty}^{\infty} [v_1(t - \tau) h_2(\tau) + v_2(t - \tau) h_1(\tau)]\, dt \tag{8.26}$$

The channel characteristics are reflected in the form of $p(t)$ and $q(t)$. These functions can rarely be expressed more explicitly than in Eq. (8.26). However, good estimates can be obtained with the use of a digital computer. Since $v_1(t)$ and $v_2(t)$ are dependent on the entire data sequence, numerical convolutions must be carried out for all possible data sequences of a given length. The choice of the length of the sequence is dictated by the channel memory and by computation limitations.

The added noise at the detector is also represented in terms of in-phase noise $n_1(t)$ and quadrature noise $n_2(t)$. This time it is convenient to represent the noise components about the center frequency ω_c. The combined wave is written in terms of resulting envelope and phase as

$$v_r(t) = R(t) \cos [\omega_c t + \Psi(t)] \tag{8.27}$$

where $$R^2(t) = [p(t) + n_1(t)]^2 + [q(t) + n_2(t)]^2$$

and $$\Psi(t) = \tan^{-1} \frac{q(t) + n_2(t)}{p(t) + n_1(t)}$$

The frequency detector, as before, differentiates the phase of Eq. (8.27). The function of the postdetection filter in the present analysis is to separate the wave representing the frequency variation from the high-frequency detection products. Differentiating $\Psi(t)$ we obtain the detected frequency from midband

$$\frac{d\Psi}{dt} = \dot{\Psi}(t) = \frac{[p(t) + n_1(t)][\dot{q}(t) + \dot{n}_2(t)] - [q(t) + n_2(t)][\dot{p}(t) + \dot{n}_1(t)]}{R^2(t)} \tag{8.28}$$

The detector decides which transmitted symbol was sent by synchronously sampling $\dot{\Psi}(t)$ and comparing the result with a set of thresholds. We now consider some special examples.

Binary FM. In this mode of operation the symbols $a_n = \pm 1$ with equal probability. Assuming that the system does not make errors in the absence of noise, we can express the probability of error in a given sample of instantaneous frequency taken at $t = t_0 + nT$ as the probability that $\dot{\Psi}(t_0 + nT)$ is positive when $\dot{\Psi}(t_0 + nT)$ is negative in the absence of noise or vice versa. Since the system has memory the parameters p, $\dot{p}$, q, and $\dot{q}$ at any sampling instant depend on the entire past history of the FM signal. Our procedure is first to show how the error probability can be evaluated at any sampling instant for any sequence. Then we calculate the error probability for specific sequences and establish bounds for most and least vulnerable sequences.

Since the denominator of Eq. (8.28) is positive, decisions are made entirely on the sign of the numerator. Thus the probability of error may be expressed as

$$P_e = P\left[(p + n_1)(\dot{q} + \dot{n}_2) + (-q - n_2)(\dot{p} + \dot{n}_1) < 0 \text{ given that } p\dot{q} - q\dot{p} > 0\right] \tag{8.29}$$

When the added noise spectrum is symmetrical about ω_c, n_1, $\dot{n}_1$, n_2, and $\dot{n}_2$ are mutually independent, gaussian random variables. Even so the calculation of the error rate is still complicated. It can be shown[6] that P_e can be represented in terms of a single integral of the form

$$P_e = \frac{1}{2} - \frac{c}{2\sqrt{\pi}} \int_{-1}^{1} e^{-c^2x^2} \operatorname{erf}\left[ac(1 - x^2)^{\frac{1}{2}} - bx\right] dx \tag{8.30}$$

where $a^2 = \dfrac{\sigma_0^2}{\sigma_1^2}\left[\dfrac{p\dot{q} - q\dot{p}}{p^2 + q^2}\right]^2 \qquad b^2 = \dfrac{[p\dot{p} + q\dot{q}]^2}{2\sigma_1^2(p^2 + q^2)} \qquad c^2 = \dfrac{p^2 + q^2}{2\sigma_0^2}$

Here σ_0^2 and σ_1^2 are the variances of the noise and differentiated noise at the output of the front-end filter. They are given explicitly by

$$\sigma_0^2 = \frac{1}{2\pi}\int_{-\infty}^{\infty} N(\omega)\,|H(\omega)|^2\,d\omega$$

$$\sigma_1^2 = \frac{1}{2\pi}\int_{-\infty}^{\infty} \omega^2 N(\omega)\,|H(\omega)|^2\,d\omega \tag{8.31}$$

where $N(\omega)$ is the double-sided power spectrum of the input noise and $H(\omega)$ is the transfer characteristic of the front-end filter.

The physical interpretation of the parameters appearing in the probability of error formula is as follows: c^2 is proportional to the signal-to-noise ratio[‡] at the output of the front-end filter. The parameter a is interpreted as a bandwidth ratio since σ_1^2/σ_0^2 is a measure of input noise bandwidth according to Gabor[7] and $p\dot{q} - q\dot{p}/p^2 + q^2$ is proportional to the signal bandwidth. (For a well-designed system, a is of the order of unity.) The parameter b is a measure of channel dispersion. It is proportional to the derivative of the received-signal envelope. For an ideal, infinite bandwidth channel the envelope is constant. It is instructive to examine the behavior of Eq. (8.30) as the parameter b is varied. The derivative of P_e with respect to b is found to vanish at $b = 0$. The second derivative of P_e with respect to b is found to be positive at $b = 0$. Thus we conclude that P_e is a minimum with respect to b when $b = 0$.

‡ The quantity c^2 is not the actual signal-to-noise ratio. However, it is within a proportionality constant since the actual average signal power at the detector input is dependent on the channel as well as transmitting- and receiving-filter characteristics.

A lower bound, therefore, on the probability of error when the other two parameters are fixed, is obtained when $b = 0$. This, of course, agrees with our physical intuition that best performance is attained with a perfect channel, meaning a channel that does not distort the FM wave. One additional observation concerning the parameter b is appropriate; namely, P_e must be monotonically increasing with $|b|$. Thus with the other two parameters fixed, the largest probability of error occurs when b has its largest possible absolute value. These observations should be of aid in selecting the data sequences which have the greatest and least probabilities of error. In Fig. 8.9 curves of P_e a a function of 20 log c are given for different values of a. It is shown[6] that the curve marked $a = \infty$ corresponds to the optimum attainable performance of a binary signaling system in additive, white, gaussian noise.

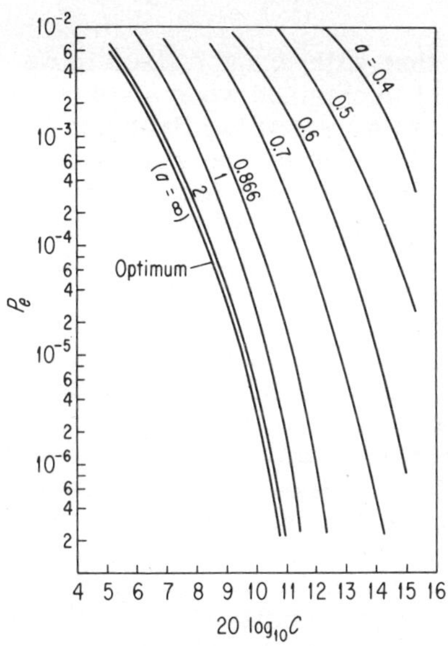

Fig. 8.9. Minimum probability of error as a function of 20 log c for different values of a in binary FM.

Observe that the integral representation for the probability of error in Eq. (8.30) is exact. It is useful, however, to examine the asymptotic behavior of P_e when c is large. As seen in Eq. (8.30), P_e is the probability that a quadratic form in gaussian variates is less than zero. This is a special case of a more general problem extensively treated in the next chapter. By making appropriate comparisons with parameters in the appendix in Chap. 9 it is possible to arrive at the following asymptotic form for P_e

$$P_e \sim e^{-\beta c^2} \tag{8.32}$$

where $$\beta(a,k) = \frac{2a^2}{k^2 + a^2 + 1 + \left[(k^2 + a^2 - 1)^2 + 4K^2\right]^{\frac{1}{2}}}$$

and $k = b/c$. These asymptotic results are also derived in Bennett and Davey[8] by approximating the integral given in Eq. (8.30).

The behavior of β as a function of the parameter a when $k = 0$ is of interest. Two cases are distinguished: $a > 1$ and $a \leq 1$. When $k = 0$ and $a > 1$, $\beta = 1$. On the other hand, when $a \leq 1$ and $k = 0$, $\beta = a^2$. Consequently the asymptotic form of P_e when $b = 0$ may be expressed as

$$P_e \sim \begin{cases} e^{-c^2} & a > 1 \\ e^{-a^2 c^2} & a \leq 1 \end{cases} \tag{8.33}$$

It can be verified that with a and c fixed in Eq. (8.32) the largest value of the exponent is obtained when $k = 0$.

Equation (8.33) tells us that for fixed but large signal-to-noise ratio, the detector degradation increases as a becomes less than unity, and no further significant improvement is gained when a is larger than unity. (This may also be seen in Fig. 8.9.) The above result is quite important in guiding the designers of binary FM systems and requires further elaboration. Recall the physical significance of the parameter a. It was interpreted as the ratio of the peak frequency deviation to noise bandwidth. As the frequency deviation is widened, the front-end filter bandwidth must grow in the same proportion to accommodate the total signal swing, thus increasing the noise bandwidth. This suggests that in well-designed systems the parameter a should be approximately unity or larger. On the other hand, how can one obtain a value of a that is less than unity? Certainly, a will be less than unity if the front-end filter is broader than necessary to accommodate the received signal. In that event excess noise is admitted to the detector, which is clearly harmful. In summary then, the front-end filter should have no wider bandwidth than absolutely necessary to accommodate the modulated signal. The bandwidth of a binary FM signal is a function of the ratio of frequency deviation to bit rate. The problem of obtaining this ratio to minimize the probability of error in binary FM systems signaling over band-limited channels will be examined in a forthcoming section.

Duobinary FM. Duobinary by FM has been described by Lender.[10] The basic principles of duobinary operation were discussed in Chap. 4. The duobinary technique utilizes controlled intersymbol interference to increase the signaling speed over binary. In linear modulation systems it is easy to see how intersymbol interference can be controlled. When this technique is applied in such nonlinear modulation methods as FM, however, the relationship between channel characteristics and baseband intersymbol interference is very complicated. The increase in speed entails a reduction in the noise immunity because of the resulting three-level eye. Whereas in the binary case decisions are made by comparing the instantaneous frequency with a threshold set at zero, in the duobinary case two thresholds are needed. One is to set midway between zero and the maximum positive instantaneous frequency; the other is midway between zero and the maximum negative instantaneous frequency.

Because of the three-level eye the performance analysis of duobinary is more complicated than binary. Definite answers may be obtained from the probability distribution of the instantaneous frequency. An exact expression for this probability distribution can be found in Salz and Stein.[11] Useful asymptotic formulas for this distribution are given in Mazo and Salz.[12] With the aid of these formulas any particular duobinary system can be evaluated.

8.2.4 The Zero-crossings Detector

A device quite often used in practice to detect FM signals is the *zero-crossing*[3] *detector*. The idea employed in this type of system is that the instantaneous frequency of a wave can be determined by measuring the number of its zero crossings per unit time. To illustrate this idea, consider a binary, continuous-phase, frequency-shift-keyed signal having the following properties: A "mark" is transmitted as one lobe of a sine wave with period $2T$ and a "space" by a complete cycle of a sine wave of period T. Because of the continuous-phase constraint, the basic signaling elements are $\pm M$ and $\pm S$, as shown in Fig. 8.10. Both $+M$ and $-M$ are needed since when a mark follows a mark, $+M$ must follow $-M$ or vice versa. The symbol $+S$ always follows $-M$ and $-S$ follows $+M$. When a steady sequence of marks is sent, the transmitted signal is a sinusoidal wave of frequency $f_M = 1/2T$; when a steady stream of spaces is sent, the signal is a sinusoidal wave of frequency $f_S = 1/T$. For a random sequence of marks and spaces the signal is a digital FM wave with signaling rate equal to $1/T$. To detect the presence of a mark or a space it is only necessary to count the number of zero crossings in the wave during T-sec intervals. If one zero crossing is measured, a mark is declared; if two, a space is declared. As can be seen from Fig. 8.10, these are the only two possible resulting measurements of zero crossings per T sec in the absence of channel impairments.

To count zero crossings the incoming wave is ideally limited to obtain a rectangular wave train. This wave then is differentiated to obtain a series of sharp pulses at the transitions. Since up crossings give positive pulses when differentiated and down crossings give negative pulses, the differentiated pulses are rectified to

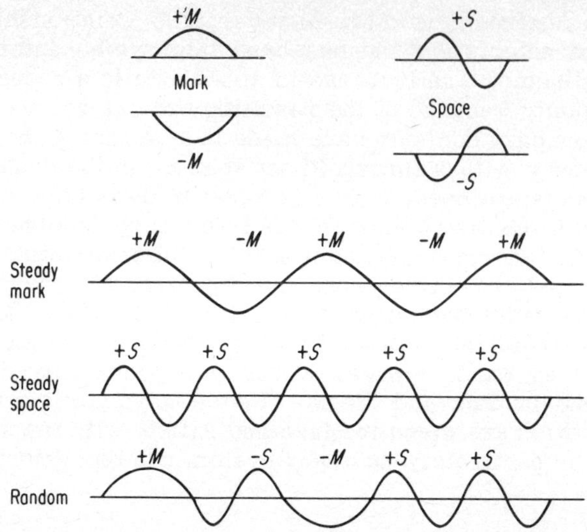

Fig. 8.10. An example of a continuous-phase frequency-shift keying signals.

obtain a pulse train of the same polarity. If the input frequency is constant, the average value of the pulses is constant. When the frequency is varied, the pulse-repetition frequency varies; consequently, the mean value of the pulse train changes. The average value therefore is proportional to the frequency of the wave and can be obtained by passing the pulse train through a low-pass filter. Figure 8.11 shows a block diagram of a zero-crossing detector with the accompanying waveforms.

The error analysis of the zero-crossing detector is extremely difficult, and no adequate theory is at present available. When the

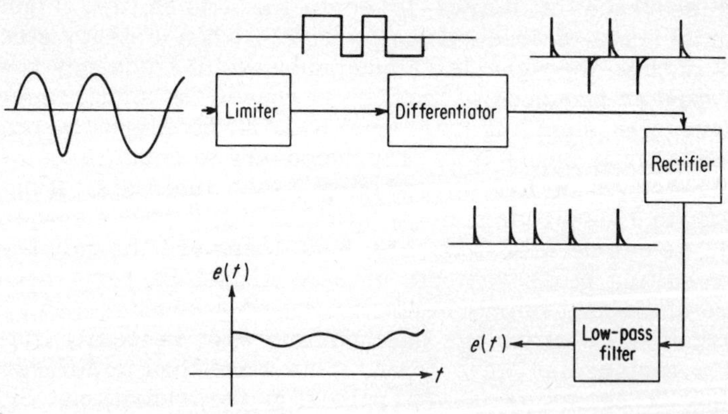

Fig. 8.11. The zero-crossing detector with accompanying waveforms.

number of cycles per bit is large, this circuit can be shown[8] to approximate the action of a limiter-discriminator and results obtained before can be applied here as well.

8.2.5 The Differential Detector

Another practical method of FM detection is called *differential detection.* When employed in a binary FM transmission system under conditions of severe delay distortion, it can be shown[9] to outperform the more conventional limiter-discriminator. Another reason for sometimes preferring this detector is that it can be synthesized with RC components and is therefore well suited for miniaturized circuitry. A block diagram of the detector is shown in Fig. 8.12. The system resembles the more familiar differentially coherent phase detector commonly employed in phase-modulation systems. The principal action of the circuit is to form the product of the incoming FM wave with its delayed replica. Because of this process, it is sometimes also referred to as a *product detector.* The device may be regarded as a particular kind of a discriminator or as a phase comparator operating on the phase changes inherent in an FM signal.

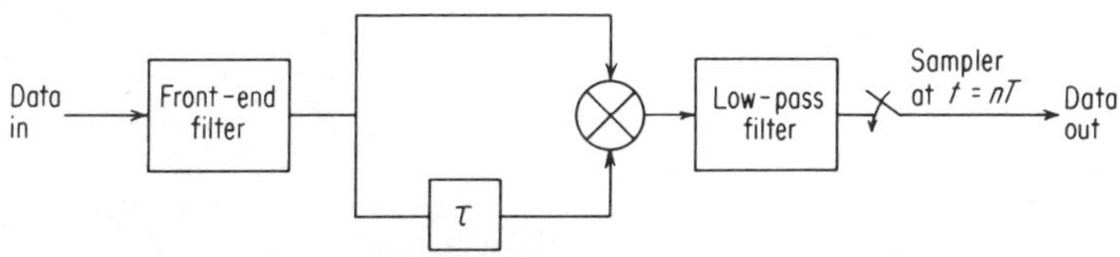

Fig. 8.12. Block diagram of a differential detector.

Recall that the discriminator is a device whose output envelope is proportional to the frequency of the input sine wave. The differential demodulator can be regarded as a discriminator from a steady-state point of view since its output may be calculated as the product of, say, $A\cos(\omega_c + \omega)t$ and its delayed replica $A\cos[(\omega_c + \omega)(t + \tau)]$:

$$d(t) = \frac{A^2}{2}\cos(\omega_c + \omega)\tau + \frac{A^2}{2}\cos[2(\omega_c + \omega)t - (\omega_c + \omega)\tau] \quad (8.34)$$

By passing this signal through a low-pass ideal filter which eliminates the double-frequency component, one obtains the response

$$d_0(t) = \frac{A^2}{2}\cos(\omega_c + \omega)\tau \quad (8.35)$$

When ω is taken as the frequency deviation from the center frequency ω_c, a discriminator characteristic is obtained as

$$d_0'(t) = \frac{A^2}{2} \sin \omega\tau \tag{8.36}$$

when $\omega_c\tau = \pi/2$. The response $d_0'(t)$ is seen to be proportional to frequency deviation over the range in which $\sin \omega\tau$ can be approximated by $\omega\tau$. When this device is used as a binary FM detector, a linear relationship is not necessary since only the sign of the deviation is important. Unambiguous noise-free results can be secured over a range of $\omega\pi$ from $-\pi$ to π. As the delay τ is made small, the linearity improves.

The analysis of this detector proceeds in essentially the same manner as in Sec. 8.2.3. The starting point is Eq. (8.25), the representation of the noise-free received FM wave after it has traversed a dispersive channel. We thus use Eq. (8.27) to represent the composite signal to be processed by the differential detector. For $0 \le \tau \le T$ the delayed version of $v_r(t)$ is

$$v_r(t-\tau) = x_1(t-\tau)\cos(\omega_c t - \omega_c\tau) - y_1(t-\tau)\sin(\omega_c t - \omega_c\tau) \tag{8.37}$$

where

$$x_1(t) = p(t) + n_1(t) \quad \text{and} \quad y_1(t) = q(t) + n_2(t)$$

Suppose the carrier frequency ω_c is chosen so as to satisfy $\omega_c\tau = 270°$. Equation (8.37) could then be written

$$v_{rd} = -x_{1d}\sin\omega_c t - y_{1d}\cos\omega_c t \tag{8.38}$$

where $f(t-\tau) = f_d$. We assume that the filter following the multiplier is ideal (refer to Fig. 8.14), and therefore write the low-frequency component $e_l(t)$ of the product $v_r(t)\,v_r(t-\tau)$ as

$$q = 2e_l(t) = y_1 x_{1d} - x_1 y_{1d} \tag{8.39}$$

The performance of this system can be derived by analyzing the probability distribution of the sign of q. This is essentially the same statistical problem treated in Sec. 8.2.3 but with an additional complication. Before, all the gaussian variables in the quadratic form were assumed to be independent. Here, allowing arbitrary delay τ, the random variables y_1, y_{1d}, x_1, and x_{1d} may become dependent since the correlation function of the noise would not be expected to vanish at $\tau < T$. The present case can be transformed mathematically into the previous solution by subjecting the four dependent random variables in Eq. (8.39) to a linear transformation.

It can be verified that the linear transformation

$$2\begin{bmatrix} z_1 \\ z_2 \\ z_3 \\ z_4 \end{bmatrix} \quad \begin{bmatrix} 1 & -1 & 1 & 1 \\ 1 & 1 & 1 & -1 \\ -1 & 1 & 1 & 1 \\ 1 & 1 & -1 & 1 \end{bmatrix} \begin{bmatrix} y_1 \\ x_1 \\ y_{1d} \\ x_{1d} \end{bmatrix} \tag{8.40}$$

renders the z variables independent while at the same time preserves the invariance of the quadratic form q. The average of the variables z is

$$2\left\langle \begin{bmatrix} z_2 \\ z_2 \\ z_3 \\ z_4 \end{bmatrix} \right\rangle = \mathrm{S} \begin{bmatrix} q(t) \\ -p(t) \\ q(t-\tau) \\ p(t-\tau) \end{bmatrix} \tag{8.41}$$

where S is the 4×4 matrix shown in Eq. (8.40). When the covariance matrix of the original variables is

$$R = \sigma_0^2 \begin{bmatrix} 1 & 0 & r & 0 \\ 0 & 1 & 0 & -r \\ r & 0 & 1 & 0 \\ 0 & -r & 0 & 0 \end{bmatrix} \tag{8.42}$$

the covariance of the z variables can be verified to be diagonal with elements

$$R_z = \sigma_0^2 \begin{bmatrix} 1+r & 0 & 0 & 0 \\ 0 & 1+r & 0 & 0 \\ 0 & 0 & 1-r & 0 \\ 0 & 0 & 0 & 1-r \end{bmatrix} \tag{8.43}$$

where

$$r\sigma_0^2 = \operatorname{cov}(y_1, y_{1d}) \quad \text{and} \quad -r\sigma_0^2 = -\operatorname{cov}(x_1, x_{1d})$$

Here σ_0^2 is the total average power measured at the output of the front-end filter and is given explicitly by Eq. (8.31).

Since now $q = z_1 z_4 + z_2 z_3$, with the z's independent, the exact probability of error for the binary mode is given by Eq. (8.30). However, in the present case the parameters c, a, and k are given by

$$c^2 = \frac{\overline{z_1}^2 + \overline{z_2}^2}{2\sigma_0^2} \qquad a = \frac{\overline{z_1}\,\overline{z_4} + \overline{z_2}\,\overline{z_3}}{2\sigma_1\sigma_2 c^2} \qquad k = \frac{b}{c} = \frac{\overline{z_2}\,\overline{z_4} - \overline{z_1}\,\overline{z_3}}{2c^2\sigma_1\sigma_2} \tag{8.44}$$

where $\sigma_1^2 = (1-r)\sigma_0^2$, $\sigma_2^2 = (1+r)\sigma_0^2$, and the bar over the z's denotes average values.

It is instructive to show that when the delay τ approaches zero, the differential detector approaches a phase differentiator $d\psi/dt$, or a discriminator. To show this we approximate $x_1(t-\tau)$ and $y_1(t-\tau)$ by

$$x_{1d} = x_1(t-\tau) \sim x_1(t) - \tau\dot{x}_1(t)$$

and

$$x_{1d} = y_1(t-\tau) \sim y_1(t) - \tau\dot{y}_1(t) \tag{8.45}$$

valid for small values of τ. Using the approximations in Eq. (8.39), we find that

$$q \sim \tau(x_1\dot{y}_1 - y_1\dot{x}_1) \tag{8.46}$$

The amplitude of the detected signal approaches zero as τ becomes small; if τ is not exactly zero, however, the small output can be amplified. The form of the random variable q is Eq. (8.46), except for a scale factor, is identical to the one appearing in Eq. (8.29) characterizing the discriminator observable. We can thus conclude that the performance of this detector for small values of τ is identical to the discriminator.

Detailed calculations show[8] that for values of $0 < \tau \le T$, this method of detection results in poorer performance than the discriminator in the presence of only gaussian noise. However, its performance is better than the discriminator in the presence of a highly dispersive channel.

8.3 Narrowband, Multilevel, FM Data-Transmission Systems

Thus far we have been primarily concerned with the theory governing the operation of various FM detectors. This section will be concerned with incorporating these devices into data-transmission systems. First we treat the narrowband case. By this we mean that the required modulation bandwidth is of the same order as that of AM. In a subsequent section we discuss transmission systems employing wideband FM requiring modulation bandwidths much wider than AM. Our principal aim throughout is to indicate optimum design procedures and to investigate various parameter tradeoffs.

8.3.1 System Analysis

A block diagram of the data-transmission system under consideration is shown in Fig. 8.13. The data source may consist of L different symbols. If the source is binary, we shall assume that a sequence of m binary digits, such that $L = 2^m$, is grouped to produce one of the L symbols. After this grouping has been achieved, we associate with each symbol a particular voltage or current and form the baseband data as shown in Eq. (8.2).

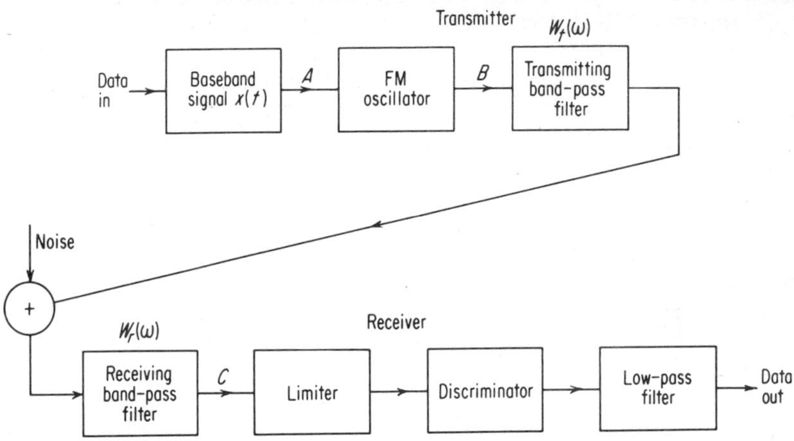

Fig. 8.13. Block diagram of an FM transmission system.

In Fig. 8.13 the baseband signal $x(t)$ appears at point A and controls the FM oscillator. The output is a constant-envelope continuous-phase FM wave appearing at point B. To eliminate frequency spillover, a transmitting bandpass filter is provided. At the receiver a bandpass filter preceds the detector. As explained in earlier sections, its function is to filter as much of the noise as possible without distorting the signal.

To facilitate analysis we combine the channel characteristics with those of the transmitting and receiving filters into a single composite filter. This filter is assumed to possess a flat-amplitude–frequency-characteristic as well as linear phase over the range of frequencies necessary to pass the FM wave.

The composite signal plus noise to be processed by the receiver is written as

$$z(t) = A \cos \left[\omega_c t + \omega_d \int_0^t x(t')\, dt' + \phi \right] + \eta(t) \tag{8.47}$$

where $\eta(t)$ is the added noise and the signal part is the same as Eq. (8.1).

The FM transmitter is completely specified by the set of allowable numbers $\{a_n\}$ and the parameter $\omega_d T$ as can be seen from the definition of $x(t)$ in Eq. (8.2). From practical considerations it is desirable that the separation between adjacent levels be a constant independence of the level. We therefore assume that the data symbols $\{a_n\}$ take on the positive and negative integers with equal probability. Thus the only remaining parameter to be specified is $\omega_d T$. We saw in Secs. 8.1.2 and 8.2.2 that the spectral shaping, bandwidth, and probability of error are critically dependent on this parameter. How should this parameter and the filter shapes be chosen?

It is possible to formulate a joint optimization problem subject to certain constraints. The general problem is to choose transmitting- and receiving-filter characteristics for arbitrary $\omega_d T$ that minimize the probability of error. The constraints governing the optimization may be the average or peak signal power at the receiver and the transmission bandwidth. This is a formidable problem which will not be pursued from this general point of view. However, in the next section we shall pose a less general optimization problem and show that it is possible to obtain near-optimum performance.

Requiring the L-level FM system to be narrowband restricts the choice of the parameter $\omega_d T$ to a small range. Let us then choose the value of this parameter on the basis of its effect on the bandwidth and spectral shaping. Figure 8.4 reveals that a bandwidth of 1.5 $1/T$ and a smooth spectral shaping can be attained when $\Delta = 1/L \rightarrow \omega_d T = \pi/L$. Furthermore, the spectral shapes and bandwidth are invariant with L. This in many applications is a very desirable feature since receiving and transmitting filters can be identical when operating at different rates. We see that the spectrum of the transmitted FM signal with this choice of parameters resembles the spectrum of a DSBSC signal.

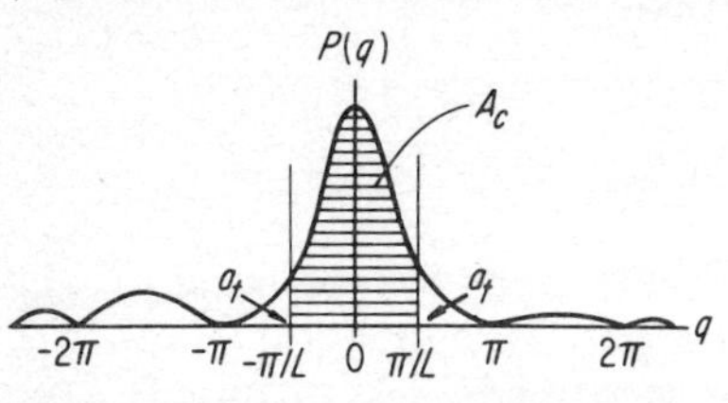

Fig. 8.14. A sketch of the probability density $P(q)$ of q.

When $\omega_d T = \pi/L$ the maximum steady angular frequency $\omega_{\max}$ and the minimum angular frequency $\omega_{\min}$ in the multilevel FM signal are

$$\omega_{\max} = \omega_c + \frac{\pi}{T}\frac{L-1}{L} \quad \text{and} \quad \omega_{\min} = \omega_c - \frac{\pi}{T}\frac{L-1}{L} \tag{8.48}$$

The frequency separation $\Delta\omega$ between the levels is

$$\Delta\omega = \frac{\pi}{LT}(a_{k-1} - a_k) = \frac{2\pi}{LT} \tag{8.49}$$

We again model the FM receiver as an ideal limiter-discriminator followed by an integrate-and-dump circuit. When the added noise is expressed in terms of quadrature and in-phase components about an angular frequency $v_k = \omega_c + a_k\omega_c$, $a_k > 0$, the response of the limiter-discriminator is given in Eq. (8.14). The output of the integrator at $t = T$ is, from Eq. (8.15),

$$q' = q + \omega_c T + a_0 \omega_d T \tag{8.50}$$

Since $\omega_c T$ is an added bias term common to all received samples, it may be subtracted from q'. Decisions as to which symbol was transmitted are made by comparing the quantity $q + a_0\pi/L$ with a set of thresholds. This is conceptually the same problem treated in Chap. 4 dealing with multilevel baseband transmission. Mathematically the problem here is more involved since the random variable q is not gaussian and the probability density is not symmetrical about its mean value.

The probability of error in multilevel data transmission can be expressed as

$$P_e = 1 - \frac{1}{L}\sum_i P(i/i) \tag{8.51}$$

where $i = 1, 2, \ldots, L$ and $P(i/i)$ is the probability of correctly decoding symbol a_i given that this symbol was transmitted. In the application here, q is independent of a_0, and the thresholds are equally spaced about each symbol (except for the extreme ones which have their upper thresholds at infinity). Therefore, the transition probabilities in Eq. (8.51) are independent of i, that is, $P(i/i) = A_c$. This permits Eq. (8.51) to be written as

$$P_e = 1 - A_c + \frac{2}{L}A_c - \frac{2}{L}A_\infty \tag{8.52}$$

where A_∞ is the probability of the extreme levels being decoded correctly.

A more explicit evaluation of Eq. (8.52) is possible by using results developed in Sec. 8.2.2. Figure 8.14 shows a sketch of the probability density of q and a pair of thresholds set at $q = \pm\pi/L$. The shaded area between these two thresholds is equal to A_c, while the area from $-\pi/L$ to ∞ is equal to A_∞. A closed-form expression for these quantities is not known. However, a good approximation can be obtained when $\rho = A^2/2\sigma^2$, the signal-to-noise ratio, is large. In this case the area to the right of π can be neglected [from Eq. (8.22)], and consequently the area to the left of $-\pi$ is just unity minus the area of the major lobe. Also from Eq. (8.22) the area of the major lobe is equal to the probability of getting exactly zero clicks. This is asymptotically equal to

$$1 - \frac{T|a_k|\omega_d}{2\pi} e^{-\rho}$$

Thus the area from $-\pi$ to $-\infty$ denoted by c is equal to

$$\frac{|a_k|\omega_d}{2\pi} e^{-\rho}$$

In terms of c and the areas marked in the figure by a_t, the probability of error becomes

$$P_e \sim 2a_t\left(1 - \frac{1}{L}\right) + c$$

The area marked a_t is asymptotically equal to $P(\pi/L)$ in Eq. (8.21) times the probability of getting zero clicks in the interval $[0, T]$. The probability of obtaining zero clicks is asymptotically equal to unity; therefore

$$P_e \sim \left(1 - \frac{1}{L}\right)\frac{\cot(\pi/2L)}{\sqrt{\cos(\pi/L)}} e^{-2\rho\sin^2(\pi/2L)} \tag{8.54}$$

since c is negligibly small compared with a_t. This implies that in a narrowband FM system the effect of clicks on the probability of error is negligibly small.

As seen from Eq. (8.54) the probability of error depends on the parameter ρ and the number of levels L. For fixed L, the probability of error decreases monotonically with ρ but is functionally dependent on the transmitting- and receiving-filter shapes. The noise variance σ^2 can be made arbitrarily small by making the receiving filter arbitrarily narrow. As this filter is made arbitrarily narrow, however, the signal to be detected becomes badly distorted. The possibility of an optimum design is investigated next.

8.3.2 *System Optimization and Evaluation*

Here we consider the problem of selection of a receiving-filter characteristic $W_r(\omega)$ which minimizes σ^2 subject to the constraint that the average signal power at the receiver be fixed. Recall that the overall characteristic of the receiving and transmitting filter was assumed to be flat. Therefore if $W_t(\omega)$ denotes the transmitting-filter characteristic,§ the relation

$$W_r(\omega)\,W_t(\omega) = 1 \qquad \omega_c - \frac{3}{4}\frac{2\pi}{T} < \omega \leq \omega_c + \frac{3}{4}\frac{2\pi}{T} \tag{8.55}$$

must be satisfied.

If the spectral density of the FM signal at point B in Fig. 8.13 is denoted by $G(\omega)$, the average signal power at the input to the detector is

$$P_s = \frac{1}{2\pi}\int_{-\infty}^{\infty} \frac{G(\omega)}{|W_r(\omega)|^2}\,d\omega \tag{8.56}$$

and the average noise power at point C is

$$\sigma^2 = \frac{1}{2\pi}\int_{-\infty}^{\infty} N(\omega)\,|W_r(\omega)|^2\,d\omega \tag{8.57}$$

where $N(\omega)$ is the two-sided spectral density of the added noise. In order to carry out the minimization we introduce a Lagrange multiplier λ and write the functional equation

$$\min_{W_r} I[W_r] = \frac{1}{2\pi}\int_{-\infty}^{\infty} N(\omega)\,|W_{r0}(\omega)|^2\,d\omega + \frac{\lambda}{2\pi}\int_{-\infty}^{\infty} \frac{G(\omega)\,d\omega}{|W_{r0}(\omega)|^2} \tag{8.58}$$

where $W_{r0}(\omega)$ is the solution of the variational equation

$$\delta I[Y] = \int_{-\infty}^{\infty}\left[N(\omega) - \lambda\frac{G_s(\omega)}{Y^2(\omega)}\right]\delta Y(\omega)\,d\omega = 0 \tag{8.59}$$

where

$$Y(\omega) = |W_r(\omega)|^2$$

Solving Eq. (8.59) for the optimum filter yields

$$Y^2(\omega) = \lambda\frac{G(\omega)}{N(\omega)} \tag{8.60}$$

which can be demonstrated to be a minimum solution.

In summary, the optimum receiving-filter characteristic is proportional to the fourth root of the ratio of the spectral-density function of the constant-envelope FM wave to the spectral density

---

§ The channel characteristics may be included in $W_t(\omega)$ without loss of generality.

of the input noise. The phase characteristic therefore can be chosen arbitrarily since it does not enter the optimization equations.

It is convenient to compare systems in terms of decibel degradation, which is the increase in signal-to-noise ratio in decibels required to match the error rate of the optimum binary communication system signaling over the white gaussian infinite-bandwidth channel. In most systems the asymptotic-error rate can be expressed in the form $f(M)e^{-\alpha M}$, where M is the average signal power at the input to the receiver divided by the average white gaussian noise power measured in a bandwidth equal to the baud. The coefficient $f(M)$ is of minor interest when M is large because of its slowly varying nature compared with the exponential. Consequently the decibel degradation can be calculated as $\beta = -10 \log \alpha$. The relationship between ρ defined in Eq. (8.54) and M can then be determined when the transmitting and receiving filters are specified.

The decibel degradation of the optimized L-level FM system is calculated from Eq. (8.54) as

$$\beta = d_L + 10 \log \left[2 \sin \frac{\pi}{2L} \right] \tag{8.61}$$

The value of d_L for $L = 2, 4, 8, 16, 32$, and 64 has been calculated for this system in Salz.[13] The values range from $d_2 = 0.25$ db to $d_{64} = 0.71$ db. These values indicate, for instance, that the performance of the optimized binary FM system suffers only 0.25 db degradation over the best-attainable binary system.

The optimized binary FM system merits further discussion since it is widely used in practice. We see from Eq. (8.49) that this system has a peak-to-peak frequency deviation equal to 50 percent of the bit rate. It was shown by Kotel'nikov[14] that in an optimum binary FM system signaling over the infinite-bandwidth, white gaussian channel the optimum peak-to-peak frequency deviation is 70 percent of the bit rate. However, the system considered in this analysis does not allow transmitting filters; consequently the received signal has a constant envelope. In contrast our optimum transmitting filter distorts the constant-envelope FM signal resulting in a received wave with a fluctuating envelope. The received signal in our case can be thought of as possessing both frequency and amplitude modulation; therein lies its improved performance over pure FM considered by Kotel'nikov. Several publications [6,15-17] have appeared which indicate that the idealized performance can be closely attained through the use of transmitting and receiving filters.

8.4 Wideband FM Systems

During the course of this chapter we have discussed the implementation of FM systems. In particular we have seen that FM techniques provide a simple means of generating orthogonal signal sets. In this section we investigate the possibility of employing such signal sets in data-transmission systems and demonstrate that orthogonal systems result in power efficiency at the expense of bandwidth utilization. First we consider coherent and incoherent, digital, FM orthogonal signaling systems over the additive gaussian channel. Then we analyze the respective optimum detectors and compare their performance with conventional discrimination detection.

8.4.1 The Optimum Coherent Orthogonal FM Detector

Let us consider an L-ary, phase-coherent digital communications system. The transmitter produces L equienergy waveforms

$$s_n(t) = A \cos \omega_n t \tag{8.62}$$

where $n = 1, 2, \ldots, L$ over the interval $[0, T]$ satisfying the orthogonality conditions

$$\int_0^T s_n(t)\, s_m(t)\, dt = \frac{TA^2}{2}\, \delta_{nm}$$

The minimum bandwidth of the signal set is approximately equal to $L/2T$. In order to achieve this bandwidth, the closest frequency separation

$$\omega_d = |\omega_n - \omega_m| \qquad m \neq n$$

must satisfy $\omega_d T \geq \pi$.

The optimum receiver for an arbitrary equienergy signal set was established in Chap. 5. It consists of a bank of L matched filters; i.e., the filters are matched to the signal set $\{s_n(t)\}$. An analogous procedure is to employ a bank of integrate-and-dump circuits, giving rise to the following L quantities:

$$x_n = \int_0^T v(t)\, s_n(t)\, dt \tag{8.63}$$

where $n = 1, 2, \ldots, L$ and $v(t)$ is the transmitted signal plus white gaussian noise with double-sided spectral density N_0.

The optimum receiver makes decisions based on the set $\{x_n\}$. It decides that $s_j(t)$ was transmitted when

$$x_j = \max_n x_n \tag{8.64}$$

Since the random variables appearing in Eq. (8.64) are gaussian, the probability of error may be arrived at in a straightforward manner. We proceed as follows: Assume the ith signal was transmitted. Under this hypothesis the received signal is $v(t) = s_i(t) + \eta(t)$. The output from the ith filter is

$$x_i = \frac{TA^2}{2} + \int_0^T \eta(t)\, s_i(t)\, dt \tag{8.65}$$

while the output from the jth filter is

$$x_j = \int_0^T \eta(t)\, s_i(t)\, dt \qquad j \neq i \tag{8.66}$$

The respective means and variances are readily calculated to be

$$\langle x_i \rangle = \frac{TA^2}{2} \qquad \langle x_j \rangle = 0 \qquad \operatorname{var}(x_i) = \operatorname{var}(x_j) = \frac{N_0\, TA^2}{2} \tag{8.67}$$

Furthermore, we observe that the variables x_i are independent since

$$\operatorname{cov}(x_i, x_j) = 0 \qquad i \neq j$$

The probability that the receiver correctly decides that $s_i(t)$ was transmitted is the probability that x_i is larger than all the other x's. To calculate this probability we write the conditional probability that x_i is the largest as

$$\begin{aligned} P_{ci} &= P[x_1 < x_i, x_2 < x_i \cdots x_{i-1} < x_i \,|\, x_i] \\ &= \int_{-\infty}^{x_i} \cdots \int_{-\infty}^{x_i} \prod_{\substack{m \\ m \neq i}} p(x_m \,|\, x_i)\, dx_m \end{aligned} \tag{8.68}$$

The probability of correct reception is just the average value of P_{ci} with respect to x_i; namely,

$$P_c = \langle P_{ci} \rangle = \int_{-\infty}^{\infty} p(x_i)\, dx_i \int_{-\infty}^{x_i} \cdots \int_{-\infty}^{x_i} \prod_{\substack{m \\ m \neq i}} p(x_m \,|\, x_i)\, dx_m \tag{8.69}$$

Since all the $p(x_m)$'s are of the same form we can write

$$P_c = \int_{-\infty}^{\infty} p(x_i) \left[\int_{-\infty}^{x_i} p(x)\, dx \right]^{L-1} dx_i \tag{8.70}$$

where
$$p(x) = \frac{1}{\sqrt{\pi N_0 A^2 T}} \exp \frac{-x^2}{N_0 A^2 T}$$

and
$$p(x_i) = \frac{1}{\sqrt{2\pi N_0 A^2 T}} \exp \frac{-(x_i - A^2 T)^2}{N_0 A^2 T}$$

When we substitute the above probability densities into Eq. (8.70) and make the appropriate change of variables, we can express the probability of error as

$$P_e(E,L) = 1 - P_c = \frac{1}{\sqrt{2\pi}} \int_{-\infty}^{\infty} e^{-\left(\frac{v - E^{\frac{1}{2}}}{\sqrt{2}}\right)^2} \left[1 - \left(\frac{1}{\sqrt{2\pi}} \int_{-\infty}^{v} e^{-t^2/2}\, dt\right)^{L-1}\right] dv$$

where (8.71)

$$E = \frac{A^2 T}{2N_0}$$

The parameter E is the energy in a signal element divided by the spectral density of the noise. Physically E may also be interpreted as a signal-to-noise ratio where the noise power is taken in a bandwidth equal to the baud $1/T$.

Curves for the probability of P_e as a function of E in decibels for different values of L taken from Akima are shown in Fig. 8.15.

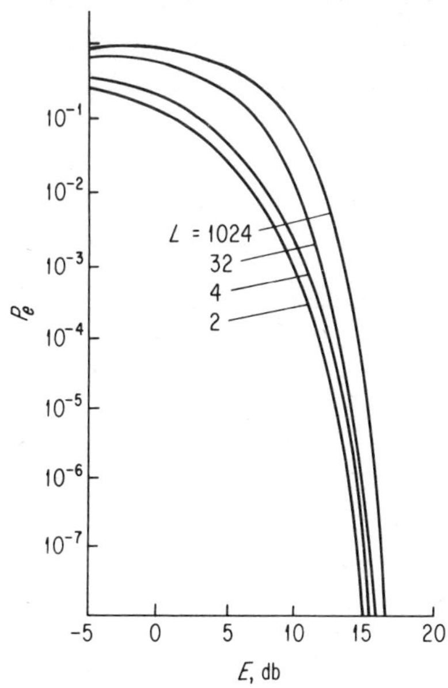

Fig. 8.15. Probability of error P_e as a function of E.

For large values of E the curves merge indicating that for only an additional 3 db of power one can increase the number of levels from 2 to 1,024. However, an obvious price in bandwidth is paid. Whereas the binary system requires roughly a bandwidth of the order of $1/2T$, the L-level system requires a bandwidth of $L/2T$. Also an additional penalty is the complex structure of the optimum detector. In a later section we shall see how equipment complexity may be reduced by the use of a conventional FM receiver. Unfortunately additional signal power must then be sacrificed.

It is interesting at this point to examine the limiting behavior of the probability of error as the number of signals L increases without bound, while the information rate remains fixed. The rate R at which information is communicated to the receiver is expressed as

$$R = \frac{\log L}{T} \text{ bps} \tag{8.72}$$

Solving for T and substituting it in the definition of E in Eq. (8.71), we obtain

$$E = \frac{A^2}{2} \frac{\log_2 L}{N_0 R} = \frac{2 \log_2 L}{(R/C)} \tag{8.73}$$

where

$$C = \frac{1}{2} \lim_{W \to \infty} \log_2 \left[1 + \frac{A^2}{2N_0 W}\right]^W = \frac{A^2}{4N_0} \log_2 e$$

is the channel capacity of the white gaussian channel. (Observe that as $L \to \infty$, $W \to \infty$.)

Substituting Eq. (8.72) into Eq. (8.71) and taking the limit as L increases without bound gives[19]

$$\lim_{L \to \infty} P_e\left(L, \frac{R}{C}\right) = \begin{cases} 0 & \dfrac{R}{C} < 1 \\ 1 & \dfrac{R}{C} > 1 \end{cases} \tag{8.74}$$

The above expression indicates that as long as $R < C$, the probability of error can be made as small as desired. Thus an orthogonal signal set may be regarded as a *constructive code* capable of signaling at a rate up to channel capacity.[20]

A useful upper bound on the probability of error may readily be deduced from Eq. (8.71) as

$$P_e \leq (L - 1)e^{-E} = e^{-TC(1/2 - R/C)} \tag{8.75}$$

This bound predicts that the error probability can be made as small as desired as long as the rate R is less than 1/2 of capacity. The factor of 1/2 is given because the bound is not very good for rates above $C/2$. The error probability can be made small by making T large, implying an exponential growth in the number of signals L.

8.4.2 The Optimum Incoherent Orthogonal FM Detector

In coherent reception the parameters specifying the transmitted waveform set are required to be known at the receiver. When all the parameters except for a uniformly distributed phase are known, the reception process is commonly referred to as *incoherent*. This is a more reasonable assumption in practice and, although its performance is somewhat inferior, results in simpler instrumentation. In this case, the transmitted signals are represented as

$$s_n(t;\theta) = A \cos [\omega_n t + \theta] \quad 0 \le t \le T \tag{8.76}$$

where $n = 1, 2, \ldots, L$ and again the orthogonality conditions are satisfied. The parameter θ is assumed random with uniform probability density on the interval $[0, 2\pi]$.

The optimum incoherent detector may be derived in the same manner as before. However, slightly more involved calculations are required because of the presence of the random angle θ. The structure of the optimum detector is shown in Fig. 8.16. Whereas in the coherent detector the filters are matched to the signal set, here the filters are matched to the respective envelopes. The

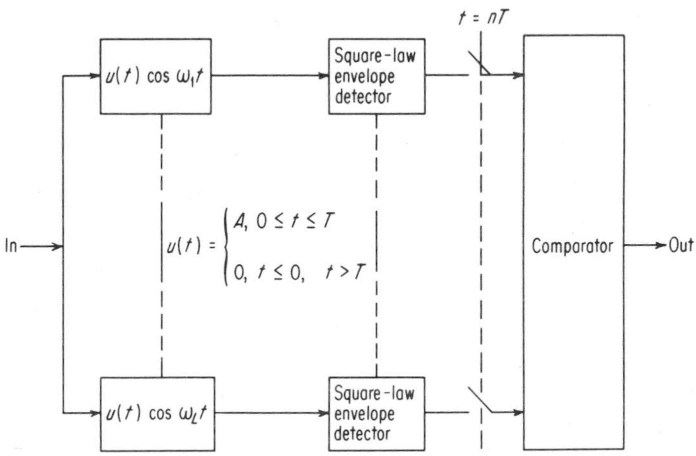

Fig. 8.16. Structure of the optimum incoherent detector.

probability of error may be derived in a manner exactly analogous to that of the coherent case and can be shown[17] to be given by

$$P_e = \int_0^{\infty} v e^{(v^2+E)/2} I_0(vE^{1/2}) \left[1 - \left(1 - e^{v^2/2}\right)^{L-1}\right] dv \tag{8.77}$$

where $I_0(v)$ is the modified Bessel function of the first kind and zero order. Figure 8.17 compares the error probabilities for incoherent and coherent reception. When E is large, the difference in performance is seen to be minor. Because of this small difference at large signal-to-noise ratios, the added complexity of achieving coherent detection is usually not warranted. It is easy to establish that the limiting form of Eq. (8.77) also satisfies Eq. (8.74).

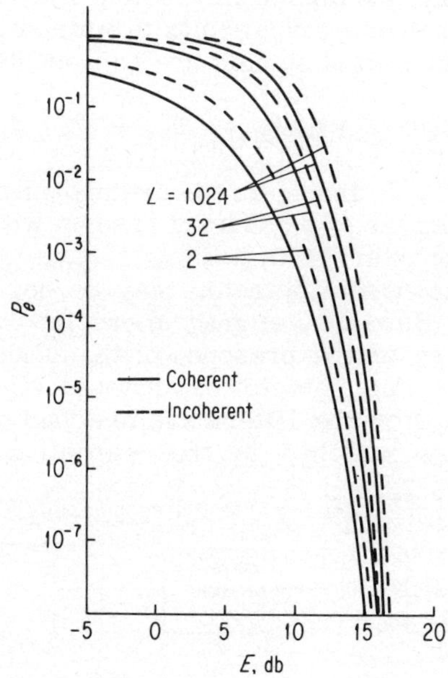

Fig. 8.17. A comparison of probability of error P_e for coherent and incoherent optimum detection.

We next investigate the performance of the discriminator when used to detect orthogonal FM signals.

8.4.3 Discrimination Detection of Wideband FM

The incoherent signal set defined in Eq. (8.76) can be generated by a voltage-controlled oscillator whose output signal is represented as a constant-envelope, continuous-phase FM wave given in Eq. (8.1).

In order to satisfy the required orthogonality conditions, the minimum frequency spacing $\Delta\omega$ must be π/T. The discriminator operates on the signal plus noise as in the narrowband case, and the output of the postdetection filter is compared with thresholds set midway between the transmitted frequencies. The decision procedures are then exactly the same as in PAM systems since the recovered baseband signal at the output of the postdetection filter is precisely a PAM signal.

The probability that the postdetection-filter output frequency is not the one transmitted can readily be calculated from the theory developed in Sec. 8.4. Referring to the output probability density curve of Fig. 8.8, it can be seen that an error in detection is committed whenever one or more clicks occur. (This is because $\omega_d T = \pi$. Using the results in Eq. (8.20), when $\alpha = \pi/2$ we see that a good estimate of the probability of error is

$$P_e \sim (L - 1)e^{-A^2/2\sigma^2} \tag{8.78}$$

To make a comparison with the ideal detector we note that $\sigma^2 \sim N_0 L/T$, and therefore the exponent in Eq. (8.78) is proportional to E/L, where E is defined in Eq. (8.71). It is thus evident that the performance of the discriminator is inferior to that of the ideal detector, as it must be. The exponent in Eq. (8.78) becomes smaller as the number of signals L increases. The physical reason for the inferior performance is because the FM receiver admits too much noise at its front end, producing clicks which cannot be eliminated by the postdetection filter. The amount of noise grows in direct proportion to the number of orthogonal signals; this gives rise to the inferior exponent. On the other hand, the optimum detector, which employs a bank of matched filters, each having bandwidth equal to approximately $1/T$, does not permit the growth in noise power. The noise power at the output of each filter does not grow with the number of signals; it is a fixed constant determined by the bandwidth of the front-end filter.

This poorer performance of conventional FM receivers as a detector of orthogonal signals can be remedied by the use of a feedback loop as described by Chaffee.[21] Without the feedback loop, the front-end or IF filter must be wide enough to pass the total swing of the incoming signal. The idea behind the feedback loop is that the incoming frequency can be tracked by the loop; therefore the IF filter could possibly be made narrower, thus admitting less noise. The analysis of FM with feedback is extremely complicated, and no adequate theory at present is available to support the above suppositions.

To pursue the performance of the discriminator futher, we suppose that it is possible to select the transmitted frequencies in a manner such that their separation is even greater than that dictated by the orthogonality conditions. It is evident from the analysis of the optimum detector that the performance cannot be improved by this selection. Indeed, if the frequencies are spaced wider than π/T, bandwidth is sacrifieed, and no additional improvement in performance is gained.

It is thus interesting to examine the performance of a discrimination detection system when $(\Delta\omega)T > \pi$, that is, a system requiring more bandwidth than an orthogonal system. When the transmitted frequencies are widely separated, only the clicks can cause errors. For these separations, a single click may not cause an error; several clicks during the detection interval may be required. Thus suppose the frequencies are spaced so that $(\Delta\omega)T = 2\pi n$, n being a positive integer. The probability of error will then be the probability of getting n or more clicks in T sec, which from Eqs. (8.20) and (8.21) behaves as

$$\frac{(f_d T)^n e^{-n\rho}}{n!} \geq \frac{(n/2)^n e^{-n\rho}}{n!} \geq \frac{1}{2} e^{-n\rho} \tag{8.79}$$

It might appear that the effect of this choice of separation has been a gain in performance over the orthogonal signal set since the exponential has changed from $e^{-\rho}$ to $e^{-n\rho}$. However, it must be realized that the ρ's in the two situations are not the same. The bandwidth for the case under consideration is essentially $2n$ times the bandwidth of an orthogonal system; therefore, for the same signal power, the nominal value of ρ has decreased by a factor of $2n$. Consequently, nothing has been gained by the wider separation. From a physical point of view it can be realized that as the frequency separation increases, the amount of noise admitted to the detector increases by a corresponding amount. The latter is predicated on the assumption that the input bandwidth is flat across the band.

It may be possible to shape the front-end filter so that increasing the frequency separation does not necessarily cause a proportionate increase in the noise power admitted. The power-spectra plots in Figs. 8.2 and 8.3 indicate that when $\omega_d T \gg \pi$, spectral peaks occur at the transmitted frequencies having widths on the order of $1/T$. Suppose we design a comb filter with transmittance peaks at the transmitted frequencies. It is then possible to conceive of a design such that as the frequency separation increases, the admitted noise power remains constant. In order not to distort the received FM signal, a filter complementary to the one at the receiver is placed at the transmitter just as in the narrow-band case. With the use of these complementary filters it might be possible to maintain constant noise power, and therefore by choosing a large-enough frequency separation, force the probability of error to be arbitrarily small, contradicting optimality considerations for reception of signals in white gaussian noise. This, of course, cannot happen and requires reexamination of the basic assumptions concerning the clicks. When we calculated the probability of getting n or more clicks, we assumed that they occurred independently. This was a reasonable assumption when the filter was essentially flat across the whole received spectrum and the correlation time out of the receiving filter was small, i.e., equal to approximately the reciprocal of the bandwidth. We saw earlier that when this situation prevailed, no contradiction was found. However, if the input noise spectrum is chopped into a few slits or notches, correlation in the noise being processed in the detector will persist for a longer time, and multiple clicks can occur with essentially the same probability that one would normally associate with a single click. These notions are made precise in Mazo and Salz[5] where it is shown that the probability of getting n or more clicks is approximately equal to $ne^{-\rho}$ rather than $e^{-n\rho}$. This resolves the contradiction.

We remark in conclusion that other wideband modulation systems exist which result in good power efficiency at the expense of bandwidth utilization. Some of these techniques are discussed and their performance compared in Jacobs.[22]

8.5 Summary

In this chapter we examined frequency-modulation and detection techniques. It was emphasized that these techniques are principally applied in low-speed, relative to available-bandwidth data-transmission systems, where they provide excellent performance with minimum complexity. Because both sidebands are needed for efficient demodulation, FM is not used in high-speed data-transmission systems. We began by indicating how digital FM is generated and by analyzing the spectral density associated with these waves. The illustrations showed how bandwidth and spectral shapes were dependent on the modulation parameters. Next we turned to the various detection methods and indicated how their performance could be evaluated. Particular attention was given to the limiter-discriminator since it is the most widely used. In Sec. 8.3 we incorporated some of the devices discussed earlier into a multilevel, narrowband, data-transmission system. This system was analyzed and the parameters optimized. The chapter concluded with a discussion of the application of FM in wideband data-transmission systems. The optimum orthogonal coherent and incoherent detector structures were derived and their performance analyzed. We showed that these systems result in good power efficiency at the expense of bandwidth efficiency. Finally, the performance of the discriminator as a detector of orthogonal FM signals was compared with the optimum and it was shown that it was considerably poorer when the number of frequencies was large.

Appendix

Spectral Density of Digital FM

A convenient definition of the spectrum of a random signal as given by Rice[23] is

$$G(f) = \lim_{\lambda \to \infty} \frac{2}{\lambda} G_{\lambda}(f) \qquad f > 0 \tag{8.80}$$

where
$$G(f) = \left\langle \left| \int_0^{\lambda} x(t)\, e^{-i2\pi ft}\, dt \right|^2 \right\rangle$$

This definition will be used here to derive digital FM spectra.

The random signal whose spectrum it is desired to calculate is representable as

$$x(t) = \exp\left[i \int_0^t \Psi(t')\, dt' \right] \qquad 0 \le t \le \infty \tag{8.81}$$

where
$$\Psi(t) = \sum_{n=0}^{n=\infty} s_n(t - nT)$$

Here $\{s_n(t)\}$, where $n = 0, 1, \ldots,$ is a sequence of independent and identically distributed random signals defined on the interval $[0, T]$ and identically zero elsewhere. The baseband signal $\Psi(t)$ is a more general representation of a digital signal than the one used in the text.

Let $\lambda = NT$, N integral in Eq. (8.80) and write

$$G(f) = \left\langle \sum_{k,s=0}^{k,s=N-1} \int_{kT}^{(k+1)T} dt_1 \int_{sT}^{(s+1)T} dt_2\, x(t_1) x^*(t_2)\, e^{i2\pi f(t_2 - t_1)} \right\rangle \tag{8.82}$$

where the asterisk (*) denotes the complex conjugate. Furthermore, it can be observed that

$$\int_{kT}^{(k+1)T} dt\, x(t)\, e^{-i2\pi ft} = \int_{kT}^{(k+1)T} dt\, e^{-i2\pi ft} \exp i \left[\sum_{n=0}^{n=\infty} \int_{-nT}^{t-nT} s_n(t')\, dt' \right]$$

$$= e^{-i2\pi fkT} \int_0^T dt\, e^{-i2\pi ft} \prod_{n=0}^{\infty} P_{n,k}(t) \tag{8.83}$$

where

$$P_{n,k}(t) = \exp\left[i\int_{-nT}^{t-(n-k)T} s_n(t')\,dt'\right] = \begin{cases} \exp\left[i\displaystyle\int_0^T s_n(t')\,dt'\right] & 0 \le n < k \\ \exp\left[i\displaystyle\int_0^t s_n(t')\,dt'\right] & n = k \\ 1 & n > k \end{cases}$$

Substituting $P_{n,k}$ as given in Eq. (8.82) yields

$$G(f) = \left\langle \sum_{k,s=0}^{N-1} F_k(t) F_s^*(t)\, e^{-2\pi i f T(k-s)} \exp i\left[\sum_{n=0}^{n=k-1} b_n - \sum_{n=0}^{n=s-1} b_n\right]\right\rangle \tag{8.84}$$

where $\quad b_n(t) = \int_0^t s_n(t')\,dt' \qquad b_n = b_n(T)$

and $\quad F_n(f) = \int_0^T e^{-i2\pi ft + ib_n(t)}\,dt$

At this point it is useful to define the following quantities:

1. $$\left\langle e^{i\alpha b_n(t)}\right\rangle = C(\alpha\,;t) \tag{8.85}$$

where $n = 0, 1, 2, \ldots,$ which is the definition of the characteristic function of $b_n(t)$.

2. $$F(f) = \langle F_n(t)\rangle = \int_0^T C(1;t)\, e^{-2\pi ift}\,dt \tag{8.86}$$

where $n = 0, 1, 2, \ldots$.

3. $$F_b^*(t) = \left\langle F_n^*(t)\, e^{ib_n}\right\rangle = \int_0^T dt\, e^{i2\pi ft}\left\langle e^{i[b_n(T) - b_n(t)]}\right\rangle$$

4. $$P(f) = \left\langle |F_n(f)|^2\right\rangle = \int_0^T\!\!\int_0^T dt_1\,dt_2\, e^{2\pi if(t_1 - t_2)}\cdot\left\langle e^{i[b_n(t_2) - b_n(t_1)]}\right\rangle \tag{8.87}$$

The above averages do not depend on the index n, since by hypothesis the random signals $s_n(t)$, $n = 0, 1, 2, \ldots,$ are identically distributed. Using this fact and the fact that the component signals are independent, it is possible, after some manipulations, to write the power spectrum in terms of the above-defined quantities as

$$G(f) = \frac{2}{T}\left\{P(f) + 2\,\mathrm{Re}\left[F(f)F_b^*(f)\,e^{-2\pi i fT}(1 + \Lambda)\,e^{2\pi i fT}\right]\right\} \tag{8.88}$$

where

$$\Lambda = \lim_{N\to\infty} \frac{1}{N} \sum_{s=0}^{s=N-3} \sum_{k=s+2}^{k=N-1} C^{k-s-1}(1;T)\,e^{-2\pi i Tf(k-s)}$$

This limit can readily be evaluated. However, care must be exercised when $|C(1,T)| = 1$ since then the series does not converge in the ordinary sense and can only be interpreted as a *distribution limit.*

It is shown in Barnard[24] that

$$\Lambda = \begin{cases} e^{-4\pi i Tf}\dfrac{C(1,T)}{1 - C(1,T)\,e^{-2\pi i Tf}} & |C(1,T)| < 1 \\ \left[\dfrac{-i}{2}\cot\dfrac{\gamma}{2} - \dfrac{1}{2} + \pi\sum_n \delta(\gamma - 2\pi n)\right]\cdot e^{-2\pi i Tf} & |C(1,T)| = 1 \end{cases} \tag{8.89}$$

where $\gamma = 2\pi fT - \arg C(1,T)$, and $\delta(\cdot)$ is the Dirac delta function. As may be seen from Eq. (8.89) discrete spectral lines appear whenever the magnitude of the characteristic function takes on values on the unit circle. It is clear that if b_n where $n = 0, 1, 2, \ldots,$ are continuous random variables, no discrete spectral lines can occur since $|C(\alpha,T)| < 1$, $\alpha \neq 0$. But if b_n where $n = 0, 1, 2, \ldots,$ are discrete random variables, the possibility exists of line spectra occurring. For a characteristic function of a discrete random variable to take on values on the unit circle the random variable defining this characteristic function must take on values on a lattice. That is,

$$b_n = 2\pi k + a \tag{8.90}$$

where $n = 0, 1, 2, \ldots,$
k = an integer
a = an arbitrary constant

When the above conditions are satisfied, the characteristic function evaluated at $\alpha = 1$ becomes e^{ia}. If Eq. (8.89) is used in (8.88) with the conditions stated in Eq. (8.90), the final results may be written as

$$\frac{TG(f)}{2} = \begin{cases} P(f) + 2\,\mathrm{Re}\, F(f) F_b^*(f) \left[\dfrac{e^{-2\pi i f T}}{1 - C(1,T)\, e^{-2\pi i f T}} \right] & |C(1,T)| < 1 \\ P(f) + |F(f)|^2 \left[2\pi \sum_n \delta(2\pi f T - a - 2\pi n) - 1 \right] & |C(1,T)| = 1 \end{cases} \tag{8.91}$$

In writing Eq. (8.91), use was made of the fact that when $C(1, T) = e^{ia}$, $F_b^*(f) = F^*(f)\, e^{ia}$.

Notes

Cohn,[3] Rice,[4] and Mazo and Salz[5] include detailed discussions of the "click" theory in FM receivers. Bennett and Davey (Ref. 9 of Chap. 9) gives a detailed discussion of the derivation of the probability of error in digital FM systems. Kotel'nikov's book, "The Theory of Optimum Noise Immunity,"[14] is a classic work on statistical communication theory where various kinds of optimum receiver structures are derived and analyzed. Sunde[25] includes some early work on digital FM. Carson and Fry,[26] Crosby,[27] Blackman,[28] and Lawson and Uhlenbeck[29] are classic works dealing with the theory of frequency modulation.

9

Digital Phase Modulation

Phase modulation (PM) is another technique available for communicating digital information over bandpass channels. This method combines some attractive features of linear and frequency modulation. It has the advantage of requiring less peak power than multilevel linear modulation and may in some applications have a reduced bandwidth requirement over FM. The digital PM signal resembles AM, while the detection process resembles that of FM.

This technique, sometimes referred to as *phase-shift keying* (PSK), has been successfully employed to transmit digital data over high-frequency channels where fading and multipath phenomena are encountered.[1,2] In these applications, digital information is conveyed in the form of phase differences between successive radio-frequency pulses. The receiver usually derives the digital information by measuring the differential phase. If path perturbations are slow compared with the duration of the pulses, both pulses are affected in the same manner, thus preserving the digital information contained in the phase differential.

In this chapter we shall examine some of the important properties of digital phase modulation and compare its performance with the other modulation techniques examined earlier.

9.1 Digital Phase-modulated Signals

Phase modulation, like FM, is a nonlinear modulation technique. When the modulating signals are analog, no real distinction between the two methods can be made. The only difference between PM and FM is the spectral shaping of the phase function and the shape of the postdetection filter. In digital data transmission, however, quite different instrumentations are used for PM and FM; therefore separate mathematical treatments are required.

In this section we first discuss the representation and generation of digital PM and then examine a method of phase coding commonly used in practice.

9.1.1 Generation and Representation

A digital PM signal may be represented as the real part of the complex signal

$$w(t) = A \sum_n g(t - nT)\, e^{i(\phi_n + \omega_c t)} \tag{9.1}$$

where for clarity $g(t)$ is a T-sec rectangular pulse with amplitude equal to 1. The sequence of phases $\{\phi_n\}$, which is the carrier of the digital information, takes on discrete values on the interval $[0, 2\pi]$; ω_c, as before, is the carrier frequency.

The real part of $w(t)$ exhibits the same structure as double-sideband AM in quadrature treated in Chap. 7. To see this, write the real part of Eq. (9.1) as

$$\text{Re}\, w(t) = A \cos \omega_c t \sum_n g(t - nT) \cos \phi_n + A \sin \omega_c t \sum_n g(t - nT) \sin \phi_n \tag{9.2}$$

By associating $\cos \phi_n$ with a_n and $\sin \phi_n$ with b_n, we see the similarities. Thus a digital PM signal possesses the same basic properties as a linearly modulated signal and consequently may be generated in the same manner. For example, binary PM is structured identically as a DSBSC signal since in this case $\{\phi_n = \pm\pi/2\}$ and

$$\text{Re}\, w(t) = s(t) = A \sin \omega_c t \sum_n a_n g(t - nT) \tag{9.3}$$

where $$a_n = \sin \frac{\pm\pi}{2} = \pm 1$$

The essentially linear structure of the digital PM signal greatly simplifies the analysis of spectral properties and response behavior to linear systems. As can be observed, the spectral density of Eq. (9.1) is determined by the Fourier transform of $g(t)$. Since the carrier phase is not continuously varied during a signaling interval T, no band spreading can be achieved as in FM.

9.1.2 Differential Phase Coding

It is often desired to encode the digital information in the phase changes rather than in the absolute phases. In that event, the sequence of phases $\{\phi_n\}$ appearing in Eq. (9.1) cannot be chosen independently. What is usually meant by differential phase modulation is that the value of $\phi_n - \phi_{n-1} = \Delta_n$ carries the digital information. More specifically, suppose it is desired to encode the source information into one of L equally spaced differential phases. One possible assignment might be

$$\Delta = \frac{(2l - 1)\pi}{L} \tag{9.4}$$

where $l = 1, 2, \ldots, L$. In order to achieve these phase changes, the set of absolute phases associated with the modulated signal in Eq. (9.1) must be computed from

$$\phi_n = \Delta_n + \phi_{n-1} \tag{9.5}$$

That is, the absolute phase transmitted in the nth time interval is computed by adding the absolute phase transmitted in the $(n-1)$th time interval and the desired phase differential at the nth time interval Δ_n. An example as to how this coding is achieved in a four-phase differential modulator is shown in Fig. 9.1.

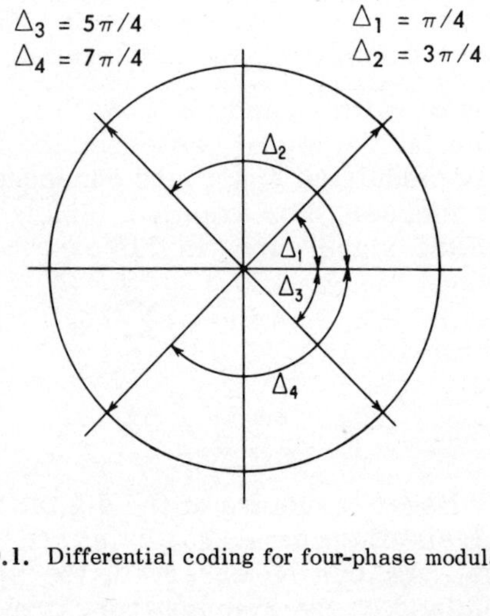

Fig. 9.1. Differential coding for four-phase modulation.

It can be verified that the spectral properties of the digital-phase-modulation signal with differential coding are unaltered, provided the discrete phase changes occur with equal probability.

9.2 Digital Phase Detection

Two generally recognized methods of digital phase detection are *coherent detection* and *comparison*, or *differential*, *detection*.

We first treat coherent detection, where a phase reference at the receiver is assumed to be available. We then examine the more practical method of comparison detection, where a phase reference is derived from past signaling intervals.

9.2.1 Coherent Detection of Multiphase Modulation

For coherent detection of multiphase modulation, an absolute phase reference must be stored at the receiver. The mode of detection is to compare the received phase with the stored reference. This method is only useful when the channel is stable so that the transmitted phasors are not perturbed in transit to the receiver. In most situations this is a very severe requirement, and therefore stored-reference phase detection is seldom used.

We study the performance of the coherent phase detector since it can be shown[3] to be optimum in the presence of gaussian noise and provides an absolute performance reference. These results will permit us later to estimate the degradation suffered when suboptimum but more practical detection systems are used.

The ideal, coherent, digital phase detector observes a sine wave with one of L phases in additive gaussian noise. The detector may be characterized as a device that performs a phase measurement during a T-sec interval. To proceed with the mathematical formulation of the problem, write the received wave plus noise as

$$V_r(t) = A\cos(\omega_c t + \phi) + \xi(t)\cos(\omega_c t + \phi) + \eta(t)\sin(\omega_c t + \phi) \qquad 0 \le t \le T \tag{9.8}$$

where $\phi = l(2\pi/L)$, $l = 1, 2, \ldots, L$, and $\xi(t)$ and $\eta(t)$ are in-phase and quadrature gaussian noise components, respectively, with common average power σ^2. Since the transmitted phases are constant over a T-sec interval, the composite phase of $V_r(t)$ may be expressed as

$$\text{Phase } V_r(t) = \phi + \tan^{-1}\frac{\eta(t)}{A + \xi(t)} \tag{9.9}$$

In the absence of noise, the ideal phase detector measures ϕ at the end of the signaling interval. Because of noise, errors are committed whenever the device measures a phase outside of the region

$$\phi - \frac{\pi}{L} \le \theta < \phi + \frac{\pi}{L}$$

(Of course, we are dealing with mod-2π arithmetic throughout.) The probability of error can be calculated by integrating the mod-2π probability density of the phase angle $\theta = \tan^{-1}\eta/(\xi + A)$ over an appropriate region. The probability density of the phase in Eq. (9.9) is well known and is given by

$$p(\theta) = \frac{1}{2\pi}e^{-\rho}\left[1 + \sqrt{4\pi\rho}\cos\theta\, e^{\rho\cos^2\theta}\, Q(\sqrt{2\rho}\cos\theta)\right] \qquad -\pi \le \theta \le \pi \tag{9.10}$$

where
$$\rho = \frac{A^2}{2\sigma^2}$$

is the signal-to-noise ratio, and

$$Q(v) = \frac{1}{\sqrt{2\pi}} \int_v^\infty e^{-t^2/2}\, dt$$

The probability distribution, which is the integral of Eq. (9.10), is not known in closed form but has been tabulated by Cahn.[5]

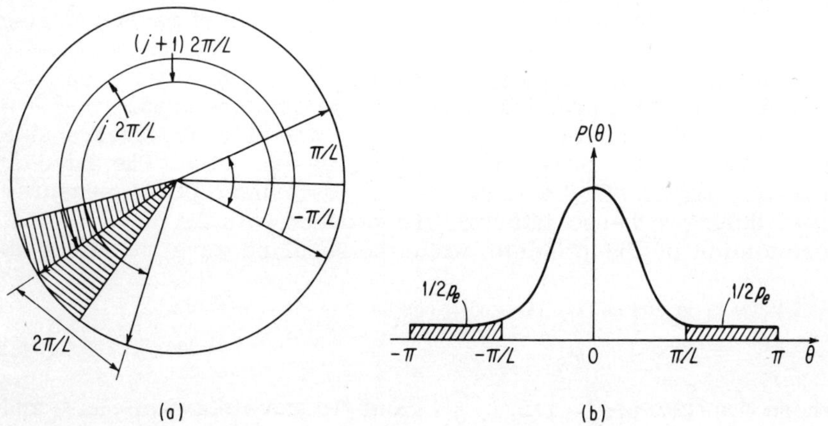

Fig. 9.2. (a) Phasor diagram showing the region of correct reception. (b) probability density of phase.

The L allowed transmitted phases are shown in the phasor diagram in Fig. 9.2(a), and the resulting probability density of the phase θ is shown in Fig. 9.2(b). The shaded area in Fig. 9.2(a) is the region where correct coding is attained when phase j is transmitted and the blanked out area under the probability density is equal to the error rate. More explicitly, the probability distribution of the phase angle θ is

$$\Phi(\alpha) = \int_{-\pi}^{\alpha} p(\theta)\, d\theta \tag{9.11}$$

Therefore, the probability of error for a digital phase-modulation system with L equally spaced phases is simply

$$P_e = 2\left[1 - \Phi\left(\frac{\pi}{L}\right)\right] \tag{9.12}$$

It is shown that Eq. (9.12) may be expressed in closed form only when $L = 2$ and $L = 4$. For these values of L, Eq. (9.12) becomes

$$P_{e2} = Q(\sqrt{2\rho}) \tag{9.13}$$

and

$$P_{e4} = 1 - [1 - Q(\sqrt{\rho})]^2$$

Fig. 9.3 is a plot of P_e as a function of a normalized signal-to-noise ratio

$$\alpha = 10 \log \left(\rho \sin^2 \frac{\pi}{L}\right) \tag{9.14}$$

for different values of L.

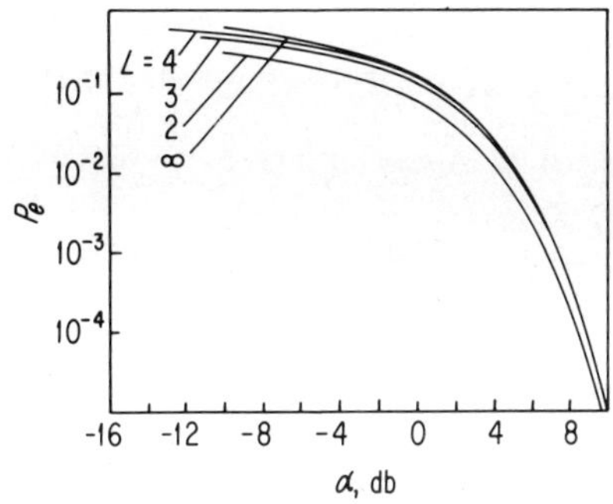

Fig. 9.3. Error probability P_e as a function of normalized signal-to-noise ratio α.

A useful estimate of the probability of error is provided by the leading terms of the asymptotic expansion of Eq. (9.12). It is also shown in Cahn[5] that for large ρ

$$P_{eL} \sim e^{-\rho \sin^2 \pi/L} \tag{9.15}$$

This essentially completes the analysis of ideal coherent phase detection in the presence of additive gaussian noise. Clearly any practical detection system can perform no better than predicted by formula (9.15). An alternative approach to digital phase modulation and detection is examined next.

9.2.2 Comparison Detection

As mentioned, it is sometimes not practical to store at the receiver an absolute phase reference so as to achieve coherent detection. Communication channels and transmitting equipment are seldom stable enough to preserve the phase of a carrier wave. A more reasonable requirement might be to ask that the phase change between two signaling intervals be preserved, and let the absolute phase of the carrier in each signaling interval take on arbitrary values. Under these conditions a suitable approach is to use differential coding and differentially coherent (comparison detection). In this type of detection system the phase reference is derived from past signaling intervals.

The question we wish to examine in this section is what penalty we must pay for using comparison detection instead of coherent detection. Here we discuss conceptually the phase-comparison technique, while in a subsequent section we analyze physical models which realize the process of comparison detection.

The ideal comparison detector measures the phase difference between two successive signaling intervals. Thus, if the signal plus noise in, say, the first time interval is

$$s_1(t) = A\cos[\omega_c t + \phi_1] + \eta_1(t)$$

and

$$s_2(t) = A\cos[\omega_c t + \phi_2] + \eta_2(t) \tag{9.16}$$

in the succeeding interval the phase difference may be expressed as

$$\text{Phase } s_2(t) - \text{phase } s_1(t) = \phi_2 - \phi_1 - \phi_n + \tan^{-1}\frac{y(2T)}{A + x(2T)} - \tan^{-1}\frac{y(T)}{A + x(T)} \tag{9.17}$$

Here, the discrete phase differences $\Delta = \phi_{n+1} - \phi_n$ take on one of L values equally spaced around a circle. For example, in binary-differential phase-shift keying, no change in phase between successive signals denotes a binary 1, while a π-rad change a binary 0. An error is committed when the differential phase due to noise,

$$\tan^{-1}\frac{y_2}{A + x_2} - \tan^{-1}\frac{y_1}{A + x_1}$$

exceeds thresholds above and below $\pm\pi/L$. Error rates can be calculated from the probability distribution of the phase differential.

Making the reasonable assumption that the additive noise samples in the two successive time intervals are independent, the probability density of the phase difference can be obtained by convolving the density function given in Eq. (9.10) with itself; namely,

$$p(\Delta\theta) = \int_{-\pi}^{\pi} p(\theta)\, p(\Delta\theta + \theta)\, d\theta \tag{9.18}$$

where the random variable $\Delta\theta$ is defined in terms of the quadrature and in-phase noise components in the two successive time intervals as

$$\Delta\theta = \tan^{-1}\frac{y_2}{A + x_2} - \tan^{-1}\frac{y_1}{A + x_1} \tag{9.19}$$

The desired probability of error is just

$$P_{eL} = 2\left[1 - \int_{-\pi}^{\pi/L} p(\Delta\theta)\, d\Delta\theta\right] \tag{9.20}$$

As is typical with these calculations, a closed-form answer is not often attainable. Fortunately, excellent approximations can again be obtained when the signal-to-noise ratio ρ is large. One such approximation is given in Cahn[5] as

$$P_{eL} \sim e^{-2\rho \sin^2 \pi/2L} \tag{9.21}$$

A direct comparison between coherent and comparison detection can now be made by taking the ratio of the exponents of Eqs. (9.15) and (9.21). This ratio is

$$\gamma = \frac{\sin^2 \pi/L}{2 \sin^2 \pi/2L} \tag{9.22}$$

and is plotted in decibels in Fig. 9.4 as a function of L. It is seen that when L becomes large, the degradation suffered in comparison detection approaches 3 db over coherent detection.

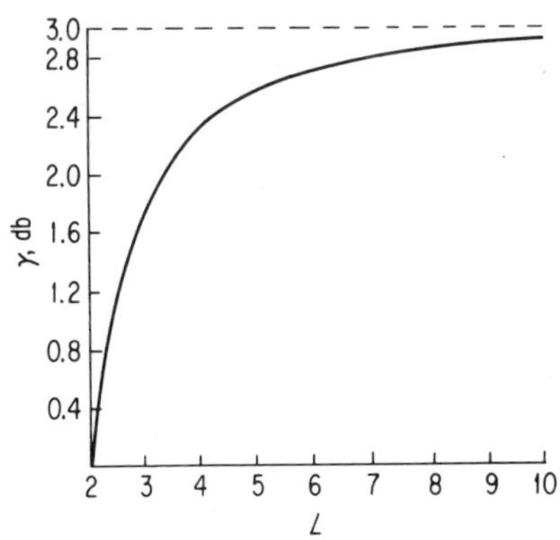

Fig. 9.4. Degradation γ db in comparison detection over coherent detection as a function of the number of phases L.

In particular we observe from Fig. 9.4 that for $L = 2$, corresponding to binary-phase-shift keying, no asymptotoc degradation is indicated. Nevertheless, some degradation does occur when the signal-to-noise ratio is not arbitrarily large. In order to evaluate this degradation, one must investigate the exact expression for P_e, in spite of its unwieldy form. Using a digital computer it is possible to plot the expression; the results appear in Fig. 9.5, where the difference in performance is indicated. Four-phase modulation with comparison detection is widely used; therefore we note in particular from Fig. 9.4 that when $L = 4$, the asymptotic degradation of the comparison detector is 2.3 db above the ideal four-phase coherent detector.

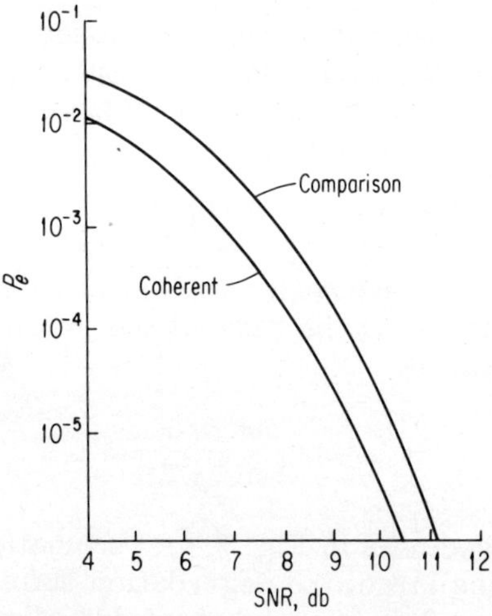

Fig. 9.5. Probability of error P_e in binary PM with coherent and comparison detection as a function of signal-to-noise ratio.

We remark at this point that, asymptotically, formula (9.21) is the same as Eq. (8.54), which expresses the probability of error in multilevel narrowband FM. This is so because an FM receiver, using discrimination or differential detection as modeled in Chap. 8, actually observes the phase changes of the received signal and is a particular realization of a phase comparison detector. In the subsequent section, we shall investigate still other physical realizations of phase comparison detectors.

9.2.3 *Physical Realization of Comparison Detection*

In the last section, we considered some conceptual aspects of digital-phase-comparison detection and derived the ultimate attainable error probability. In this section, we shall study specific methods of realizing the conceptual process of determining phase differences. We begin by considering some simple examples and then discuss general multiphase comparison techniques. We conclude with an analysis of the quadratic detector, a general type of incoherent detector.

Binary-phase modulation. The structure of the binary-phase comparison detector is shown in Fig. 9.6. The delay line and the multiplier are assumed to be ideal, and the postdetection filter is such that it eliminates the double-frequency components from the product signal and leaves the resulting baseband signal intact. Decisions are made by synchronously sampling, at $t = mT$, the output of the ideal low-pass filter. In order to investigate the mathematical behavior of this detector we use the representation of the binary, digital, phase-modulated signal as given in Eq. (9.3) plus additive gaussian noise and calculate the response as follows.

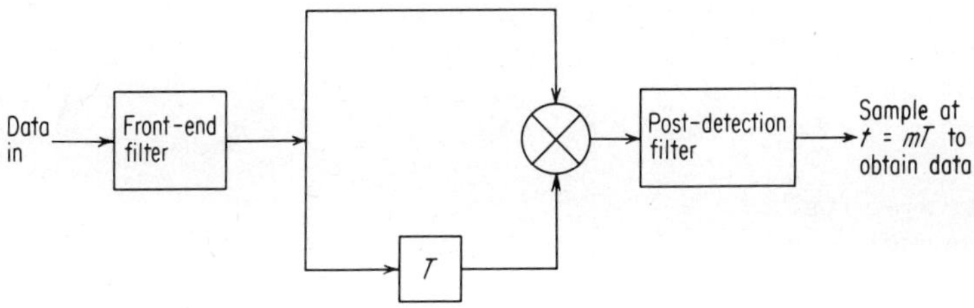

Fig. 9.6. Block diagram of a binary phase comparison detector.

The input signal plus noise is

$$s'(t) = A \sin \omega_c t \left[\sum_n a_n g(t - nT) + \xi(t) \right] + \eta(t) \cos \omega_c t \tag{9.23}$$

where $\xi(t)$ and $\eta(t)$ represent in-phase and quadrature noise components, respectively, and $a_n = \sin \phi_n = \pm 1$. The pulse shaping is appropriately chosen for optimum performance over an equivalent baseband-transmission medium as in linear modulation. For instance, we may assume for purposes of analysis that $g(t)$ is band-limited, with a 100 percent raised-cosine Fourier transform*

$$G(\omega) = \begin{cases} \dfrac{T}{2}\left[1 + \cos \dfrac{T\omega}{2}\right] & -W \leq \omega \leq W \\ 0 & |\omega| > W \end{cases} \tag{9.24}$$

where $W = 2\pi/T$. The pulse $g(t)$ then satisfies $g[T(n - m)] = \delta_{nm}$. For this analysis we assume a perfect channel (we shall discuss later a more general model including the effects of intersymbol interference). When the output of the ideal delay line is multiplied by $s'(t)$, the result is

* Tighter roll-offs can be used and even be more desirable in practice in order to conserve bandwidth.

$$z_0(t) = s'(t)\,s'(t - T)$$

$$= \frac{1}{2}\left\{\left[A\sum_n a_n g(t - nT) + \xi(t)\right]\left[A\sum_n a_n g(t - nT - T) + \xi(t - T)\right]\right\}$$
$$+ \frac{1}{2}\,\eta(t)\,\eta(t - T) + \text{terms centered at } 2\omega_c \qquad (9.25)$$

where we assumed that $\omega_c T = 2\pi$.

The wave $z_0(t)$ is passed through the ideal low-pass post-detection filter whose only action in this present model is to eliminate the terms centered at twice the carrier frequency. The result is Eq. (9.25) minus the terms center at $2\omega_c$. A sample of this wave taken at $t = mT$ yields

$$2q_m = [Aa_m + \xi_m][Aa_{m-1} + \xi_{m-1}] + \eta_m \eta_{m-1} \qquad (9.26)$$

where $f_m = f(mT)$.

Decisions are made by observing the quadratic form $2q_m$. The decision rule is as follows: When $2q_m$ is negative, the binary digit 1 is chosen; when it is positive, the binary digit 0 is chosen. If a_m and a_{m-1} differ in sign and $2q_m$ is positive, or vice versa, an error is committed.

The observable $2q_m$ is a quadratic form in gaussian variables. The probability of error may be evaluated by making use of results developed in the Appendix at the end of this chapter. It can be shown[6,7] that an exact expression for the probability of error in this case is

$$P_e = \frac{1}{2}\,e^{-(A^2/2\sigma^2)} \qquad (9.27)$$

where $\sigma^2 = \operatorname{var}(\xi_m) = \operatorname{var}(\xi_{m-1}) = \operatorname{var}(\eta_m) = \operatorname{var}(\eta_{m-1})$

and $\operatorname{var} x = \left\langle (x - \langle x\rangle)^2 \right\rangle$

Equation (9.27) holds when the noise samples taken T sec apart are statistically independent. This, however, is a very reasonable assumption since the noise bandwidth need not be larger than $1/T$, and consequently the noise correlation function evaluated at T is expected to vanish. We see that Eq. (9.27), a result based on a physical model, has an exponent identical to that in Eq. (9.21) when $L = 2$.

Four-phase modulation. Four-phase data-transmission systems using comparison detection are widely used in the field (see Baker) and have proved very reliable. The four-phase modulated signal occupies the same bandwidth as the biphase signal but carries twice as much information. Whereas in biphase modulation two phase changes are permitted, in four-phase modulation four possible phase changes are allowed. Each phase change is uniquely related to a dibit (pairs of bits are termed *dibits*). The four-phase modulated signal in the absence of channel distortion is represented as

$$z(t) = A \sum_n g(t - nT) \cos(\omega_c t + \phi_n) \tag{9.28}$$

where, for example, the four possible phase changes Δ are $\pm\pi/4$ and $\pm 3\pi/4$, and $g(t)$ is an ideal pulse.

The four-phase comparison detector is composed of essentially two biphase comparison detectors and is shown in block diagram form in Fig. 9.7. One detector computes the cosine of the phase difference, while the other computes the sine of the phase difference. If the signs of the outputs of the two detectors are compared, a unique resolution of one of the four transmitted phases is made. We now analyze the performance of this four-phase detector in the presence of gaussian noise with independent quadrature and in-phase components $\xi(t)$ and $\eta(t)$, respectivel, and common variance σ^2.

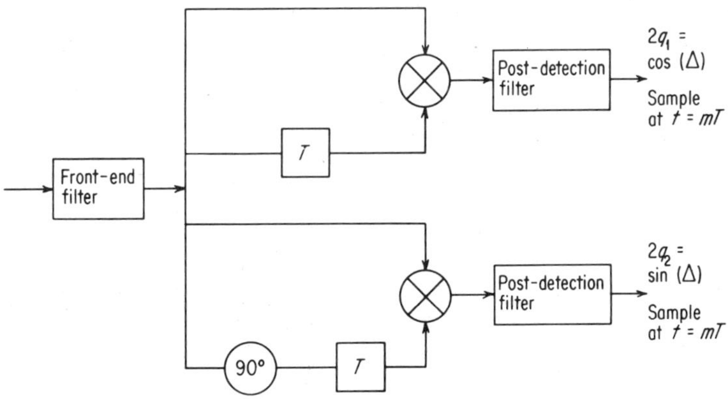

Fig. 9.7. Block diagram of a four-phase comparison detection system.

We write the signal plus noise as

$$w_1(t) = [\xi(t) + x(t)] \cos \omega_c t + [\eta(t) + y(t)] \sin \omega_c t \tag{9.29}$$

where
$$x(t) = A \sum_n g(t - nT) \cos \phi_n$$

and
$$y(t) = A \sum_n g(t - nT) \sin \phi_n$$

When we multiply $w_1(t)$ by its delayed replica $w_1(t - T)$, pass the result through the ideal low-pass filter, and take a sample at $t = mT$, we obtain at the output of the first detector the quadratic form

$$2q_1 = (\xi_m + x_m)(\xi_{m-1} + x_{m-1}) + (\eta_m + y_m)(\eta_{m-1} + y_{m-1}) \tag{9.30}$$

by again assuming that $\omega_c T = 2\pi$. In the second biphase detector an ideal 90° phase shifter is inserted in the forward path of the delay line. This phase-shifting operation converts $w_1(t - T)$ to

$$w_{1\theta}(t - T) = [\eta(t - T) + y(t - T)] \cos \omega_c t - [\xi(t - T) + x(t - T)] \sin \omega_c t \tag{9.31}$$

Performing the same operation with $w_{1\theta}(t - T)$ as with $w_1(t - T)$ in the first detector yields a quadratic form for the output of the second detector

$$2q_2 = (\xi_m + x_m)(\eta_{m-1} + y_{m-1}) - (\eta_m + y_m)(\xi_{m-1} + x_{m-1}) \tag{9.32}$$

The noise-free quadratic forms are determined by setting the noise components in Eqs. (9.30) and (9.32) to zero:

$$\begin{aligned} 2q_1\,{}_{\text{noise} \Rightarrow 0} &= x_m x_{m-1} + y_m y_{m-1} \\ 2q_2\,{}_{\text{noise} \Rightarrow 0} &= x_m y_{m-1} - y_m x_{m-1} \end{aligned} \tag{9.33}$$

From Eq. (9.29) we determine

$$x_m = x(mT) = A \cos \phi_m$$

and

$$y_m = y(mT) = A \sin \phi_m$$

and when we insert these equations into Eq. (9.33), we obtain

$$\begin{aligned} 2q_1\,{}_{\text{noise} \Rightarrow 0} &= A^2 \cos(\phi_m - \phi_{m-1}) = \pm \frac{A^2}{\sqrt{2}} \\ \text{and} \quad 2q_2\,{}_{\text{noise} \Rightarrow 0} &= A^2 \sin(\phi_m - \phi_{m-1}) = \pm \frac{A^2}{\sqrt{2}} \end{aligned} \tag{9.34}$$

Since there are four possible phase changes, the signs of Eq. (9.34) provide a unique resolution in the absence of noise. An error in decoding a dibit is made if an error is made in either detector. The probability of error in either detector may readily be determined as in the biphase case. However, the errors made in the two detectors are not independent, since the quadratic forms Eqs. (9.30) and (9.32) are not statistically independent. Excellent bounds on the probability of error may be established by use of the well-known union bound. Denote the event of an error committed in the first and second detector by E_1 and E_2, respectively. The probability of either E_1 or E_2 occurring is

$$P(E_1 \cup E_2) = P(E_1) + P(E_2) - P(E_1 E_2) \leq P(E_1) + P(E_2) \tag{9.35}$$

A lower bound is also obtained by max $[P(E_1), P(E_2)]$. Since no statistical mechanism is at work to make $P(E_1)$ and $P(E_2)$ different

$$P(E_1) \leq P(E_1 \cup E_2) = \text{probability of error} \leq 2P(E) \tag{9.36}$$

We see than, that the probability of error of the four-phase detector can at most be off by a factor of 2 from the individual biphase detectors shown in Fig. 9.7. The probability of error in one of the biphase detectors is

$$P_e = P[zz_d + pp_d < 0] \tag{9.37}$$

subject to the condition that

$$\langle zz_d + pp_d \rangle = A \cos(\phi_m - \phi_{m-1}) > 0$$

where we set for notational convenience

$$z = \xi_m + x_m \qquad z_d = \xi_{m-1} + x_{m-1} \tag{9.38}$$

and

$$p = \eta_m + y_m \qquad p_d = \eta_{m-1} + y_{m-1}$$

Since the gaussian random variables appearing in the quadratic form of Eq. (9.37) are all independent with the same variance σ^2, it is possible to evaluate an asymptotic form for the probability of error by making use of the results developed in the appendix to this chapter. Using the notation in the appendix at the end of this chapter, we write

$$P_e \sim \frac{1}{2\sqrt{\pi}} \frac{u+v}{u-v} \frac{1}{\sqrt{uv}} e^{-(u-v)^2/4\sigma^2} \tag{9.39}$$

where u and v in this application are the lengths of the two-dimensional vectors

$$\begin{aligned} \mathbf{u} &= \frac{1}{\sqrt{2}} \left(\langle z + z_d \rangle, \ \langle p + p_d \rangle \right) \\ &= \frac{A}{\sqrt{2}} \left(\cos\phi_m + \cos\phi_{m-1}, \ \sin\phi_m + \sin\phi_{m-1} \right) \end{aligned} \tag{9.40}$$

$$\mathrm{v} = \frac{1}{\sqrt{2}}\Big(\langle z - z_d\rangle, \ \langle p - p_d\rangle\Big)$$

$$= \frac{A}{\sqrt{2}}\Big((\cos\phi_m - \cos\phi_{m-1}, \ \sin\phi_m - \sin\phi_{m-1})\Big) \tag{9.40 (cont'd.)}$$

Use of Eq. (9.34) allows the vector magnitudes to be calculated as

$$u = A\left(1 + \frac{1}{\sqrt{2}}\right)^{1/2}$$

and

$$v = A\left(1 - \frac{1}{\sqrt{2}}\right)^{1/2} \tag{9.41}$$

Substituting Eq. (9.41) into (9.39), we obtain the asymptotic behavior of the probability of error as

$$P_e \sim \exp\left[-\frac{A^2}{2\sigma^2}\left(1 - \frac{1}{\sqrt{2}}\right)\right] \tag{9.42}$$

The factor $1 - (1/\sqrt{2})$ appearing in the exponent represents a degradation suffered by the four-phase detector over the biphase detector. The value of the degradation is calculated to be 5.3 db of which 2.3 db results from the use of comparison detection instead of coherent detection. This value of 2.3 db is the value of γ in Eq. (9.22) when $L = 4$. The 3 db that remains results from our use of four phases instead of two. This factor may be accounted for from geometrical considerations as follows: The distance between differential signal phasors in biphase modulation is $2A$ since the angle differences are 0 and π. In four-phase modulation the possible values of the phase changes are $\pi/4, 3\pi/4, 5\pi/4$, and $7\pi/4$. Since the length of each phasor is still A, the minimum distance between phasors becomes $\sqrt{2}\,A$ instead of $2A$; hence the 3 db additional degradation.

L-phase comparison detection. The implementations given in the last section for biphase and four-phase detection may be generalized to L phases. The idea exploited in the L-phase comparison detector is that, by a series of phase splitting actions on the received wave, resolution of the particular one of L-phase differences can be effected. We have seen that the four-phase comparison detector computes the quantities $\cos\Delta$ and $\sin\Delta$, where Δ is the transmitted phase change, and bases its decisions on the signs of these quantities. In general, for efficient decoding an L-phase comparison detector must compute $\log_2 L$ quantities. Although the differential angle Δ may be determined by making threshold decisions on both $\cos\Delta$ and $\sin\Delta$, this method yields a rather poor probability of error. An efficient method of detecting the L-phase changes is to compute the following $L = 2^m$ quantities:

$$\Lambda = \begin{cases} \sin \Delta \\ \cos \Delta \\ \cos 2\Delta \\ \cos 2^2 \Delta \\ \vdots \\ \cos 2^{m-2} \Delta \end{cases} \tag{9.43}$$

and observe their signs. This particular method is discussed in Khvorostenko,[8] while still other techniques are presented in Bennett and Davey.[9] For representational convenience we have taken L to be a power of 2. The first two entries in Λ are computed by using the same instrumentation as in the four-phase detector. The remaining $m-2$ quantities may be derived from the first two. The sign of each entry in Λ isolates the transmitted phase change to a smaller region in the phase plane than the previous entry. For example, the sign of $\cos \Delta$ indicates that the transmitted phase change is in either the right- or the left-hand plane. The sign of $\sin \Delta$ gives an indication as to the upper- or lower-half plane, thus narrowing the possibility to one of the four quadrants. To make further resolutions, the sign of $\cos 2\Delta$ divides the respective quadrant into 45° segments, and so forth.

When it is desired to transmit data from m parallel binary sources via L-phase differential phase modulation, this form of detection can provide at its m outputs the original binary sequences in a one-to-one manner, provided the proper correspondence is made between binary sequences and differential phases. To illustrate this point, consider the m-digit binary sequence $(a_1 a_2 \cdots a_m)$. The choice of the 2^m differential phases is made as follows: The first digit determines whether it should be in the right or the left plane. For example, if $a_1 = 1$, the phase must be in the right plane; if $a_1 = 0$, the phase is assigned to the left plane. This choice determines the sign of $\cos \Delta$ that decodes a_1 directly. The next binary digit a_2 determines whether the differential phase should be in the upper or the lower left plane. Thus the sign of $\sin \Delta$ directly decodes the secondary binary digit. The third binary digit determines a particular 45° segment, etc. Consequently, each parallel output of the detector corresponds directly to a digit in the serial binary sequence.

The probability of error of the L-phase comparison detector may be bounded by again using the union bound. Thus a good indication of the average error rate may be had by the knowledge of the error rate in a single output. Suppose we concentrate on the output indicated by $\cos \Delta$ in Eq. (9.43). The analysis proceeds in exactly the same manner as in the four-phase case, except that the quantity $\cos \Delta = \cos (\phi_{m-1} - \phi_m)$ must be modified to reflect that we are dealing with L phase changes rather than only four. Thus to evaluate Eq. (9.33), we need to determine the squared difference of the vector lengths $\mathbf{u}$ and $\mathbf{v}$. It can be seen from the definitions of $\mathbf{v}$ and $\mathbf{u}$ in Eq. (9.40) that when $\Delta = \phi_m - \phi_{m-1}$ takes on L values equally distributed around a circle,

$$u = A(1 + \cos \Delta)^{1/2}$$

and

$$v = A(1 - \cos \Delta)^{1/2} \tag{9.44}$$

The value of Δ in L-phase modulation is the maximum angle that makes $\cos \Delta$ positive. The minimum phase change is π/L; and therefore $\Delta = \pi/2 - \pi/L$. With this value of Δ

$$u = A\left(1 + \sin \frac{\pi}{L}\right)^{1/2}$$

$$v = A\left(1 - \sin \frac{\pi}{L}\right)^{1/2} \tag{9.45}$$

and

$$(u - v)^2 = 2A^2\left(1 - \cos \frac{\pi}{L}\right)$$

Substituting the above quantities into Eq. (9.33), we obtain for the asymptotic form of the probability of error

$$P_e \sim \exp\left[-\frac{A}{2\sigma^2}\left(2 \sin^2 \frac{\pi}{2L}\right)\right] \tag{9.46}$$

This expression agrees with Eq. (9.21), indicating that the bound is very sharp when the signal-to-noise ratio is large.

9.2.4 Double Error Rates in Comparison Detection

In comparison detection, successive decisions are not independent as in coherent detection, and errors tend to bunch. The frequency of double errors may be as high as 25 percent of that of single errors. This phenomenon seems physically reasonable since in this mode of detection the reference signal is obtained from the previous signaling interval. One may reason that in a well-designed system and in low-noise conditions, a large phase deviation from the one intended is very rare. However, when such a deviation does occur, an error will be committed with a high probability since the last bit was correctly decoded with a high probability. In the same manner, since the next phase is very likely to be correct, an error in the detection of that bit will occur with high probability. Thus the probability of two successive errors may be large. On the other hand single errors do occur, for example, when successive deviations have opposite signs but the amount of each deviation is only half large enough to cause an error.

The probability of occurrence of double errors can be computed for the models studied in the previous sections by calculating the joint probability of two successive errors. It can be shown[10] that the joint probability of two successive errors in biphase comparison detection is

$$P(e_1,e_2) = \frac{1}{4\pi} e^{-\rho} \int_0^{\pi} \Big\{[1 - \operatorname{erf}(\sqrt{\rho}\cos\theta)]^2 + \sqrt{\rho\pi}\cos\theta\, e^{\rho\cos^2\theta}[1 - \operatorname{erf}(\sqrt{\rho}\cos\theta)]^2[1 + \operatorname{erf}(\sqrt{\rho}\cos\theta)]\Big\}\, d\theta \tag{9.47}$$

where $\rho = A^2/2\sigma^2$. Since the probability of a single error is [from Eq. (9.27)] $1/2\, e^{-\rho}$, one can write for the conditional probability of error

$$P(e_2 | e_1) = 2e^{\rho} P(e_2, e_1) \tag{9.48}$$

This conditional probability of error as a function of ρ is plotted in Fig. 9.8. It can be observed that the ratio of double errors to single errors tends to zero as the signal-to-noise ratio ρ tends to infinity. Since the biphase error rates serve as bounds on L-phase error rates, Eq. (9.48) is a typical indication of what might be expected in L-phase comparison detection.

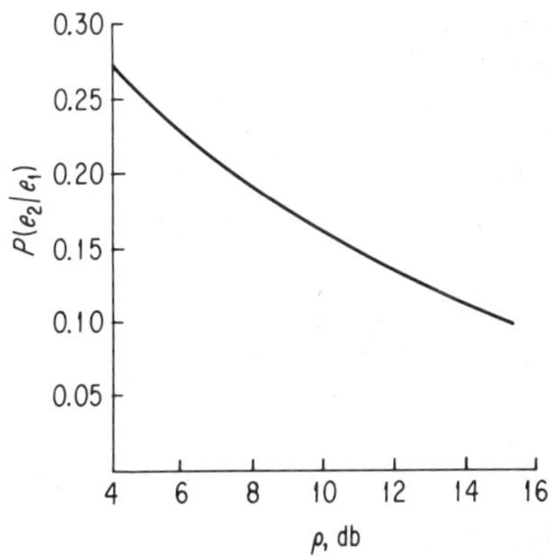

Fig. 9.8. Double error rate to single error rate in binary PM using comparison detection.

9.2.5 The Quadratic Detector

In the preceding section we analyzed the effect of added noise on the performance of comparison detectors and neglected the effect of an imperfect channel. We saw that the performance of L-phase comparison detection could be determined by modeling the detectors as a contiguous set of biphase detectors where each biphase detector essentially computed the correlation function of the incoming signal plus noise. In this section we provide a more practical model for this correlator and develop formulas taking into account channel dispersion, from which error rates may be calculated.

Here we analyze a general class of *quadratic* detectors. They are called quadratic because they base their decision logic on a random variable that can be represented by a quadratic form in signal-plus-noise components. The quadratic-form representation arises because these detectors filter the products of signal plus noise and their delayed replicas. Because of these actions, the detection process is sometimes referred to as *autocorrelation detection.*

We saw in the preceding sections that phase comparison detection can be accomplished with a special kind of a quadratic detector. These were modeled as multipliers of the baseband incoming signal with its delayed replica. The analysis was restricted to the inclusion of postdetection filters that only accept low-frequency difference products and reject high-frequency sum products. Furthermore, channel imperfections were ignored. The analysis of some simple models provides sufficient indication of actual performance as well as physical insight. However, in many applications the postdetection filter does significant selective processing of the detected low-frequency signal and noise components and cannot be ignored.

The process of delaying a signal is most often achieved with linear filters. Since perfect delay action cannot be physically realized, filters approximating delay lines must be included in the model. Although our emphasis in this section is on comparison detection, quadratic detectors are not restricted in use to phase modulation. Many other communications devices such as noncoherent frequency detectors, energy measuring devices, and various diversity-combining receivers base their decision logic on the sign of quadratic forms. In detection theory, for instance, optimum discrimination (in the sense of minimum error probability) of two gaussian signals consists in comparing a quadratic form in gaussian variables with a threshold.

The performance evaluation of these detectors requires knowledge of the probability distribution of quadratic forms in statistical variates. The determination of the probability distribution of a quadratic form in arbitrary statistical variates is a very difficult problem and is usually not mathematically tractable. Even with added white gaussian noise the exact probability distribution is too complicated to be useful. Because of the analytically unwieldy form of the exact expression for the probability of error, we shall treat only its asymptotic form. This kind of treatment usually yields sharp estimates of performance.

The schematic representation of a general quadratic detector is shown in Fig. 9.9. The three filters shown are arbitrary, and the multiplier is ideal. This general model provides for the presence of a dispersive channel and imperfect filtering in the delay lines. The quadratic detector is conveniently characterized by a system function $g(u,v)$. The output $e_0(t)$ is then expressed as a quadratic integral with $g(u,v)$ as the kernel. Reference to Fig. 9.9 can verify that

$$e_0(t) = \iint_{-\infty}^{\infty} e_1(t-u)\,g(u,v)\,e_1(t-v)\,du\,dv \tag{9.49}$$

where the system function $g(u,v)$ is represented in terms of the three filter impulse responses as

$$g(u,v) = \int_{-\infty}^{\infty} f_1(u-z)\,f_3(z)\,f_2(v-z)\,dz \tag{9.50}$$

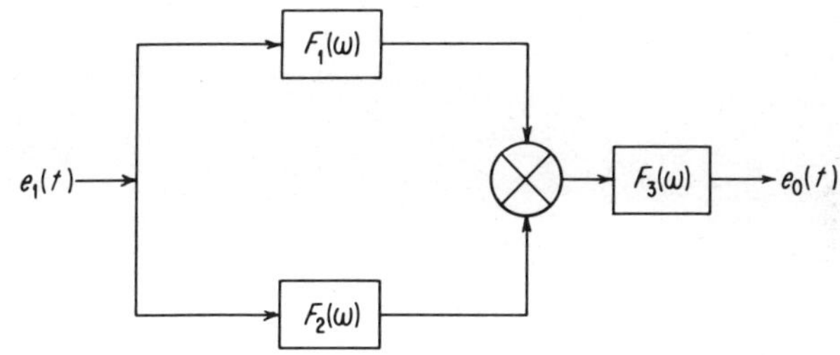

Fig. 9.9. Schematic representation of a Quadratic detector.

For analytical purposes it is often desirable to have a representation such as indicated in Eq. (9.49) in terms of a system function that is a symmetric function in the variables u and v. If the filters f_1 and f_2 are not identical, $g(u,v) \neq g(v,u)$. It is possible, however, to form a new system function $G(u,v)$ that is symmetric. It may be ascertained that the new system function

$$G(u,v) = \frac{1}{2}[g(u,v) + g(v,u)] \tag{9.51}$$

is symmetric and renders the representation of $e_0(t)$ invariant when $g(u,v)$ is replaced by $G(u,v)$ in Eq. (9.49).

The integral representation in Eq. (9.49) is still not in a convenient form for error-rate analysis. We wish to represent $e_0(t)$ as a finite quadratic form such that, as the number of terms in the limit approaches infinity, it approaches the integral representation. To accomplish this we use a theorem well-known in analysis which states that if $G(u,v)$ is square-integrable, a condition always satisfied in practical situations, it may be represented as

$$G(u,v) = \sum_n \lambda_n \phi_n(u)\,\phi_n(v) \tag{9.52}$$

on the rectangle

$$0 \le u \le T \quad 0 \le v \le T$$

where the set of functions $\{\phi_n(\cdot)\}$ are solutions of the integral equation

$$\int_0^T G(u,v)\,\phi_n(v)\,dv = \lambda_n \phi_n(u) \qquad 0 \le u \le T \tag{9.53}$$

The functions $\{\phi_n(\cdot)\}$ can be shown to be orthonormal. With each function $\varphi_n(\cdot)$ there is associated a real number λ_n called its *eigenvalue*. Upon substituting Eq. (9.52) into (9.49), we obtain a useful representation; namely,

$$\begin{aligned} e_0(t) &= \sum_n \lambda_n \int_{-\infty}^{\infty}\!\!\int_{-\infty}^{\infty} du\,dv\,e_1(t-u)\,\phi_n(u)\,e_1(t-v)\,\phi_n(v) \\ &= \sum_n \lambda_n e_{1n}{}^2(t) \end{aligned} \tag{9.54}$$

where
$$e_{1n}(t) = \int_{-\infty}^{\infty} \phi_n(\xi)\,e_1(t-\xi)\,d\xi$$

In most cases of interest the eigenvalues decrease quite rapidly with increasing n; thus when carrying out practical calculation, only a few terms in the series need be considered. This fact makes the representation indicated in Eq. (9.54) amenable to analysis.

We now write the input wave $e_1(t)$ as the sum of signal plus noise

$$e_1(t) = s(t) + \xi(t) \tag{9.55}$$

and assume that $\xi(t)$ is a stationary, white gaussian noise process with correlation function

$$R(\tau) = \langle \xi(t)\,\xi(t+\tau)\rangle = N_0\,\delta(\tau) \tag{9.56}$$

Observe that the particular choice for the noise correlation function is not a severe restriction since the inclusion of the arbitrary filters F_1 and F_2 permits analysis with an arbitrary noise spectrum. The white-noise assumption is made strictly for mathematical convenience. If the input noise were actually not white, the filter characteristics could suitably be modified to account for a specified input-noise spectrum.

Using the representation for $e_{1n}(t)$ in Eq. (9.54), we can verify that

$$\langle e_{1n}{}^2(t)\rangle = N_0 \tag{9.57}$$

Very often, the probability of error may be estimated from the probability that the output $e_0(t)$ at one instant is greater or less than zero, given that in the absence of noise $e_0(t)$ is less or greater than zero. The mathematical details of this calculation are worked out in the Appendix at the end of the chapter. By making suitable identifications with formula (9.71) in the appendix to this chapter, it can be established that

$$P_e = P\left[e_0(t_0) < 0 \,\middle|\, \sum_n \lambda_n s_n^2(t) > 0\right]$$

$$\sim \frac{\sqrt{2N_0}}{\sqrt{2\pi}} \frac{1}{y_0} \frac{1}{\left[\prod_n (1 + y_0\lambda_n)\right]^{1/2}} \frac{e^{-(1/2N_0)f(y_0)}}{\sqrt{-f''(y_0)}} \tag{9.58}$$

where
$$s_n(t_0) = \int_{-\infty}^{\infty} \phi_n(t)\, s(t_0 - t)\, dt$$

and
$$f(y) = \sum_n \frac{y s_n^2(t_0)}{1 + y\lambda_n}$$

and y_0 is determined by

$$f'(y_0) = 0$$

The system function $G(u,v)$ plays an important role in determining the asymptotic behavior of the probability of error. As can be seen from the above equations, the degradation, i.e., the exponential behavior of p_e for a fixed signal-to-noise ratio, is determined by $f(y_0)$. Knowing the system function, one could conceivably calculate $f(y)$ and search for a maximum. In general this is a formidable task, and approximation methods must be used.

We now cite an example illustrating some of the ideas. Consider the differential detector treated in the previous section and assume that the two filters F_1 and F_2 are identical except that one of the filters has a delay of T sec. Further assume that these filters are ideal-low-pass with cutoff frequency W rads per sec. Finally the postdetection filter has a rectangular impulse response with unit height and T-sec duration.

The system function for this detector is readily calculated as

$$g_d(u,v) = \int_{-T/2}^{T/2} \frac{\sin W(u - z)}{\pi(u - z)} \frac{\sin W(v - z)}{\pi(v - z)}\, dz \tag{9.59}$$

The symmetric system function as defined in Eq. (9.51) thus becomes

$$G(u,v) = \frac{1}{2}\sum_n \lambda_n \psi_n(u + T)\psi_n(v) + \frac{1}{2}\sum_n \lambda_n \psi_n(u)\psi_n(v + T) \tag{9.60}$$

where it can be shown[11] that the set of functions $\{\psi_n(t)\}$ satisfying the integral equation (9.53) are the prolate spheroidal functions orthogonal on the interval $[-T/2, T/2]$. Eq. (9.60) can be approximated by only a few terms in the infinite series because of the rapidity with which the eigenvalues λ_n associated with the prolate spheroidal functions decrease when n is large. The rapidity with which they decrease depends on the parameter $WT/2$, which is proportional to the ratio of input bandwidth to output bandwidth. This parameter is sometimes referred to as the *processing gain*. As a practical example, suppose $WT/2 = \pi$. The first six eigenvalues corresponding to this example are,[12] in decreasing order, 0.981, 0.749, 0.243, 0.025, and 0.001, $\sim 10^{-5}$. Thus the first few eigenvalues should be sufficient when making a practical calculation.

We remark in closing that in any particular situation a detailed study of the system function must be made to establish the degree of allowed approximations. The general problem treated in this section is sufficiently complicated to exclude the possibility of giving a closed-form solution to all possible choices of parameters. We merely intended to indicate methods of attack and provide the reader with some useful techniques.

9.3 Comparisons with Other Modulation Systems

We wish to emphasize again that the choice of a modulation method depends on the particular application. The choice may be based on the relative noise immunity, immunity to other channel impairments (such as delay and amplitude distortion, nonlinearities, phase jitter, and frequency offset), or simplicity of instrumentation and compatibility with other equipment. Clearly it would not be realistic to compare modulation systems under all the circumstances cited above. Our intent is to provide the reader with physical insight and analytical techniques useful to him in evaluating different systems. However, some fundamental comparisons based on ideal situations can be made. For instance, if one is interested in high-speed data transmission over a noisy, band-limited medium, vestigial-sideband modulation may be a better choice than PM or FM. As a concrete example suppose it is desired to send 7,200 bps over a usable 2,400-Hz bandwidth. One could use 50 percent roll-off vestigial-sideband signaling at 2,400 bauds with eight levels. To achieve the same speed with phase modulation, one could only signal at 1,200 bauds requiring 16 phases, since phase modulation is basically a double-sideband technique.

Setting aside all other considerations except such gross parameters as flat bandwidth, rate, signal-to-noise ratio (gaussian noise), and probability of error, is it then possible to make some meaningful comparisons? Even here the answer is not clear. Shall we compare systems operating in the same bandwidth but at different speeds or vice versa?

We have developed formulas in the preceding chapters which relate all the above-mentioned parameters for each particular modulation system. We saw that the bandwidth required to transmit a digital phase-modulated wave is the same as that of double-sideband suppressed carrier AM. Because of the similar structure of the PM and AM signals, identical transmitting and receiving filters can be used, and these can be chosen in an optimum manner to minimize the noise variance for a fixed average signal power at the receiver. Thus the only remaining parameter distinguishing the performance of the two systems signaling at the same rate, and hence at the same number of levels, is the degradation suffered as a result of the increase in the number of levels. This degradation in decibels for coherent AM is (from Chap. 4), because of the baseband equivalence,

$$D_A = 10 \log \frac{3}{L^2 - 1} \sim 4.8 - 20 \log L \qquad L \gg 2 \tag{9.61}$$

and for PM employing comparison detection from Eq. (9.21)

$$D_\phi = 10 \log \left(2 \sin \frac{2\pi}{2L}\right) \sim 7 - 20 \log L \qquad L \gg 2 \tag{9.62}$$

Thus for large L, AM suffers approximately 2 db degradation over PM employing comparison detection. This means that if the same probability of error is desired, AM requires 2 db more average signal-to-noise ratio than PM.

We saw in Chap. 8 that the probability of error for narrowband FM employing the same bandwidth as PM and double-sideband AM is of the same form as Eq. (9.62). Thus we conclude that the optimum narrowband FM system is identical in performance, under these conditions, to PM using comparison detection. Additional comparisons of modulation systems based on some other performance indices are given in Bennett and Davey.[9]

In concluding the discussion of phase modulation, we wish to remark at this point that for large values of L, say, $L > 8$, combined amplitude and phase modulation may yield better performance than phase modulation alone. We do not treat combined modulation schemes since, for these large values of L where they are theoretically more efficient, the modulation technique becomes impractical. Combined AM and PM modulation is discussed in Hancock and Lucky.[3]

9.4 Summary

In this chapter we discussed methods of digital phase modulation as a vehicle for digital data transmission. We briefly examined the structure of the modulated signal and showed that its spectral properties were the same as that of AM. We emphasized that the modulation method is double-sideband and therefore is not efficient from the point of view of bandwidth utilization. Two detection methods were investigated. Coherent detection was discussed first. This is an optimum procedure and the derived error-rate results therefore provide an absolute performance reference. We next turned to the more practical method of comparison detection and showed the degradation suffered over the coherent detector. The discussion was concluded by analyzing the performance of particular PM systems using differential coding and comparison detection. In Sec. 9.2.5 we treated a general class of differential detectors called quadratic detectors. The model for this class of detector provides for the presence of a dispersive channel as well as nonideal postdetection filters. Analytical techniques were developed for studying system performance. The chapter was concluded with a comparison of the various modulation methods.

Appendix

Probability of Error in Quadratic Detectors

In this Appendix, we derive an asymptotic expansion for the probability that a quadratic form in gaussian variates with arbitrary means and variances is less (greater) than zero. Using these results, we demonstrate how the probability of error of some quadratic detectors may be estimated. Let $\mathrm{w} = (w_1 \cdots w_n)$ be a gaussian n vector with a mean vector $\overline{\mathrm{w}} = (\overline{w}_1 \cdots \overline{w}_n)$ and correlation coefficients $M_{ij} = \overline{w_i w_j}$, where $i, j = 1, 2, \ldots, n$. The desired probability is

$$p = P[q < 0] \tag{9.63}$$

where
$$q = \sum_{i,j}^{n} Q_{ij} w_i w_j$$

and $\{Q_{ij}\}$, where $i, j = 1, 2, \ldots, n$ is a set of real numbers. Without loss of generality we may assume $Q_{ij} = Q_{ji}$.

The characteristic function of gaussian quadratic forms is well known[13] and is given in matrix notation by

$$C(\omega) = \langle e^{i\omega q} \rangle_q = \frac{\exp \frac{1}{2} [\overline{\mathrm{w}}^T \mathrm{M}^{-1} (\mathrm{I} - 2i\omega \mathrm{MQ}^{-1}) \overline{\mathrm{w}} - \overline{\mathrm{w}}^T \mathrm{M}^{-1} \overline{\mathrm{w}}]}{\sqrt{\det(\mathrm{I} - 2i\omega\mathrm{MQ})}} \tag{9.64}$$

where $\overline{\mathrm{w}}$ is the n-dimensional mean column vector with elements $\overline{w}_i$, Q is an $n \times n$ real symmetric matrix with elements Q_{ij}, and M is an $n \times n$ real, symmetric, and positive-definite covariance matrix with elements M_{ij}. The notation $\mathbf{M}^{-1}$, $\overline{\mathbf{w}}^T$, $\mathbf{I}$ and det, stand for inverse, transpose, identity matrix and determinant, respectively.

It is convenient at this point to introduce the following transformations (it can be shown that this is always possible for the problem at hand):

$$\mathbf{M} = \sigma^2 \mathbf{S}\mathbf{S}^T \quad \text{and} \quad \mathbf{Q} = (\mathbf{S}^T)^{-1} \mathbf{D}\mathbf{S}^{-1} \tag{9.65}$$

where $\mathbf{S}$ is a real matrix, $\mathbf{D}$ is a diagonal matrix, and σ^2 is a positive parameter introduced for later convenience.

By defining a new gaussian vector

$$\mathbf{y} = \mathbf{S}^{-1}\mathbf{W} \tag{9.66}$$

and substituting it into Eq. (9.64), one obtains for the characteristic function a more suitable representation:

$$C(\omega) = \frac{\exp \dfrac{1}{\sigma^2}\left[\dfrac{1}{2}\overline{\mathbf{y}}^T(\mathbf{I} - 2i\omega\sigma^2\mathbf{D})^{-1}\overline{\mathbf{y}} - \dfrac{1}{2}\overline{\mathbf{y}}^T\overline{\mathbf{y}}\right]}{\sqrt{\det(\mathbf{I} - 2i\omega\sigma^2\mathbf{D})}} \tag{9.670}$$

The principal simplification now is that $\mathbf{D}$ is a diagonal matrix. Note that

$$\overline{\mathbf{w}}^T\mathbf{Q}\overline{\mathbf{w}} = \overline{\mathbf{y}}^T\mathbf{D}\overline{\mathbf{y}} \quad \text{and} \quad \overline{\mathbf{w}}^T\mathbf{Q}\overline{\mathbf{w}} = \overline{\mathbf{y}}^T\mathbf{D}\overline{\mathbf{y}} \tag{9.68}$$

The desired probability, Eq. (9.63), is now expressed, via the well-known inversion formula, as

$$p = -\frac{1}{2\pi i}\int_{-\infty}^{\infty}\frac{C(\omega)}{\omega + i\epsilon}\,d\omega \tag{9.69}$$

The $i\epsilon\,(\epsilon > 0)$ appearing in the denominator of the integrand is used to signify that in the complex ω plane, the contour of integration implied in Eq. (9.69) goes above the singularity at $\omega = 0$. By using Eqs. (9.67) and (9.68), Eq. (9.69) may be expressed in more detail as

$$p = -\frac{1}{2\pi i}\int_{-\infty}^{\infty}\frac{d\omega}{\omega + i\epsilon}\left[\prod_{j=1}^{n}(1 - i\omega\, d_j)\right]^{-1/2} \exp\left[-\frac{\omega}{2\sigma^2}\sum_{j=1}^{n}\frac{\overline{y}_j^{\,2}}{\omega + i/d_j}\right] \tag{9.70}$$

where d_j is the jth element of $\mathbf{D}$.

We see from Eq. (9.70) that $1/\sigma^2$ appears only in the exponent; therefore this representation is in a suitable form for an asymptotic expansion. In specific applications the quantity $1/\sigma^2$ corresponds to the signal-to-noise ratio.

Proceeding with the analysis of Eq. (9.70), we remark that the integrand falls off so rapidly at infinity that one can close or distort the contour at infinity without changing the value of the integral. We further note the singularities of the integrand. There is a simple pole at $\omega = 0$ which has already been discussed. In addition to the simple pole at $\omega = 0$, the exponent has simple poles at $\omega = -i/d_i$; these all lie along the imaginary axis. Also at these points the denominator of the integrand has, in general, branch points due to the square root. For doubly degenerate eigenvalues, the branch points become single poles.

We now focus our attention on performing the integration in Eq. (9.70) for small σ^2 by the saddle-point method. To locate the saddle points, let $\omega = i\mu$ and consider the solutions of

$$-\frac{d}{d\mu}\,\mu\sum_{i=1}^{n}\frac{\bar{y}_i^{\,2}}{\mu + (1/d_i)} = 0 \tag{9.71}$$

or

$$F(\mu) \equiv -\sum_{i=1}^{n}\frac{\bar{y}_i^{\,2}/d_i}{[\mu + (1/d_i)]^2} = 0 \tag{9.72}$$

Consider $F(\mu)$ defined by Eq. (9.72) and let μ be real. We shall show that a saddle point always exist for positive μ (ω positive-imaginary), which occurs before the first singularity of the exponent on the positive-imaginary axis.† We separate the sum of Eq. (9.72) into two parts:

$$F(\mu) = -\sum_{d_i\,\mathrm{pos}}\frac{\bar{y}_i^{\,2}/d_i}{[\mu + (1/d_i)]^2} - \sum_{d_i\,\mathrm{neg}}\frac{\bar{y}_i^{\,2}/d_i}{[\mu + (1/d_i)]^2} \tag{9.73}$$

Note that when $\mu \approx -(1/d_i)$, $F(\mu)$ is large and positive for $d_i < 0$ and $F(\mu)$ is large and negative for $d_i > 0$. Also note that

$$F(0) = -\sum_{i=1}^{n} d_i\bar{y}_i^{\,2} \le 0$$

by Eq. (9.68) and our assumption that the noise-free signal is positive. Hence by continuity, there must be at least one $\mu > 0$ for which $F(\mu) = 0$, which locates a saddle point. By the monotonicity of the two parts of expression (9.73) there can be only one such saddle point between the origin and the first singularity of the exponent on the positive-imaginary axis.

---

† This assumes that at least one of the d_i is negative with nonvanishing residue $\bar{y}_i^{\,2}$. The quadratic form q in the absence of noise is, of course, also assumed positive.

The relations in the complex ω plane described above are illustrated in Fig. 9.10. Dashed line L_1 is the original contour of integration; the small circle labeled "SP" is the (unique) saddle point on the positive imaginary axis lying before the nearest singularity $\omega = -i/d_i$, $d_i < 0$, which has nonvanishing residue $\bar{y}_i^2$ in the exponent of the integrand. Also drawn on the imaginary axis in Fig. 9.10 are the branch cuts of the denominator, indicated by the hatched bars in the figure. It is assumed that the saddle point occurs before the nearest singularity $\omega = -i/d'$ on the positive imaginary axis.‡ This is guaranteed to be the case if only $(\bar{y}')^2 \neq 0$.§

If the saddle point is situated as shown in Fig. 9.10, the contour may then be shifted from the real axis (line L_1 in Fig. 9.10) to a contour which is a straight line parallel to the real axis and passing through the saddle point (the line L_2 in Fig. 9.10). The integrand drops off sufficiently rapidly for large $|\omega|$ so that the ends of the contour connecting L_1 and L_2 (in accordance with Cauchy's theorem) give no contribution. We shall now, for large signal-to-noise ratios (small σ^2), approximate the integral along the contour L_2 by the contribution in the immediate neighborhood of the saddle point. It can be verified that the magnitude of the exponential term in the integrand of Eq. (9.70) is a monotonically decreasing function of ω as one recedes from the saddle point on the contour L_2; therefore this saddle-point evaluation is asymptotically correct. It can also be shown that the contour L_2 is in fact tangent to the path of the steepest descent at the saddle point.

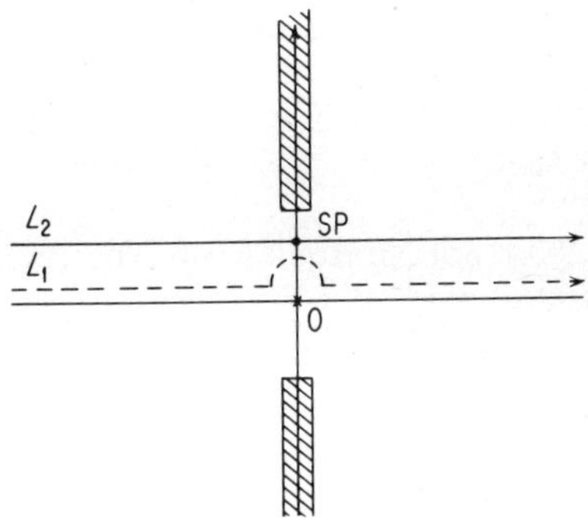

Fig. 9.10. Contours and singularities in the ω plane.

To obtain an explicit formula for our asymptotic evaluation of p under these conditions, write Eq. (9.70) as

$$p = -\frac{1}{2\pi i}\int_{L_2} g(z)\, e^{-f(z)/\sigma^2}\, dz \tag{9.74}$$

with

$$g(z) = \frac{1}{z}\,\frac{1}{[\Pi(1 - iz\,d_i)]^{1/2}} \tag{9.75}$$

---

‡ For definiteness we have based our analysis on the assumption that the noiseless signal is positive. We emphasize that the role of upper- and lower-half planes would be interchanged if one assumed the noiseless signal were negative.

§ We have used d' to denote that particular d_i which corresponds to the nearest singularity on the positive imaginary axis; $\bar{y}'$ is the associated $\bar{y}_i$.

and
$$f(z) = \frac{z}{2} \sum_{j=1}^{n} \frac{\bar{y}_j^2}{z + (i/d_j)} \tag{9.76}$$

Using a standard saddle-point technique, we finally write

$$p \sim \frac{1}{\sqrt{2\pi}} \frac{1}{\Gamma[\prod_j (1 + \Gamma d_j)]^{1/2}} \frac{\sigma}{[f''(i\Gamma)]^{1/2}} \exp\left[-\frac{\Gamma}{2\sigma^2} \sum_j \frac{\bar{y}_j^2}{\Gamma + (1/d_j)}\right] \tag{9.77}$$

where Γ is the smallest positive root of the equation

$$\left.\frac{d}{dz} f(z)\right|_{z = i\Gamma} = 0 \tag{9.78}$$

The notation $f''(i\Gamma)$ in Eq. (9.77) means the second derivative of $f(z)$ evaluated at $z = i\Gamma$. Equation (9.77) is true only when

$$\Gamma < \min_{d<0} \frac{1}{|d_i|}$$

a condition which is satisfied for all our applications. (See Mazo and Salz[13] for further discussion.)

Equation (9.77) was obtained under the condition that

$$\bar{\mathbf{w}}^T \mathbf{Q} \bar{\mathbf{w}} = \bar{\mathbf{y}}^T \mathbf{D} \bar{\mathbf{y}} = \sum_i d_i \bar{y}_i^2 > 0 \tag{9.79}$$

Physically this means that the quadratic form in the absence of noise is positive. Of course, the probability that the quadratic form q is greater than zero and that $\sum_i d_i \bar{y}_i^2$ is less than zero may be obtained analogously. Eq. (9.77) represents the area of the tail of the probability density of q when it is assumed that the bulk of its mass is concentrated at large positive values of q. For binary systems p is the actual probability of error. When the decision thresholds are set at values other than zero, the probability distribution of q, denoted by $F(\eta)$, is required in order to calculate error probabilities. However, this quantity may be calculated in an analogous manner as $F(0)$ since

$$F(\eta) = P[q < \eta] = -\frac{1}{2\pi i} \int_{-\infty}^{\infty} \frac{C(\omega)}{\omega + i\epsilon} e^{-i\omega\eta}\, d\omega \tag{9.80}$$

and the discussion in Mazo and Salz[13] of asymptotically calculating p applies to the above case as well.

Many practical error-rate calculations, as we have seen in the text, may be cast into the form where the quadratic form q is the difference of the squared lengths of two 2-dimensional gaussian vectors; that is,

$$q = |\mathbf{u}|^2 - |\mathbf{v}|^2 \tag{9.81}$$

where $\mathbf{u} = (v_1, v_2)$, and $\mathbf{v} = (u_1, u_2)$ and the variances of both components of the vector $\mathbf{u}$ are equal to σ_2^2 and of both those of $\mathbf{v}$ equal to σ_1^2. Furthermore, in these applications it was assumed that all the components are independent. We saw that the observable quantity in differential detection of FM and comparison detection of PM reduce to this case with, in general, $\sigma_1^2 \neq \sigma_2^2$.

For these applications, the characteristic function of q reduces to

$$C_q(\omega) = \frac{\exp i\omega\left[u^2/(1 - 2i\omega\sigma_2^2) - v^2/(1 + 2i\omega\sigma_1^2)\right]}{(1 + 2i\omega\sigma_2^2)\,(1 + 2i\omega\sigma_1^2)} \tag{9.82}$$

and

$$p = -\frac{1}{2\pi i}\int_{-\infty}^{\infty} \frac{d\omega}{\omega + i\epsilon}\; \frac{\exp\left(-\omega/2\sigma_1^2\right)\left[u^2/(\omega - iA) + v^2/A^2(\omega + i/A)\right]}{(\omega - iA)(\omega + i/A)} \tag{9.83}$$

where $A = \sigma_2/\sigma_1$

$u^2 = |\langle\mathbf{u}\rangle|^2$

$v^2 = |\langle\mathbf{u}\rangle|^2$

The asymptotic expansion of Eq. (9.83) is from Eqs. (9.77) and (9.78):

$$p \sim \frac{1}{\sqrt{2\pi}}\;\frac{\sigma_1}{\sqrt{1 + A^2}}\;\frac{(u - v)^2}{[u^2 + A^2 v^2 - uv(1 + A^2)]}\, e^{-(u - v)^2/2\sigma_1^2(1 + A^2)} \tag{9.84}$$

Often also necessary for evaluation purposes is the probability that the inner product q' of two independent two-dimensional gaussian vectors $\boldsymbol{\alpha}$ and $\boldsymbol{\beta}$

$$q' = 2(\boldsymbol{\alpha} \cdot \boldsymbol{\beta}) \tag{9.85}$$

is negative when the inner products of their means is positive. Here again the components of each vector are independent, and the variance of $\boldsymbol{\alpha}$ and $\boldsymbol{\beta}$ may differ. However, it is clear that multiplication of Eq. (9.85) by a positive constant leaves the desired probability invariant and can adjust the two variances to be the same with an appropriate modification of the mean values. Hence, it suffices to choose the variances to be equal to the same value σ^2.

Observe that q' in Eq. (9.85) equals q in Eq. (9.81), provided

$$u = \frac{1}{\sqrt{2}}(\alpha + \beta) \quad \text{and} \quad v = \frac{1}{\sqrt{2}}(\alpha - \beta) \tag{9.86}$$

The desired probability p now becomes, from Eq. (9.84) with $\sigma_1^2 = \sigma_2^2 = \sigma^2$ and $A = 1$,

$$p \sim \frac{\sigma}{2\sqrt{\pi}} \frac{1}{\sqrt{uv}} \frac{u + v}{u - v} e^{(u - v)^2/4\sigma^2} \tag{9.87}$$

Notes

In Cahn[5] the reader can find an extensive tabulation of the probability of error in digital phase-modulation systems. Chapter 10 in Bennett and Davey[9] gives a detailed derivation of the probability of error in digital phase modulation, while Chap. 11 compares the performance of various modulation systems. In Bussgang and Leiter the error performance of a transmitted-reference phase-shift-keying system is analyzed in detail. Kac and Siegert[15] and Emerson[16] include some early work on quadratic detectors.

10

Fundamental Properties of Error-Correcting Codes

10.1 Introduction

We have concluded that portion of the book dealing with modulation and signal processing. Attention will now be turned to the problem of coding. We saw in Chap. 3 that in order to attain virtually error-free transmission, the duration of the signals employed must be large. For a fixed rate the number of signals required increases exponentially with signal duration and even for modest durations becomes unmanageably large.

At present, digital coding is the only practical method of realizing the high reliability and efficiency known to be attainable with signals of long duration. As mentioned in Chap. 1, with this approach a relatively small set of analog signals (often two) is selected for transmission over the channel; this is the modulation process discussed in Chaps. 4 through 9. Then sequences of these basic channel signals are considered and a set of such sequences is chosen to form the information-carrying alphabet, or code.

Finding methods of selecting sequences of basic channel signals to form alphabets is the coding problem. As will be seen in this and the following chapters, several extensive classes of codes whose performance is reasonably close to optimum are known. More important, simply implemented encoding and decoding procedures have been devised for many of these codes so that the complexity of their encoders and decoders tends to increase linearly with, or at most as a small power of, the total signal duration (code length).

It has been shown (Chap. 3) that this separation of the signal-design process into the better-understood processes of modulation and coding results in some degradation in the performance attainable with a receiver of unlimited complexity and cost. However, for the real systems of today the reverse seems to be true: for a system with fixed total cost this separation usually improves performance.

This step toward reality replaces the general signal-design problem by what might be called the *channel-symbol-coding problem*. Although the latter problem is the more tractable of the two, where nonbinary channel symbols are used it is usually difficult enough to cause one to seek a further simplification even at the cost of another slight reduction in ultimately attainable performance. Such a simplification can be effected by using binary coding; thus real data-transmission systems often take the form depicted in Fig. 10.1.

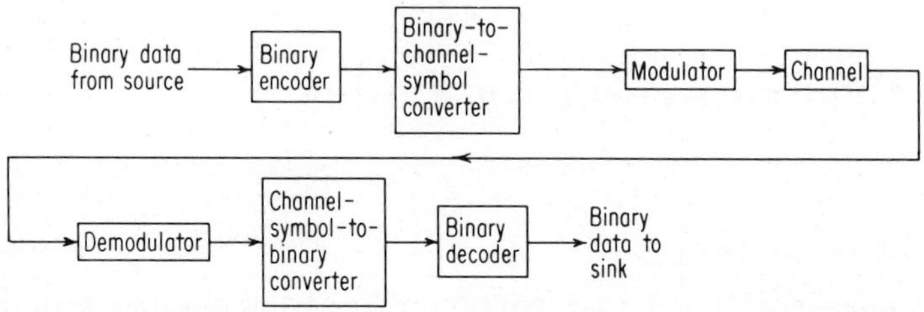

Fig. 10.1. Block diagram of data-transmission system using binary coding.

As pointed out in Chap. 3, unless the channel symbols themselves are binary, the performance of binary coding is slightly inferior to that attainable with ideal channel-symbol coding. However, because of our present limited understanding of nonbinary codes and the relative difficulty of building nonbinary machines, better performance can usually be attained with a fixed amount of coding equipment if binary coding is used. Thus in this and the following chapter we shall consider binary coding almost exclusively and shall mention other coding techniques only in passing.

In this chapter the notion of an error-correcting code is defined, and various fundamental properties of these codes are exhibited. We consider only parity check codes, and, as mentioned above, all codes are taken to be binary. The material presented in this chapter is necessary, and hopefully sufficient, background for a working understanding of the specific classes of codes and methods of implementation discussed in Chaps. 11 and 12. The scope and depth of the presentation are necessarily limited, and many interesting subjects have had to be omitted entirely. The reader interested in pursuing the general subject of coding or any of its specific applications is recommended to W. W. Peterson's book, "Error-Correcting Codes,"[1] or to the numerous technical papers on the subject in the references of Chaps. 11 and 12.

10.1.1 Methods of Correcting Errors

The essential function of a *receiver* is to classify the infinity of possible received signals into a finite number of information-carrying messages. A receiver composed of a demodulator and decoder accomplishes this classification in two steps. First the demodulator classifies sections of the received signal into a small number of classes. Then the *decoder* classifies the demodulator output sequence into a considerably larger number of classes corresponding to the possible transmitted messages. This latter classification can be, and usually is, interpreted as an error-correction process. That is, the decoder attempts to correct the noise-induced errors made by the demodulator.

If systems considerations permit, errors can be handled in an entirely different manner. Upon examining the demodulator output sequence, the decoder forwards it to the data sink if it corresponds to an acceptable message sequence. If it does not, the decoder either notifies the transmitter (via a reverse channel) that errors have occurred and that the corrupted message must be repeated, tags the erroneous message and delivers it to the sink, or simply discards it. A reverse channel is clearly unnecessary in the latter two situations. From the viewpoint of coding, these alternatives are identical and will be considered under the general heading of error detection.

These two techniques, forward-acting error correction and error detection, are usually used separately, but occasionally it may be necessary to use both in a particular system. For example, the receiver may decode a sequence only if it closely resembles an allowable combination of signals, and request a retransmission otherwise. Hybrid systems of this type are quite flexible and can sometimes be used where neither error-detection nor forward-acting correction alone will yield satisfactory performance. However, precisely because of this flexibility, analysis of such systems must be restricted to specific cases and consequently is of limited usefulness. Thus in the following discussion only passing mention is made of hybrid systems.

In most situations, systems considerations determine which, if either, of the above techniques is to be employed. For example, if data cannot be conveniently discarded and a reverse direction channel is unavailable, the first technique, forward-acting correction, must be employed. On the other hand, if a large reduction in error rate must be accomplished with a modest amount of equipment, detection-retransmission is indicated. However, in many situations either or both techniques can be employed. In these situations a detailed study of the merits of each method, in terms of the cost of implementation, average error rate, and average data rate, may be necessary to decide between them. Much of the material in this and the following chapter is directed toward helping the communications system designer-analyst make this type of decision intelligently.

Both error detection and error correction are possible because the received sequence of channel symbols contains redundancy or, in other words, because only a fraction of the possible received sequences correspond to messages and the remainder indicate that certain types of errors have occurred. The coding problem is to find ways of inserting this redundancy so as to minimize, or at least reduce substantially, the probability of delivering errors to the data sink. This problem is stated more precisely in Sec. 10.1.2.

10.1.2 Error-detecting and -correcting Codes

As noted in Chap. 2, from the point of view of coding the modulator, demodulator, and channel-symbol-to-binary converters can be taken as part of the channel. Thus the channel depicted in Fig. 10.1 reduces to that of Fig. 10.2. We denote by R_c the rate, in bits per second, at which binary digits can be transmitted through the channel. The source emits binary data at a rate of RR_c bps, $0 < R \leq 1$; redundancy is added to the data stream by the encoder at the rate of $(1-R)R_c$ bps; and the result is transmitted over the channel. The parameter R is called the *code rate* or *efficiency*.

Fig. 10.2. Channel model.

Again, as in previous chapters, it will be assumed that the data source is memoryless and emits 1s and 0s with equal probability; that is, the data is assumed to be random. In fact, this is often a pessimistic approximation to reality, but no attempt will be made to take advantage of any redundancy inherent in the source data.

The encoder can be thought of as a device which maps the bit stream emitted by the source onto another, somewhat longer, bit stream. In general no easily described set of rules for performing this mapping is necessary; rather it may be describable only by a double-entry table which relates an encoder output with each input. However, there are obvious advantages to encoding by means of a simple rule, provided this does not degrade the performance of the resulting code. For this reason only parity check codes in systematic form will be considered. This restriction usually results in little, if any, degradation in performance. In a *parity check code* the encoder accepts a block of k_0 information bits from the source, calculates a set of n_0 modulo-2* sums of various of these and possibly other past information bits, and shifts the block of n_0 sums out to the channel. The code efficiency is, of course, equal to k_0/n_0. In a *systematic parity check code* the first k_0 bits transmitted are identical to the information bits. Throughout the remainder of this

*In arithmetic modulo-2, 0 + 0 = 0, 1 + 0 = 0 + 1 = 1, and 1 + 1 = 0.

book we shall employ this convention of transmitting information digits first, then parity.

10.1.3 Types of Codes

If the n_0-k_0 parity bits check only the k_0 information bits immediately preceding them in time, the code is called a *block code*,[2] and an n_0-bit block is called a code word.

If the n_0-k_0 parity checks in a block also check information bits which appeared in preceding blocks, the code is called a *convolutional*,[3] or recurrent, code. The constraint length of the code, n, is n_0 times the number of blocks, m, checked by the n_0-k_0 parity checks. For block codes $n = n_0$; therefore the codes can be thought of as convolutional codes with $m = 1$. Briefly, a parity check code of either type with constraint length n and efficiency $R = k_0/n_0$ is usually referred to as an (n,k) *code*, where $k = Rn$.

As will be seen in Chaps. 11 and 12, the methods used to implement these two types of codes differ substantially. These differences are sometimes important enough to give one type of code a decided advantage over the other for a particular application. However, it will also be seen that both types of codes are subject to the same limitations and have, roughly speaking, the same inherent capabilities.

As desirable as it might be to unify the mathematical description of block and convolutional codes, such a description tends to be unnecessarily cumbersome. In the interest of clarity of presentation, therefore, we shall treat the two types of codes separately. Since block codes are, generally speaking, the better understood of the two, they will be described first. In describing convolutional codes an attempt will be made to relate them to block codes whenever convenient.

10.2 Block Codes

The codes under discussion here are systematic, binary, parity check codes with the property that the check bits in any block are sums of information bits in only that block, i.e., block codes. In this section various properties of these codes which follow as logical consequences of this definition are examined. Attention is also directed to a subclass of the class of parity-check codes, the cyclic codes. In Chap. 11 we shall see that the latter have several practically significant features.

10.2.1 Matrix Description

An (n,k) block code, $k \leq n$, is defined as the collection of 2^k n-tuples produced by encoding all possible k-tuples according to some predetermined set of parity check rules. Because the code is a parity check code, this encoding can be described conveniently in terms of matrices. In such a code the encoding of a data block into a code word can be represented mathematically by

$$\mathbf{c} = \mathbf{d}\mathbf{G} \tag{10.1}$$

where $\mathbf{c}$ is an n-bit code word represented as an n-place row vector (n-tuple) and $\mathbf{d}$ is a k-bit data block represented as a k-place row vector (k-tuple). The k-by-n *matrix* $\mathbf{G}$ is the *generator matrix* of the code and has the form

$$\mathbf{G} = [\mathbf{I}_k \mathbf{P}] \tag{10.2}$$

The matrix $\mathbf{I}_k$ is the identity matrix of order k, and $\mathbf{P}$ is an arbitrary k-by-$(n\text{-}k)$ matrix. Clearly the leftmost k symbols of $\mathbf{c}$ are identical to the corresponding symbols of $\mathbf{d}$, while the rightmost n-k symbols are modulo-2 sums of, or parity checks on, various symbols of $\mathbf{d}$. Specifically, the ith parity bit of $\mathbf{c}$, $1 \le i \le n\text{-}k$, equals the sum of the symbols of $\mathbf{d}$ which correspond to 1s in the ith column of $\mathbf{P}$. Alternately a code word can be considered to be the modulo-2 place-by-place sum of the rows of $\mathbf{G}$ corresponding to 1s in $\mathbf{d}$. Mathematically the code is the row space of $\mathbf{G}$.

It is easily verified that the place-by-place, modulo-2 sum of any two code words, that is, n-tuples which satisfy Eq. (10.1) is another code word. Furthermore, because the sum of any number of 0s is zero, the all-zero code word is formed by encoding the all-zero k-tuple and hence is a code word in every code. In the binary case these two facts are sufficient to guarantee that the set of code words form a *group* under the operation of place-by-place, modulo-2 addition. Hence every binary parity-check code is a *linear*, or *group*, *code* as well.[4]

Example 10-1 illustrates the generator-matrix concept.

Example 10-1. A (15,11) code is specified by the 11-by-15 generator matrix shown below:

$$\mathbf{G} = [\mathbf{I}_k \mathbf{P}] = \begin{bmatrix}
1&0&0&0&0&0&0&0&0&0&0&&1&0&0&1\\
0&1&0&0&0&0&0&0&0&0&0&&1&1&0&1\\
0&0&1&0&0&0&0&0&0&0&0&&1&1&1&1\\
0&0&0&1&0&0&0&0&0&0&0&&1&1&1&0\\
0&0&0&0&1&0&0&0&0&0&0&&0&1&1&1\\
0&0&0&0&0&1&0&0&0&0&0&&1&0&1&0\\
0&0&0&0&0&0&1&0&0&0&0&&0&1&0&1\\
0&0&0&0&0&0&0&1&0&0&0&&1&0&1&1\\
0&0&0&0&0&0&0&0&1&0&0&&1&1&0&0\\
0&0&0&0&0&0&0&0&0&1&0&&0&1&1&0\\
0&0&0&0&0&0&0&0&0&0&1&&0&0&1&1
\end{bmatrix} \tag{10.3}$$

Encoding the data sequence $\mathbf{d} = [1\ 0\ 0\ 0\ 1\ 0\ 0\ 1\ 0\ 1\ 0]$ using Eq. (10.1) gives

$$\mathbf{c} = \mathbf{dG} = [1\ 0\ 0\ 0\ 1\ 0\ 0\ 1\ 0\ 1\ 0\quad 0\ 0\ 1\ 1] \tag{10.4}$$

Note that the first 11 bits of $\mathbf{c}$ are identical to $\mathbf{d}$ and that the last four bits of $\mathbf{c}$ are the sums of the bits of $\mathbf{d}$ specified by the last four columns of $\mathbf{G}$. Thus, for example, the first parity bit is the sum of the first, second, third, fourth, sixth, eighth, and ninth bits of $\mathbf{d}$, that is, $= 1 + 0 + 0 + 0 + 0 + 1 + 0 = 0$. Equivalently, $\mathbf{c}$ can be thought of as the sum of the rows of $\mathbf{G}$ which correspond to ones in $\mathbf{P}$. Thus in the example $\mathbf{c}$ is the sum of the first, fifth, eighth, and tenth rows of $\mathbf{G}$.

As defined, a code word in a systematic linear code is an n-tuple with the property that the n-k subsets of the k information bits of the word specified by the rightmost n-k columns of the generator matrix add to the corresponding parity check bit. This can be stated more elegantly as follows: $\mathbf{c}$ is a code word in the code generated by the matrix $\mathbf{G}$ if and only if

$$\mathbf{cH}^T = \mathbf{0} \tag{10.5)*}$$

where $\mathbf{H}^T$ is the transpose of the $(n\text{-}k)$-by-n *parity-check matrix*

$$\mathbf{H} = [\mathbf{P}^T \mathbf{I}_{n-k}] \tag{10.6}$$

Since

$$\mathbf{GH}^T = [\mathbf{I}_k \mathbf{P}]\begin{bmatrix}\mathbf{P} \\ \mathbf{I}_{n-k}\end{bmatrix} = \mathbf{0} \tag{10.7}$$

it follows that the inner, or dot, product of any row or sum of rows of $\mathbf{G}$, that is, any code word, and any row or sum of rows of $\mathbf{H}$ must equal zero. On the other hand, if Eq. (10.5) holds, the n-k parity-check equations specified by $\mathbf{P}$ are satisfied, and $\mathbf{c}$ must be a code word.

As a consequence of this result, for any n-tuple $\mathbf{e}$ that is not a code word

$$\mathbf{eH}^T = \mathbf{s} \neq \mathbf{0} \tag{10.8}$$

The $(n\text{-}k)$-tuple $\mathbf{s}$ is referred to as the *syndrome* of the n-tuple; clearly every n-tuple has one and only one syndrome. A 1 in the

---

* In matrix equations of this type, the symbol $\mathbf{0}$ will be used to denote the all-zero matrix of the proper dimensions.

ith position of the syndrome indicates that the parity-check equation specified by the ith column of $\mathbf{H}^T$ failed. Thus adding the syndrome to the parity section of $\mathbf{e}$ yields a code word.

Throughout this and the following chapter we shall denote by $\mathbf{e}_d$ and $\mathbf{e}_p$ the data and parity sections of $\mathbf{e}$, respectively. Equation (10.8) gives

$$\mathbf{s} = [\mathbf{e}_d \mathbf{e}_p]\begin{bmatrix}\mathbf{P} \\ \\ \mathbf{I}_{n-k}\end{bmatrix} = \mathbf{e}_d\mathbf{P} + \mathbf{e}_p \tag{10.9}$$

But $\mathbf{e}_d[\mathbf{P}]$ is simply the parity section of the code word into which $\mathbf{e}_d$ is mapped by Eq. (10.1). Thus the syndrome of any n-tuple can be formed by encoding the data section of the n-tuple and adding the resulting parity bits to the parity section of the n-tuple.

The notions of parity check matrix and syndrome should be clarified by Example 10-2.

Example 10-2. The parity-check matrix of the (15,11) code discussed in Example 10-1 is

$$\mathbf{H} = [\mathbf{P}^T \mathbf{I}_{n-k}] = \begin{bmatrix} 1\,1\,1\,1\,0\,1\,0\,1\,1\,0\,0 & 1\,0\,0\,0 \\ 0\,1\,1\,1\,1\,0\,1\,0\,1\,1\,0 & 0\,1\,0\,0 \\ 0\,0\,1\,1\,1\,1\,0\,1\,0\,1\,1 & 0\,0\,1\,0 \\ 1\,1\,1\,0\,1\,0\,1\,1\,0\,0\,1 & 0\,0\,0\,1 \end{bmatrix} \tag{10.10}$$

The n-tuple $\mathbf{e}$ = [0 0 1 0 0 0 0 0 0 0 0 1 0 0 0] is not a code word for its syndrome

$$\mathbf{e}\mathbf{H}^T = 0\,1\,1\,1 \neq 0\,0\,0\,0 \tag{10.11}$$

That adding this syndrome to the n-tuple $\mathbf{e}$ produces a code word should be clear. If it is not, the reader should perform the necessary matrix multiplication.

Encoding the data section of $\mathbf{e}$ by means of Eq. (10.3) yields the code word [0 0 1 0 0 0 0 0 0 0 0 1 1 1 1]. Adding the parity section of $\mathbf{e}$ to the parity section of this code word produces the syndrome given in Eq. (10.11), as it must.

10.2.2 Hamming Weight and Distance

At this point we introduce some of the basic terminology which will be used in the sequel. The *(Hamming) weight* of an n-tuple is the number of zero elements it contains. A concept closely related to weight is that of distance. The *(Hamming) distance* between two

n-tuples is the weight of their sum, or in other words, the number of positions in which they differ.

The minimum weight of any nonzero word in a linear code, commonly referred to as the minimum weight of the code, has special significance in coding theory. First of all, since the all-zero n-tuple is a code word, it is clear that the minimum distance between code words cannot be greater than the weight of the minimum-weight nonzero code word. But if it is less, the sum of the two closest code words is a word of weight lower than the minimum weight, which is impossible. This yields Theorem 10-1.

Theorem 10-1. In a linear block code the weight of the minimum-weight nonzero code word equals the minimum distance between code words; i.e., minimum weight equals minimum distance.

From the point of view of analysis this result is quite important. For one thing, it guarantees that no loss of generality is incurred by considering the all-zero code word, rather than an arbitrary code word, when making deductions about many of the properties of linear codes.

Another significant aspect of the minimum weight of a code is the following. If some nonzero word $\mathbf{e}_w$ in a linear code has weight exactly w and there are no words of lower weight, by Eq. (10.5)

$$\mathbf{e}_w \mathbf{H}^T = 0 \tag{10.12}$$

In other words the w columns of $\mathbf{H}$ (rows of $\mathbf{H}^T$) that correspond to the w 1s of $\mathbf{e}_w$ add to zero. Similarly no fewer than w colums can add to zero since this would imply that a nonzero code word with weight less than w existed. For emphasis, this is stated in Theorem 10-2.

Theorem 10-2. In a code with minimum weight exactly w there exists at least one set of w columns, and no set of fewer than w columns, of the parity check matrix of the code which add to ezro.

This result relates the important parameter minimum distance to a reasonably tractable mathematical property of the parity-check matrix. We shall use this result in Chap. 11 to prove that certain cyclic codes have relatively large minimum distance.

Note the similarity and difference between the minimum distance associated with the matrix $\mathbf{H}$ and the rank of matrix. Minimum distance is the smallest number of columns which add to zero; rank is the largest number of columns such that no subset of the set adds to zero. Intuitively one would expect the two quantities to be related. We shall see that they are and that large minimum distance can in general be attained only by increasing the rank of H, i.e., the number of check bits per code word.

10.2.3 The Standard Array

Consider the set of n-tuples which have the same syndrome, say $\mathbf{e}\mathbf{H}^T$, where $\mathbf{e}$ is an arbitrary member of the set. No two n-tuples formed by adding different code words $\mathbf{c}$ and $\mathbf{c}'$ to $\mathbf{e}$ can be identical because if $(\mathbf{e}+\mathbf{c}) = (\mathbf{e}+\mathbf{c}')$, then $\mathbf{c} = \mathbf{c}'$, which is impossible. Each of these n-tuples must be in the set of n-tuples having the same syndrome as $\mathbf{e}$ since

$$(\mathbf{e} + \mathbf{c})\mathbf{H}^T = \mathbf{e}\mathbf{H}^T \tag{10.13}$$

and consequently at least 2^k n-tuples must be in the set. But if two n-tuples $\mathbf{e}'$ and $\mathbf{e}$ have the same syndrome, that is, $\mathbf{e}'\mathbf{H}^T = \mathbf{e}\mathbf{H}^T$, $(\mathbf{e}'+\mathbf{e})\mathbf{H}^T = 0$ and $\mathbf{e}+\mathbf{e}'$ is a code word. Thus exactly 2^k n-tuples have the same syndrome. This set of n-tuples forms a *coset* of the (group) code. Now every $(n\text{-}k)$-tuple is a syndrome; in particular it is the syndrome of the n-tuple whose first k positions are zero and whose last $n\text{-}k$ positions match the $(n\text{-}k)$-tuple. Then since every n-tuple has exactly one syndrome and since there are 2^{n-k} syndromes, the cosets are disjoint.

That the collection of n-tuples can be broken down into 2^{n-k} disjoint subsets as above suggests visualizing the collection as a rectangular array with the following properties. The first row of the array consists of the 2^k code words; the leftmost element of this row is customarily taken to be the all-zero word. Each other row consists of the coset of n-tuples formed by adding any n-tuple which is not a word to each code word. The leftmost element of the row formed in this manner is the original n-tuple and is referred to as the *coset leader*. By choosing different n-tuples for coset leaders, it is possible to construct numerous different arrays. This type of array, usually referred to as a *standard array*[4] or *coset decomposition* of the code, is a powerful tool for analyzing codes. The example below explicitly illustrates the properties of a standard array.

Example 10-3. The eight code words in the (6,3) linear block code generated by

$$\mathbf{G} = \begin{bmatrix} 1\,0\,0 & 1\,1\,0 \\ 0\,1\,0 & 0\,1\,1 \\ 0\,0\,1 & 1\,0\,1 \end{bmatrix} \tag{10.14}$$

are (000 000), (100 110), (010 011), (001 101), (110 101), (011 110), (101 011), and (111 000). One coset decomposition of this code is

Syndrome	Coset leader							
000	000000	100110	010011	001101	110101	011110	101011	111000
110	100000	000110	110011	101101	010101	111110	001011	011000
011	010000	110110	000011	011101	100101	001110	111011	101000
101	001000	101110	011011	000101	111101	010110	100011	110000
100	000100	100010	010111	001001	110001	011010	101111	111100
010	000010	100100	010001	001111	110111	011100	101001	111010
001	000001	100111	010010	001100	110100	011111	101010	111001
111	100001	000111	110010	101100	010100	111111	001010	011001

Note that each of the sixty-four 6-tuples appears in the array once and only once and that the syndromes associated with the eight cosets exhaust the set of 3-tuples. In this case the coset leaders were chosen to have the lowest weight possible. If we had so desired, any other n-tuple in any coset could have been chosen as leader.

The coset decomposition can be used as a decoding table. Just as encoding consisted of associated a code word with a given data k-tuple, decoding consists of associating a code word (or k-tuple) with each received n-tuple. For channels which introduce errors independently of which code word is transmitted, the probability of erroneous decoding is minimized by associating with each code word the n-tuples formed by adding a fixed set of error patterns to that code word, provided code words are equally likely to be transmitted. (For other channels this is not always true; because it is by far the simplest decoding to implement, however, we shall always assume that this type of decoding is employed.) That is, these error patterns and no others are correctable regardless of which code word is transmitted. The collection of coset leaders in a standard array (or a subset thereof) can be regarded as such a set. Thus the standard array can be used as a decoding table by simply locating the received n-tuple in the table and associating with this n-tuple the code word heading the column in which the n-tuple appears.

A simple way of performing this association, and one that is occasionally used in practice, is to calculate the syndrome of the received n-tuple, find by a trial-and-error search the error pattern (coset leader) associated with this syndrome, and then form the code word by adding this pattern to the received n-tuple. Except for the trial-and-error step, this decoding procedure is very simply implemented. Unfortunately there are 2^{n-k} correctable error patterns, and even for modest values of n and k the number of operations involved in the search is impractically large. The several decoding algorithms presented in Chap. 11 have been devised specifically to make this search unnecessary.

Throughout the remainder of this book only decoding procedures that, as above, can be visualized in terms of a coset decomposition will be considered. One fact about this type of decoding should be clear. Every coset has only one syndrome and one coset leader, i.e., correctable error pattern. Thus if two n-tuples appear in the same coset of a particular code, at least one must be an uncorrectable error pattern. This fact will be used extensively in the sequel.

One important result which can be visualized conveniently in terms of the coset decomposition is contained in Theorem 10-3.

Theorem 10-3. The sum of any two n-tuples is a code word in an (n,k) block code if and only if the n-tuples are in the same coset.

Let $(\mathbf{e}+\mathbf{c})$ and $(\mathbf{e}+\mathbf{c}')$ denote $\mathbf{e}$-tuples in the coset which has the pattern $\mathbf{e}$ as coset leader. Then since

$$(\mathbf{e} + \mathbf{c}) + (\mathbf{e} + \mathbf{c}') = \mathbf{c} + \mathbf{c}' \tag{10.15}$$

the sum of any two n-tuples in a coset is a code word. Similarly if two patterns $\mathbf{e}$ and $\mathbf{e}'$ add to a code word $\mathbf{c}$, they must be in the same coset. Thus if $\mathbf{e} + \mathbf{e}' = \mathbf{c}$, $\mathbf{e}' = \mathbf{c} + e$; by Eq. (10.13)

$$(\mathbf{e} + \mathbf{c})\mathbf{H}^T = \mathbf{e}'\mathbf{H}^T = \mathbf{e}\mathbf{H}^T \tag{10.16}$$

This theorem has several immediate and interesting consequences. For example, in a code with minimum distance d, no two n-tuples of weight $t = [(d-1)/2]$‡ or less can be in the same coset, and in at least one coset two n-tuples of weight not greater than $t+1$ appear. Thus it is possible to choose as coset leaders all error patterns of weight t or less. With this choice the code is capable of correcting all error patterns of weight t, but not all of weight $t + 1$. In brief, the code is t*-error-correcting*.

As another application of Theorem 10-3 we define a burst of length x to be a set of x bits, the first and last of which are 1s. Obviously, every code word in a linear code has all its 1s confined to a burst of length n or less. Let l denote the length of the shortest burst that in a code word. Now consider the coset which contains the n-tuple which matches this code word in the first $[(l+1/2]$ positions and is 0 elsewhere. By Theorem 10-3 this coset contains two n-tuples, each of whose 1s are confined to a burst of length not greater than $[(l+1)/2]$. Since only one can be decoded correctly, an uncorrectable error pattern exists in which errors are confined to $[(l+1)/2]$ consecutive positions. That is,

$$b \leq \frac{l-1}{2} \tag{10.17}$$

where b denotes the maximum guaranteed-correctable burst length or, more simply, the *burst-correcting ability* of the code.

‡ The symbol $[x]$ denotes the largest integer not greater than x.

These two results will prove quite useful in Chap. 11, where the random-error and burst-correcting abilities of block codes are considered in detail.

We have pointed out that error patterns which are coset leaders are correctable; all others are not. Consequently it is desirable to construct the coset decomposition so as to make the 2^{n-k} most likely error patterns the coset leaders. Although it is sometimes impossible to do this, usually it is not. For example, if two of the most likely error patterns happen to add to a code word, by Theorem 10-3 they are in the same coset; thus one must be uncorrectable. For certain mathematical channels, such as the binary symmetric channel, codes can be designed for which the probability of an error pattern occurring that is not a coset leader is extremely low. However, on real channels it is often difficult to order the error patterns and even more difficult to design codes to take advantage of any but the simplest orderings. As a consequence, usually no real attempt is made to match codes to channels exactly. Instead, codes are used which are known to perform well on simple mathematical models, models that in some sense approximate the real channels. Several specialized classes of codes that do this are subclasses of the class of cyclic codes, which we now discuss.

10.2.4 Cyclic Codes

Almost all of the research in block-coding theory has been directed toward the problem of choosing the matrix P [see Eq. (10.2)] so as to ensure that the code determined by P and generated by G has certain desirable properties. Among the properties which have been sought are general ease of implementation, ability to correct large numbers of random errors, ability to correct long bursts of errors, ability to correct losses of synchronization, etc. Much of this research has been concentrated on a small subclass of the class of linear codes, the cyclic codes.[6] These codes have two important advantages over most other types of codes. First of all, the codes possess a fair amount of mathematical "structure." This places several powerful tools in the hands of the coding theorist and allows him to design codes with useful error-correcting properties. Secondly, the cyclic nature of the codes usually simplifies the required encoding and decoding equipment considerably. With few exceptions all block codes considered hereafter will be cyclic codes or codes simply derived from them.

A *cyclic* (n,k) *code*, a linear block code of length n having k information symbols, has the property that every cyclic shift of a code word is another code word. That is, if

$$\mathbf{c} = \{c_{n-1}, c_{n-2}, \ldots, c_0\} \tag{10.18}$$

is a code word, so are

$$\{c_{n-2}, c_{n-3}, \ldots, c_{n-1}\}$$
$$\{c_{n-3}, c_{n-4}, \ldots, c_{n-2}\}$$
$$\vdots$$
$$\{c_0, c_{n-1}, \ldots, c_1\} \tag{10.19*}$$

This formulation of cyclic codes suggests treating the elements of each code word as coefficients of a polynomial of degree $n-1$. With this convention, the code word c can be represented as a code polynomial; that is,

$$c(x) = c_{n-1}x^{n-1} + c_{n-2}x^{n-2} + \cdots + c_1 x + c_0 \tag{10.20}$$

Hereafter we shall use the terms *code word* and *code polynomial* interchangeably. The condition that every cyclic shift of a code word be another code word can be expressed concisely as follows. That $c(x)$ is a code word implies that $x^i c(x)$ modulo-(x^n+1) is also a code word for all i. The reader may wish to verify that multiplication by x-modulo-(x^n+1) results in a cyclic shift and that the coefficients of $x^i c(x)$-modulo-(x^n+1) are, in fact,

$$\{c_{n-i-1}, c_{n-i-2}, \ldots, c_1, c_0, c_{n-1}, \ldots, c_{n-i}\} \tag{10.21}$$

Several interesting properties of the generator matrix of a cyclic code can be deduced quite simply from the above definition. Consider the kth row of the matrix G defined previously and assume that the code generated by G is cyclic. This row, which has 0s in its first $k-1$ positions and a 1 in the kth position, must also have a 1 in the nth position. Otherwise, a single shift to the right would produce a code word with an all-zero information section and a nonzero parity section, and this is impossible. Hence, this polynomial, denoted by $g(x)$ and referred to as the *generator polynomial* of the code, has two terms with coefficient 1 separated by $n-k-1$ terms with undetermined coefficients and consequently has degree n-k. This is the only code polynomial of degree n-k because if there were another, when added to $g(x)$ it would produce a code word of degree less than n-k, which is impossible.

The $(k-1)$th row of the generator matrix of a cyclic code is formed by cyclically shifting the kth row once to the left, and, if the kth entry in the result is 1, adding the kth row to it. That is, the $(k-1)$th row equals $xg(x)$ if the coefficient of x^{n-k-1} in $g(x)$ is 0, and $xg(x) + g(x) = (x+1)g(x)$ if it is 1. In like manner the $(k-j)$th row can be formed from the row below it, and assuming that the row was a

---

*It is not obvious, at this point, that such codes exist. The reader seeking reassurance is referred to Example 10-1 for an example of a parity-check code with the cyclic property.

multiple of $g(x)$, the $(k-j)$th row will be also. Hence, every row of G, when represented as a polynomial of degree less than n, is divisible by $g(x)$, the polynomial representation of the kth row. However, since every code word is the sum of rows of G, every code polynomial must be divisible by $g(x)$; consequently $g(x)$ completely specifies every word in the code.

Since $g(x)$ has degree n-k, there are 2^k polynomials, i.e., those of degree less than k, which can be multiplied by $g(x)$ to yield a polynomial of degree less than n. Clearly there is a 1-to-1 correspondence between these polynomials and the 2^k code words in the code.

Example 10-4 is intended to illuminate the preceding discussion.

Example 10-4. The (15,11) code examined in Examples 10-1 and 10-2 is a cyclic code. Written in polynomial form, its generator matrix is

$$
\mathrm{G} = [\mathrm{I}_k\mathrm{P}] = \begin{bmatrix}
x^{14} & & & & & & & & & & & x^3 & & & 1 \\
& x^{13} & & & & & & & & & & x^3 & x^2 & & 1 \\
& & x^{12} & & & & & & & & & x^3 & x^2 & x & 1 \\
& & & x^{11} & & & & & & & & x^3 & x^2 & x & \\
& & & & x^{10} & & & & & & & & x^2 & x & 1 \\
& & & & & x^9 & & & & & & x^3 & & x & \\
& & & & & & x^8 & & & & & & x^2 & & 1 \\
& & & & & & & x^7 & & & & x^3 & & x & 1 \\
& & & & & & & & x^6 & & & x^3 & x^2 & & \\
& & & & & & & & & x^5 & & & x^2 & x & \\
& & & & & & & & & & x^4 & & & x & 1
\end{bmatrix}
$$

$$
= \begin{bmatrix}
(x^{10} + x^7 + x^6 + x^4 + x^2 + x + 1)\, g(x) \\
(x^9 + x^6 + x^5 + x^3 + x + 1)\, g(x) \\
(x^8 + x^5 + x^4 + x^2 + 1)\, g(x) \\
(x^7 + x^4 + x^3 + x)\, g(x) \\
(x^6 + x^3 + x^2 + 1)\, g(x) \\
(x^5 + x^2 + x)\, g(x) \\
(x^4 + x + 1)\, g(x) \\
(x^3 + 1)\, g(x) \\
x^2\, g(x) \\
x g(x) \\
g(x)
\end{bmatrix} \tag{10.22}
$$

where $g(x) = x^4 + x + 1$. (To simplify the writing, terms with coefficient 0 are usually omitted in matrices of this type.)

Encoding a k-bit data block by multiplying it by the generator matrix G of Eq. (10.1) is equivalent to the following polynomial operation. The polynomial representation of the information block, denoted by $d(x)$, has degree less than k; therefore $x^{n-k}d(x)$ has degree less than n. Also

$$\frac{x^{n-k}d(x)}{g(x)} = q(x) + \frac{r(x)}{g(x)} \tag{10.23}$$

where $q(x)$ has degree less than k and $r(x)$ has degree less than n-k, the degree of $g(x)$. Thus the polynomial $x^{n-k}d(x) + r(x)$ is divisible by $g(x)$ and is a code word in the code generated by $g(x)$.

This word consists of the unaltered k-bit information block followed by n-k linear combinations of the information bits. It must be identical to the word formed by multiplying the k-place vector d by the matrix G since, as we have seen, there is only one code polynomial of degree n-k in a cyclic (n,k) code.

The syndrome associated with a given n-tuple can be calculated in a similar manner. Let $e(x) = e_d(x) + e_p(x)$ be the polynomial representation of such an n-tuple. As pointed out earlier, the syndrome is obtained by encoding the information section of an n-tuple and adding the resulting check bits to the corresponding bits of the parity section of the n-tuple. That is, the syndrome $s(x)$ is given by

$$s(x) = e_p(x) + r(x) \tag{10.24}$$

where $r(x)$ is the remainder obtained by dividing $e_d(x)$ by $g(x)$. But since $e_p(x)$ has degree less than $g(x)$, this is identical to the remainder obtained by dividing $e(x) = e_d(x) + e_p(x)$ by $g(x)$. That is,

$$s(x) = \text{rem}\,\frac{e(x)}{g(x)} \tag{10.25}$$

Of course, since $g(x)$ divides every code word $c(x)$, the syndrome is identical to that of $e(x) + c(x)$. As will be seen in Chap. 11, encoding and syndrome calculation can be implemented in a remarkably simple manner as a consequence of Eqs. (10.23) and (10.25).

Examples 10-1 and 10-4 are continued here to clarify the encoding and syndrome calculation procedures just described for cyclic codes.

Example 10-5. If the cyclic properties of the (15,11) code discussed earlier are ignored, an 11-bit data block can be encoded by multiplying it by the generator matrix of the code (Example 10-1). However, because the code is cyclic (Example 10-4), encoding can also be performed by using Eq. (10.23). Since $g(x) = x^4 + x + 1$,

$$c(x) = x^4 d(x) + r(x) \tag{10.26}$$

where
$$r(x) = \text{rem}\ \frac{x^4 d(x)}{x^4 + x + 1} \tag{10.27}$$

As we did earlier [Eq. (10.4)], let us encode the data sequence [10001001010]. From Eq. (10.27)

$$r(x) = \text{rem}\ \frac{x^{14} + x^{10} + x^7 + x^5}{x^4 + x + 1} = x + 1 \tag{10.28}$$

and
$$c(x) = x^{14} + x^{10} + x^7 + x^5 + x + 1 \tag{10.29}$$

Note that this result is identical to that obtained earlier when the cyclic properties of the code were ignored.

The syndrome of a given n-tuple, say, e = [00100000000 1000], can be calculated using Eq. (10.25). Thus

$$s(x) = \text{rem}\ \frac{e(x)}{g(x)} = \frac{x^{12} + x^3}{x^4 + x + 1} = x^2 + x + 1 \tag{10.30}$$

which is the identical result obtained by matrix multiplication in Example 10-2.

Another interesting property of cyclic codes can be exhibited as follows. The generator matrix of a cyclic code has the form

$$\mathbf{G} = [\mathbf{I}_k \mathbf{P}] \tag{10.31}$$

while its parity check matrix is

$$\mathbf{H} = [\mathbf{P}^T \mathbf{I}_{n-k}] \tag{10.32}$$

As we pointed out in defining the parity-check matrix, the dot product of any row of H and any row or sum of rows of G is 0. In particular the dot product of the top row of the matrix H and every cyclic shift of the bottom row of G is 0. But this is a necessary and sufficient condition that

$$g(x)h(x) \equiv 0 \bmod x^n + 1 \tag{10.33}$$

where $h(x)$ is the polynomial representation of the top row of H with the order of the components reversed; i.e., the ith component replaces the $(n\text{-}i)$th. Observe that $g(x)$ has degree $n\text{-}k$ and $h(x)$ had degree k; thus

$$g(x)h(x) = x^n + 1 \tag{10.34}$$

and $g(x)$ divides $x^n + 1$.

This argument can be reversed in a straightforward manner. Let $g(x)$ represent a divisor of x^n+1 of degree n-k. Then the 2^k multiples of $g(x)$ of degree less than n form a group under the operation polynomial addition. (This is exactly equivalent to the statement made earlier that the code words of a linear code form a group under the operation place-by-place modulo-2 addition.) Also, if $c(x) = i(x)g(x)$ denotes one such multiple, since multiplication by x is equivalent to a cyclic shift, and since

$$\begin{aligned} xc(x) &= xi(x)g(x) && \text{degree } c(x) \leq n-2 \\ &= xi(x)g(x) + (x^n+1) && \text{degree } c(x) = n-1 \end{aligned} \qquad (10.35)$$

it follows that $g(x)$ divides $xc(x)$. In other words every cyclic shift of a multiple of $g(x)$ is another multiple. But this and the group property are all that are needed for the set of multiples to form a cyclic (n,k) code. Thus we have proved Theorem 10-4, which is as follows.

Theorem 10-4. Every cyclic (n,k) code is generated by a divisor of x^n+1 of degree n-k; conversely, every such divisor generates a cyclic (n,k) code.[1,6]

If $g(x)$ divides x^m+1 as well as x^n+1 (where $m < n$), x^m+1 is a code word in the cyclic code of length n generated by $g(x)$. To eliminate this obviously undesirable situation (minimum weight 2) n will henceforth be taken as the smallest integer such that $g(x)$ divides x^n+1, except for the case of $g(x) = x+1$. This guarantees that every code under consideration having more than one parity check has minimum weight 3 and hence by Theorem 10-3 is at least single-error-correcting.

Theorem 10-4 states that the divisors of x^n+1 specify all of the cyclic codes of length n. A considerable body of knowledge of the properties of these polynomials exists. For example, divisors have been tabulated for all values of n of interest to us here.[1] With such a table the generator polynomial of almost any interesting cyclic code can be determined readily. By selecting appropriate divisors to form $g(x)$, it is possible to produce codes capable of correcting random errors, bursts of errors, losses of synchronization, or combinations of these types of errors. Also, the implementation of these codes is often reasonably simple as a consequence of their cyclic structure. Both these points will be explored in more detail in Chap. 11.

In previous examples we considered a cyclic (15,11) code. In the following one we investigate the class of cyclic codes of length 15.

Example 10-6. Theorem 10-4 states that every cyclic code of length 15 is generated by a divisor of $x^{15}+1$. In Table 11.3 we find that $x^{15}+1$ has five nontrivial factors, that is,

$$x^{15} + 1 = (x + 1)(x^2 + x + 1)(x^4 + x + 1)(x^4 + x^3 + 1)(x^4 + x^3 + x^2 + x + 1) \tag{10.36}$$

Hence there are at most

$$\binom{5}{1}^{*} + \binom{5}{2} + \binom{5}{3} + \binom{5}{4} = 30$$

cyclic codes of length 15. The reader should verify that because of the restriction that $g(x)$ does not divide $x^m + 1$, $m < n$, except when $g(x) = x + 1$, there are actually only 26 nontrivial codes of length 15. The generator polynomial and minimum distance of some of these codes are listed in Table 10.1.

Table 10-1. Important Parameters of Several Cyclic Codes of Length 15.

Number of information bits, k	Minimum distance, d	Generator polynomial, $g(x)$
14	2	(1, 0)[a]
11	3	(4, 1, 0)[a]
10	4	(1, 0) (4, 1, 0)
7	5	(4, 1, 0) (4, 3, 2, 1, 0)
6	6	(1, 0) (4, 1, 0) (4, 3, 2, 1, 0)
5	7	(4, 1, 0) (4, 3, 2, 1, 0) (2, 1, 0)
4	8	(1, 0) (4, 1, 0) (4, 3, 2, 1, 0) (2, 1, 0)
2	10	(1, 0) (4, 1, 0) (4, 3, 2, 1, 0) (4, 3, 0)
1	15	(4, 1, 0) (4, 3, 2, 1, 0) (4, 3, 0) (2, 1, 0)

[a] The symbol (1, 0) (4, 1, 0) represents the polynomial $(x + 1)(x^4 + x + 1) = x^5 + x^4 + x^2 + 1$.

Note that $d \geq 3$ provided that $g(x)$ does not divide $x^m + 1$, $m < 15$, but that the code generated by $x + 1$ (which divides $x + 1, x^2 + 1, \ldots, x^{14} + 1$) has minimum distance 2.

The polynomial $h(x)$ defined in Eq. (10.34) is referred to as the *parity check polynomial* of the code generated by $g(x)$. It is related to the matrix H with the order of its columns reversed in exactly the same manner as $g(x)$ is to G. In fact, if $h(x)$ is considered a generator polynomial, the $(n, n\text{-}k)$ cyclic code generated by $h(x)$ is referred to as the *dual code* of the (n, k) code generated by $g(x)$. Thus in Table 10.1 the first code is the dual of the last, the second the dual of the seventh, and the third the dual of the sixth. That is,

---

* $\binom{n}{i} = n!/(i!(n - 1)!)$.

the parity check matrix of the cyclic code geerated by $g(x)$ can be formed by reversing the column positions of the generator matrix of the dual code generated by $h(x) = (x^n + 1)/g(x)$. These points are illustrated in Example 10-7.

Example 10-7. The parity-check polynomial of the (15,11) cyclic code discussed in Example 10-5 is

$$h(x) = \frac{x^n + 1}{g(x)} = \frac{x^{15} + 1}{x^4 + x + 1} = x^{11} + x^8 + x^7 + x^5 + x^3 + x^2 + x + 1$$

$$= (x + 1)(x^2 + x + 1)(x^4 + x^3 + 1)(x^4 + x^3 + x^2 + x + 1) \tag{10.37}$$

If $h(x)$ is considered a generator polynomial, the corresponding generator matrix is

$$\begin{bmatrix} x^{14} & & & & x^{10} & & & x^7 & x^6 & & x^4 & & x^2 & x & 1 \\ & x^{13} & & & x^{10} & x^9 & & x^7 & & x^5 & x^4 & x^3 & x^2 & & \\ & & x^{12} & & & x^9 & x^8 & & x^6 & & x^4 & x^3 & x^2 & x & \\ & & & x^{11} & & & x^8 & x^7 & & x^5 & & x^3 & x^2 & x & 1 \end{bmatrix}$$

$$= \begin{bmatrix} (x^3 + 1)\,h(x) \\ x^2\,h(x) \\ xh(x) \\ h(x) \end{bmatrix} = \begin{bmatrix} \mathbf{I}_{n-k}\,\overline{\mathbf{P}}^T \end{bmatrix} \tag{10.38}$$

where $\overline{\mathbf{P}}^T$ denotes the mirror image of the matrix $\mathbf{P}^T$ about horizontal and vertical lines through its center. Reversing the matrix of Eq. (10.38) and rearranging rows yields the parity check matrix $\mathbf{H} = [\mathbf{P}^T\,\mathbf{I}_{n-k}]$ discussed in Example 10-2.

The properties of the foregoing cyclic codes are direct consequences of the first of the two main advantages of cyclic codes, namely, their mathematical structure. We shall continue this investigation and shall also demonstrate their second advantage, ease of implementation, in the succeeding chapter. First, however, it is worthwhile to enlarge the class of codes possessing these valuable properties.

10.2.5 Shortened Cyclic Codes

In general, the polynomial $x^n + 1$ has relatively few divisors; as a result there are usually very few cyclic codes of a given length. To circumvent this difficulty and increase the number of pairs (n,k)

for which useful codes are known, cyclic codes are often used in a shortened form. That is, the encoder and decoder agree that the first i information bits will always be zero and consequently will not be transmitted. The decoder for the original cyclic code can decode the shortened code words simply by prefixing each received $(n\text{-}i)$-tuple by i zeros. Thus given an (n,k) cyclic code, it is always possible to construct a $(n\text{-}i, k\text{-}i)$ shortened cyclic code for all i, $0 \leq i \leq k$. Furthermore, because the latter code is a subset of the cyclic code, its minimum distance, and hence its guaranteed error-correcting ability, is at least as great as that of the cyclic code. More generally, any error pattern correctable with a cyclic code is correctable with a shortened version of it.

Since every code word in a shortened cyclic $(n\text{-}i, k\text{-}i)$ code is also a code word in the original cyclic code, the generator polynomial divides every such word, and all polynomials of degree less than $n - i$ divisible by $g(x)$ are code polynomials. As a result, shortened cyclic codes inherit nearly all of the implementation advantages and much of the mathematical structure of cyclic codes. In particular, encoding and syndrome calculation for shortened cyclic codes can be performed by means of Eqs. (10.23) and (10.25), respectively.

10.3 Convolutional Codes

In the linear block codes described in the preceding section, the parity bits in a particular n_0-bit block check only the k_0 information bits in that block. In other words, the block length of the code, n_0, equals its constraint length, n. In a convolutional code, on the other hand, the parity bits in a given block check information bits in other blocks as well, and the constraint length exceeds the block length n_0. The constraint length of the code has been defined as the block length times m, the number of blocks between the first and last blocks (inclusive) checked by the parity bits in a block. That is,

$$n = mn_0 \tag{10.39}$$

where m is an integer. As with block codes, the $n_0 - k_0$ parity bits in a particular block can check information bits which either precede or follow them on the channel. As before, all codes under consideration will be of the former type, and blocks will be transmitted data first, then parity. Also, only systematic codes will be considered so that all information bits are transmitted without alteration. These restrictions usually simplify the encoding and decoding processes without degrading the performance of the codes.

In block codes, the parity-check matrix

$$\mathrm{H} = \left[\mathrm{P}^T \mathrm{I}_{n-k}\right] \tag{10.40}$$

specifies which information bits in the present block are checked by which parity bits. Since the parity bits in a convolutional code also check information symbols in blocks preceding the present block, the basic parity-check matrix has the form

$$\mathbf{h} = \left[\mathbf{P}_{m-1}{}^{T}\ \mathbf{0}\mathbf{P}_{m-2}{}^{T}\ \mathbf{0} \ldots \mathbf{P}_{0}{}^{T}\ \mathbf{I}\right] \tag{10.41}$$

where the $(n_0\text{-}k_0)$-by-k_0 matrices $\mathbf{P}_l{}^T$ are arbitrary, and $\mathbf{0}$ and $\mathbf{I}$ represent, respectively, the all-zero and identity matrices of order $n_0 - k_0$. If $p_l{}^T(i,j) = 1$; that is, if the element in the ith row, jth column of the matrix $\mathbf{P}_l{}^T$ is 1, the ith parity bit in each block checks the jth information bit in the block which precedes it in time by l blocks.

Most of the convolutional codes which are known to have worthwhile error-correcting properties and which can be implemented reasonably simply have either k_0 or $(n_0 - k_0) = 1$. In other words, for these codes, $R = \ldots, 1/5, 1/4, 1/3, 1/2, 2/3, 3/4, 4/5, \ldots$. A code of this type is illustrated in Example 10-8.

Example 10-8. A (12,9) convolutional code with parameters $k_0 = 3$, $n_0 = 4$, and $m = 3$ is specified by the basic parity-check matrix

$$\mathbf{h} = [1100 \quad 1010 \quad 1111] \tag{10.42}$$

This matrix states that every parity bit is the sum of the information bits which precede it by 11, 10, 7, 5, 3, 2, and 1 bit.

The encoder for a convolutional code calculates a set of $n_0 - k_0$ parity checks, as specified by the matrix $\mathbf{h}$, once every n_0 channel-bit times. It appends these checks to data block numbered 0 and shifts the resulting n_0-tuple out to the channel. Syndrome calculation differs only in that the received parity bits are added to the checks calculated on the received data bits.

Decoding consists of associating an error pattern with the syndrome. However, because of the nature of convolutional codes, this syndrome is semi-infinite. For obvious reasons, it is possible to consider only a finite section of the syndrome at a time. Two different approaches to this problem have been proposed and investigated.

1. In the *algebraic decoding*[7,8] approach the decoder decodes the n_0-bit blocks one at a time and bases its decision only on the (n,k)-bit section of the syndrome directly involved in the decision, i.e., on the check digits within a constraint length of the block. At any particular time the block numbered $m - 1$ is being decoded. This method has two advantages. First, it lends itself to analysis by the algebraic techniques employed in connection with block codes. Second, the implementation of this type of decoder often tends to be

relatively simple. Unfortunately this method of decoding suffers from the fact that since it ignores all but a short section of the syndrome, its performance is decidedly inferior to that of the probabilistic decoding.

2. In the *probabilistic decoding*[9] approach the decoder bases every decoding decision on a section of the syndrome that is much longer than the constraint length of the code. This improves error-correcting performance substantially. Although the decoding process is inherently much more complex than algebraic decoding, by taking advantage of the probabilistic nature of the perturbing noise, the complexity and cost of the decoder can be reduced to the point where this technique may be attractive in certain situations.

Although both these techniques employ convolutional codes, they do so in entirely different ways. As a consequence, most of the basic concepts relevant to the coding-decoding problems (i.e., code word, minimum distance, parity-check matrix, burst-correcting ability, etc., which are defined in terms of the number of bits on which the decoder bases its decisions) have different meanings, depending on how the code is decoded. For this reason, and also because at present algebraic decoding appears to be more attractive for most real data-transmission systems, we usually assume that each decoding decision is based on exactly n bits. Probabilistic decoding will only be described briefly; the interested reader should see the references of Chap. 12.

Before proceeding with a description of convolutional codes it is necessary to mention a point on which these codes differ basically from block codes. After calculating the syndrome, a convolutional decoder uses it to decode the oldest block in the decoder. Since all past blocks are known except for decoding errors, the effects of past bit errors on the parity-check equations involved in the decoding of the present block can usually be removed. Whether he wishes to use this feedback information or not is up to the system designer. However, because fewer possibly erroneous digits are involved in decoding when the information is used, performance is usually improved by so doing. Also, most analyses of convolutional codes have assumed that feedback is used. For these reasons this assumption will also be made in what follows, unless otherwise stated.

10.3.1 Matrix Description

A convolutional code can be described conveniently in terms of parity check and generator matrices similar to those used to describe block codes.[7,8] An n-tuple $\mathbf{c}$ is a code word in the convolutional code encoded by Eq. (10.41) if and only if

$$cH^T = 0 \tag{10.43}$$

where H^T denotes the transpose of the parity-check matrix

$$H = \begin{bmatrix} P_0{}^T I & & & \\ P_1{}^T 0 P_0{}^T I & & & \\ \cdot & & & \\ \cdot & & & \\ \cdot & & & \\ P_{m-1}{}^T 0 P_{m-2}{}^T 0 & \cdots & P_0{}^T I \end{bmatrix} \tag{10.44}$$

where $P_l{}^T$, 0 and I are defined in Eq. (10.41). As usual, blank spaces in the matrix represent zeros.

The matrix H has been constructed by taking the left-shifted, truncated versions of the basic parity-check matrix h for its rows. Some authors have employed a slightly different matrix:

$$B = \begin{bmatrix} P_0{}^T I \\ P_1{}^T 0 \\ \cdot \\ \cdot \\ \cdot \\ P_{m-1}{}^T 0 \end{bmatrix} \tag{10.45}$$

and formed H by taking down-shifted, truncated versions of B for its columns. Each of these two approaches can be used to exhibit various properties of convolutional codes which would be considerably more difficult to exhibit using the other approach. Consequently, in the discussion of convolutional codes in Chap. 12 both characterizations of H will be employed.

The n-tuple c is a code word if and only if Eq. (10.43) holds. This is equivalent to stating that c is a code word if and only if it is obtained from a $k = mk_0$-bit data sequence

$$d = d_{m-1} \cdots d_1 d_0 \tag{10.46}$$

by multiplying the sequence by the generator matrix

$$
G = \begin{bmatrix} IP_0\,0P_1 & \cdots & 0P_{m-1} \\ & IP_0 \;\ldots & 0P_{m-2} \\ & \cdot & \cdot \\ & \quad\cdot & \cdot \\ & \qquad\cdot & \cdot \\ & & IP_0 \end{bmatrix} \tag{10.47}
$$

In this matrix 0 and I represent, respectively, the all-zero and identity matrices of order k_0. It can be verified directly that, as with block codes,

$$
GH^T = 0 \tag{10.48}
$$

The syndrome of an n-tuple r is defined as

$$
rH^T = s \tag{10.49}
$$

Clearly, all n-tuples which are not code words have nonzero syndromes.

The ideas presented in the preceding paragraphs are illustrated by Example 10-9.

Example 10-9. The (12,9) convolutional code discussed in the preceding illustration is specified by the parity-check matrix

$$
H = \begin{bmatrix} 1111 & & \\ 1010 & 1111 & \\ 1100 & 1010 & 1111 \end{bmatrix} \tag{10.50}
$$

Note that H can also be constructed from down-shifted, truncated versions of the matrix

$$
B = \begin{bmatrix} 1111 \\ 1010 \\ 1100 \end{bmatrix} \tag{10.51}
$$

The sequence

$$
c = 1001 \quad 1011 \quad 1101 \tag{10.52}
$$

is a code word since $cH^T = 0$. It was obtained by multiplying the data sequence

$$\mathbf{d} = 100 \quad 101 \quad 110 \tag{10.53}$$

by the generator matrix

$$\mathbf{G} = \begin{bmatrix} 1001 & 0001 & 0001 \\ 0101 & 0000 & 0001 \\ 0011 & 0001 & 0000 \\ & 1001 & 0001 \\ & 0101 & 0000 \\ & 0011 & 0001 \\ & & 1001 \\ & & 0101 \\ & & 0011 \end{bmatrix} \tag{10.54}$$

10.3.2 The Standard Array

Decoding can again be performed by associating an error pattern with each syndrome. This association can be visualized in terms of a standard array in which the code words form the first row and each other row is composed of the 2^k n-tuples which have the same syndrome. Many, but not all, of the properties of the standard array for block codes carry over. For example, every n-tuple appears in the array once; the sum of any two n-tuples is a code word if and only if they are in the same coset; etc.

This array can be used as a decoding table but in a substantially different manner from the way in which it was used for block codes. This difference is directly attributable to the fact that here only an n_0-bit section of the received sequence, rather than the entire n-bit block, is to be decoded. Thus, received sequences decoded into code words identical in their $(m-1)$th blocks are equivalent for decoding purposes. Similarly nothing is gained by distinguishing between syndromes whose coset leaders are identical in the first n_0 positions. In order to use the standard array as a decoding table it must be modified as follows. The 2^k columns of the array corresponding to the 2^k code words are grouped into 2^{k_0} sets. Each set contains the 2^{k-k_0} columns headed by code words identical in their $(m-1)$th blocks. Next the 2^{n-k} rows of the array are grouped into sets with the property that the coset leaders in a given set are identical in their first n_0 positions. Clearly there must be not more than 2^{n_0} of these sets. Decoding can be accomplished by calculating the syndrome, determining the first n_0 bits of the coset leader associated with the syndrome, and adding this error pattern to the $(m-1)$th block of the received n-tuple.

The set of 2^{k_0} code words which match the transmitted code word in the $(m-1)$th block has been referred to as the *correct subset*; the *incorrect subset* is composed of the other $2^k - 2^{k_0}$ code words. For a given set the decoding table should be ideally constructed to minimize the probability that the received sequence will be decoded incorrectly. However, practical considerations usually dictate that a somewhat inferior, but more simply implemented, arrangement be employed. This point will be considered further in Chap. 12.

10.3.3 Other Properties

Although it is of limited usefulness in decoding, the standard array of Sec. 10.3.3 makes apparent several interesting properties of convolutional codes. For example, the *minimum distance* of the code d can be defined as the minimum distance between code words in different sets, i.e., code words which differ in the first n_0 positions. Similarly, the *minimum weight* w can be defined as the weight of the minimum-weight code word in the incorrect subset. Then by reasoning similar to that used for block codes, because the code is linear minimum weight equals minimum distance. Also all patterns of weight $t = [(d-1)/2]$ or less are correctable if and only if the code word has minimum distance d.

With this definition of minimum weight, a slightly modified version of Theorem 10-2 holds for linear convolutional codes. Thus in a convolutional code with minimum weight w there exists at least one set of w columns of the parity check matrix of the code which add to zero, provided at least one column in the set is chosen from among the first n_0 columns of the matrix. Conversely, no such set of fewer than w columns exists.

Let l denote the length of the shortest burst that is a code word in a convolutional code and that has its first 1 in the first n_0 positions of the code word. Then since the code is linear, two bursts of length not greater than $[(l+1)/2]$ have the same syndrome. Only one can be corrected so

$$b \leq \frac{l-1}{2} \tag{10.55}$$

where b again denotes the burst-correcting ability of the code.

In the context of convolutional codes the concept of *guard space* is important. A code is said to have guard-space requirement g if at least g correct bits must precede and follow a burst of length not greater than b to assure that the burst is correctable. Clearly for a convolutional (n,k) code

$$g \leq n - 1 \tag{10.56}$$

This follows directly from the definition of constraint length.

In the next two chapters various error-control techniques that employ the codes presented in this chapter will be discussed. It will be seen that although the fundamental concepts of coding theory presented in this chapter are seemingly very abstract, they are nevertheless quite useful in the design of efficient, reliable data-communications systems.

10.4 Summary

We have seen that coding is a mathematically convenient means of communicating with large time-bandwidth-product signals. It can be used either to correct noise-induced errors made by the demodulator or to detect them. Because of the relative ease with which they can be implemented, only binary codes were considered.

There are basically two types of codes: (1) block codes, in which data are grouped into independent blocks and parity checks computed only on the data bits of the block, and (2) convolutional codes, in which the parity checks in a given block check data bits in other blocks as well. Both types of codes are conveniently described in terms of matrices, although they are treated in somewhat different ways. With block codes, we restricted our attention to a subclass of the class of parity check codes, the cyclic codes. These latter have two very desirable features: they use a great deal of mathematical structure and consequently readily lend themselves to analysis, and they are relatively easy to encode and decode. With respect to convolutional codes, the algebraic rather than probabilistic aspects were emphasized.

Notes

Interest in coding theory was sparked by Shannon's[2] famous noisy-channel theorem. Golay[11] and Hamming[5] devised the first nontrivial error-correcting codes and defined several of the concepts basic to coding theory. The concise description of linear codes presented in Secs. 10.2.1 and 10.2.3 is from Slepian.[4] Abramson and Elspas[10] were the first to use linear codes for burst-error correction. Cyclic codes were discovered by Prange.[6] Peterson[1] unified the theory of error-correcting block codes, and Sec. 10.2 draws heavily on his work.

Convolutional codes were invented by Elias.[3] Wyner and Ash[8] organized the algebraic theory of these codes and Sec. 10.3 is based primarily on their work. Massey's[7] approach to these codes is couched in slightly different terminology but is essentially equivalent to that of Wyner and Ash. Nearly all of the work on the probabilistic aspects of these codes has followed from the original work of Wozencraft.[9]

11

Error Control Using Block Codes

In this chapter the problem of correcting errors with block codes is considered. In the past decade this topic has been the subject of literally hundreds of technical journal articles. Clearly it would be impossible to cover all of this material here in any meaningful depth. In our selection we have attempted to include work which is of immediate practical significance *and* which can be explained, for the most part, with the mathematics presented in Chap. 10. Thus a number of theoretically interesting results have been ignored completely or mentioned only in passing. Also, and perhaps more importantly, it has been necessary to treat superficially two or three topics which are of practical importance but which are also rather involved mathematically. These are pointed out, however, and it is hoped that the interested reader will seek out the material in the referenced works.

Block codes are well suited to both of the basic error-control techniques mentioned in Chap. 10, namely, error detection and forward-acting correction. However, because it is inherently more difficult to understand and implement, error correction will be discussed in considerably more detail than detection. We shall see that error detection is, in fact, the necessary first step toward correction. Because of their important advantages of mathematical tractability and ease of implementation, attention will be focused on cyclic codes and the closely related shortened cyclic codes.

It is natural to ask which of the two above-mentioned error-control techniques is better in a particular situation. Quite often the answer is so obvious as to make the question trivial. In other situations common sense produces the answer, while in still others a detailed analysis of systems of both types may be required. Unfortunately, there does not seem to be a universally applicable rule for deciding between the two techniques; usually one or the other is preferred on the basis of systems considerations such as the presence or absence of a reverse channel or the need or lack of same for continuous data flow, delay requirements, etc. For these reasons a general, detailed comparison of error detection and error correction will not be made here. Instead, the basic properties of both techniques will be described in the hopes of facilitating comparisons of specific systems by interested analysts. Comparisons of the two techniques in particular situations are available in the literature, e.g., Refs. 1 and 2.

11.1 Encoding, Syndrome Calculation, Error Detection, and Error Correction

Encoding consists of appending a set of parity checks to a data block before transmission. The syndrome is simply the modulo-2 sum of these checks after transmission and the check bits calculated on the received data block at the decoder. Error detection consists of noting whether or not the syndrome is the all-zero $(n-k)$-tuple. If it is not, transmission errors have occurred. These errors can usually be corrected by adding to the received word the error pattern associated with the syndrome. This association of error pattern with syndrome is the decoding problem and is examined in some detail in Secs. 11.2 through 11.4.

A fuller, more comprehensive treatment of the material covered in this section can be found in Chaps. 7 and 8 of Peterson.[3]

11.1.1 Encoding

In Chap. 10 we showed that a k-bit data block $d(x)$ could be encoded into an n-bit cyclic code word $x^{n-k}d(x) + r(x)$ according to the following equation:

$$\frac{x^{n-k}d(x)}{g(x)} = q(x) + \frac{r(x)}{g(x)} \tag{11.1}$$

where $g(x)$ denotes the generator polynomial of the code. That is, the remainder obtained by dividing $x^{n-k}d(x)$ by $g(x)$ is added to $x^{n-k}d(x)$; by Eq. (11.1) the result is a multiple of $g(x)$ and hence a code word.

Since the data bits are transmitted without alteration, encoding consists simply of determining $r(x)$. The necessary division can be effected with the circuit shown in Fig. 11.1(a). It is assumed that the reader has at least a passing acquaintance with linear sequential networks; however, to avoid confusion the conventions used will be stated explicitly.

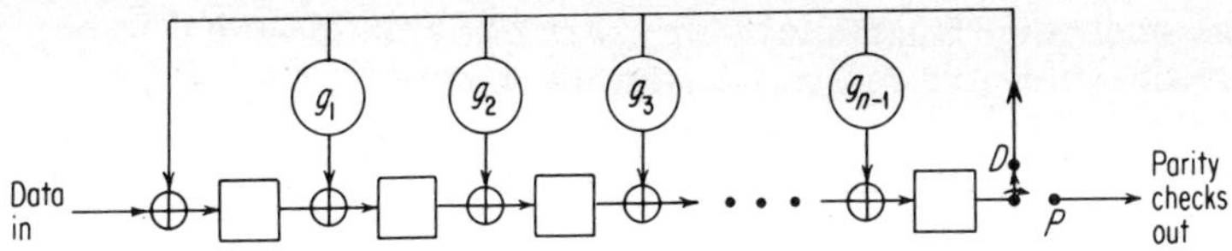

(a)

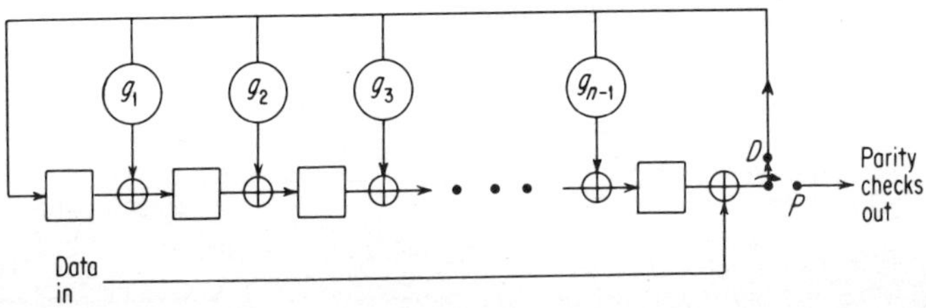

(b)

Fig. 11.1. Two $(n\text{-}k)$-stage encoding circuits for the cyclic code generated by $g(x) = x^n + g_{n-1}x^{n-1} + \cdots + g_2x^2 + g_1x + 1$.

1. The symbol —☐— denotes a single binary shift register stage which is shifted by an external synchronous clock (not shown) so that its input at a particular time appears at its output one unit of time later.

2. The symbol ⊕→ denotes an EXCLUSIVE-OR circuit or modulo-2 adder. The output of the adder at a particular time is the modulo-2 sum of its inputs at that time. No restriction is placed on the number of allowable inputs. An arrowhead will be used to distinguish the single output lead.

3. The symbol —(f_i)— simply denotes a connection or the lack of one, depending on whether f_i is 1 or 0, respectively.

Encoding is accomplished as follows: The data polynomial $d(x)$ is shifted into the feedback shift register as shown. After $n\text{-}k$ shifts, the register contains the $n\text{-}k$ high-order terms of $x^{n-k}d(x)$. After the last data bit has been fed into the register, the feedback connection is broken by throwing the switch from position D (data) to position P (parity), and the contents of the register are shifted to the channel. That this circuit actually divides $x^{n-k}d(x)$ by $g(x)$ to produce the $(n\text{-}k)$-bit remainder $r(x)$ can be seen heuristically by considering Example 11-1.

Example 11-1. To demonstrate that $r(x)$ is obtained as claimed, we examine the (15,11) code considered in the examples of Chap. 10. The generator polynomial of the code is $g(x) = x^4 + x + 1$. Suppose that the data polynomial $d(x) = x^{10} + x^6 + x^3 + x$ considered earlier is to be encoded. The remainder $r(x)$ can be obtained mathematically in a straightforward manner by dividing $x^4 d(x)$ by $g(x)$. Thus

$$
\begin{array}{r|ccccccccc}
 & & & x^{10} & & x^7 & & x^4 & & 1 \\
\hline
x^4 + x + 1 & x^{14} & & x^{10} & & x^7 & x^5 & & & \\
 & x^{14} & x^{11} & x^{10} & & & & & & \\
\cline{2-7}
 & & x^{11} & & & x^7 & x^5 & & & \\
 & & x^{11} & & x^8 & x^7 & & & & \\
\cline{3-7}
 & & & & x^8 & & x^5 & & & \\
 & & & & x^8 & & x^5 & x^4 & & \\
\cline{5-8}
 & & & & & & & x^4 & & \\
 & & & & & & & x^4 & x & 1 \\
\cline{6-10}
 & & & & & r(x) & = & & x & + 1
\end{array}
$$

Now consider the encoder for this code, shown in Fig. 11.2(a). To facilitate a comparison with the above division, the ordinary convention of signal flow from left to right has been reversed. Clearly the first four shifts merely serve to "load" the first four terms of the dividend into the register. Table 11.1 shows the contents of the register as the encoding process is performed. Also shown are the dividends in the above division which correspond to the register contents at various stages of the encoding process.

Table 11-1. Comparison of Register Contents and Dividends.

Shift number	Register contents	Bits in data source	Corresponding dividends
0	1000	1001010	x^{14} x^{10} x^7 x^5
1	0010	001010	
2	0100	01010	
3	1000	1010	x^{11} x^7 x^5
4	0010	010	
5	0100	10	
6	1001	0	x^8 x^5
7	0001		
8	0010		
9	0100		
10	1000		x^4
11	0011		x 1

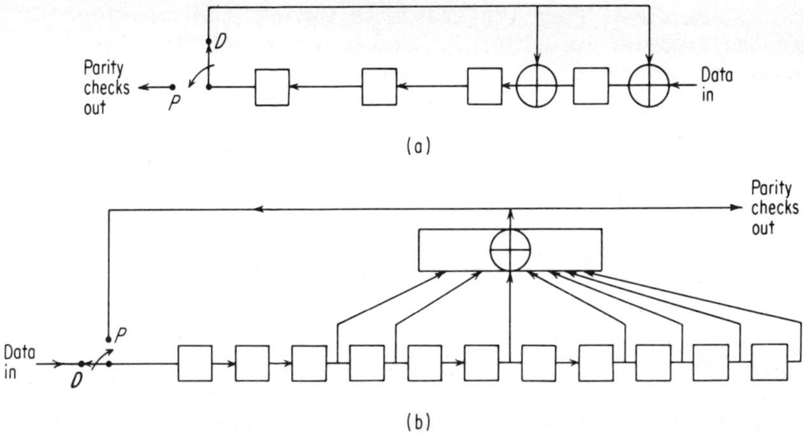

Fig. 11.2. Encoders for the (15,11) code of Example 11.1. (a) An n-k = 4-stage encoder. (b) At k = 11-stage encoder.

After the first four shifts, the four high-order bits of $x^4 d(x)$ occupy the register. On the next shift a 1, corresponding to the quotient term x^{10}, is emitted. Thus 1s are added into the stages following the feedback connections, and the register contents then correspond to the new dividend obtained by adding $x^{10} g(x)$ to the first dividend. The key point here is that when the register output is 1, the dividend is modified exactly as in the division process; likewise when it is 0, nothing is done in either case.

Since the first n-k shifts serve only to load the register, the encoder of Fig. 11.1(a) can be modified to the form of Fig. 11.1(b). In the latter figure those "load" shifts are unnecessary. Clearly the feedback information, which is simply the sum of the high-order bit of the dividend and the various past multiples of $g(x)$, is the same in either case. The contents of the register differ until after the nth shift of Fig. 11.1(a) [kth shift of Fig. 11.1(b)]. At that time the dividend has been shifted completely through the encoder, and thus both registers contain $r(x)$.

Because it requires n-k fewer shifts, the circuit of Fig. 11.1(b) is usually preferable to that of Fig. 11.1(a) for encoding serial data. Encoding is performed here by shifting the data sequence simultaneously into the register and onto the line. After the last data bit has been shifted out, the switch is thrown from D to P, and the n-k parity bits are shifted out. This leaves the register empty and ready to accept the first data bit of the next word immediately.

In a cyclic code the highest-order parity bit is the sum of the information bits specified by the top row of the matrix H or, equivalently, by the parity-check polynomial $h(x)$. This suggests the k-stage encoder shown in Fig. 11.3, which operates as follows. The k information bits are shifted simultaneously into the register and out to the line. Then the switch is thrown from position D to P, and the register is shifted n-k times. The parity bits corresponding to $r(x)$ appear at the output of the multi-input EXCLUSIVE-OR circuit and are shifted out after the data.

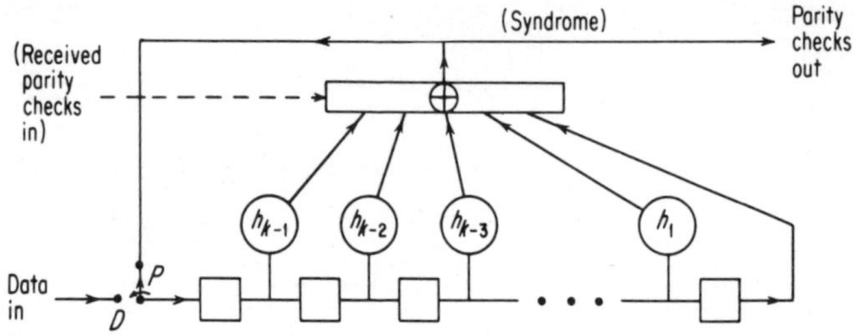

Fig. 11.3. A k-stage encoding circuit for the code whose parity check polynomial is $h(x) = x^k + h_{k-1}x^{k-1} + \cdots + h_1 x + 1$.

When the register contains the k data bits, the output of the EXCLUSIVE-OR is the sum of the data bits corresponding to nonzero terms in the first row of H; that is, it is the first parity bit. Because of the cyclic nature of the code, however, the second parity bit is simply the sum of the bits in the same positions relative to itself as the first parity bit. Thus if the register is shifted once and the first parity check replaces the last information bit, the output of the EXCLUSIVE-OR will be the second parity bit. The other parity checks are generated in like manner. The k = 11-stage encoder for the (15,11) code discussed in the various illustrations is shown in Fig. 11.2(b).

It is also possible to encode with an (n-k)-stage register of the form of Fig. 11.3, that is, a register which employs a single multi-input EXCLUSIVE-OR gate rather than a number of two-input EXCLUSIVE-ORs. In some applications these circuits yield slightly simpler encoders than those of Figs. 11.1 and 11.3. However, their operation is more difficult to explain and will not be discussed here (see Peterson[3]). In general, for codes whose efficiency is greater than one-half, the circuit of Fig. 11.1 is usually more economical; for codes whose efficiency is lower than one-half, the more economical circuit is that shown in Fig. 11.3.

11.1.2 Syndrome Calculation

Syndrome calculation differs from encoding only in that the received parity bits must be added to the checks calculated on the received data. The circuit of Fig. 11.1(a) performs this addition automatically if the received parity bits are shifted in immediately after the data. Addition of the dotted connection to Fig. 11.3 yields the same result; that is, the output is the syndrome.

One other property of the syndrome is worth mentioning. Let $s(x)$ denote the syndrome of an n-tuple $e(x)$. Then

$$s(x) = \text{rem}\ \frac{e(x)}{g(x)} \tag{11.2}$$

or

$$\frac{e(x)}{g(x)} = q(x) + \frac{s(x)}{g(x)} \tag{11.3}$$

Now let $e'(x) = x^i e(x)$ denote the polynomial $e(x)$ shifted cyclically i places to the left. The syndrome of this polynomial $s'(x)$ is given by

$$s'(x) = \text{rem}\ \frac{e'(x)}{g(x)} = \text{rem}\ \frac{x^i e(x)}{g(x)} = \text{rem}\ \frac{x^i g(x)\, q(x) + x^i s(x)}{g(x)} = \text{rem}\ \frac{x^i s(x)}{g(x)} \tag{11.4}$$

However $x^i s(x)$ is simply the syndrome of $e(x)$ shifted left i places. Thus if $s(x)$ is stored in a division circuit of the type depicted in Fig. 11.1(a), that is, a *syndrome generator*, $s'(x)$ can be obtained simply by shifting the generator i times. This result is stated in Theorem 11-1.

Theorem 11-1: *The syndrome of an i-place cyclic shift of a given n-tuple is simply an i-place cyclic shift of the syndrome of the n-tuple in its syndrome generator.*

11.1.3 Error Detection

We have shown that an n-tuple is a code word in a cyclic code if and only if its syndrome is the all-zero $(n-k)$-tuple. As a result, error detection can be implemented simply with a single additional flip-flop connected to the output of the syndrome generator. If one or more 1s appear in the syndrome, the flip-flop sets and an error has been detected; otherwise the received n-tuple is a code word.

The performance of error detection is examined in Sec. 11.5.4.

11.1.4 Error Correction

We have pointed out that error correction is simply an association of a syndrome with an error pattern. This association can be visualized conveniently in terms of the standard array introduced in Sec. 10.2.3.

A given code has exactly 2^{n-k} distinct cosets, including the code itself. Only one error pattern in each coset is correctable. Thus by choosing the most probable error pattern in each coset as the coset leader, the probability of erroneous decoding is minimized.

This type of decoding, rarely used in practice because it is usually difficult to implement, is referred to as maximum-likelihood decoding. Its importance lies in its use as a standard against which to compare practical decoding techniques.

A forward-acting, block-code error-control system, which would be optimal for a given channel, could be designed by selecting the code in the class of suitable codes for which the highest-probability set of n-tuples can be selected as coset leaders and then decoding with maximum likelihood decoding. There are several obvious difficulties here. First, detailed knowledge of channel statistics is presumed; this is often unavailable in practice. Second, except for very short codes and the few cases where analytic results can be applied, it is impractical to examine every code of a given type in detail, even with the aid of a high-speed digital computer. Usually the codes are far too large, and there are too many of them. Finally, even if an optimum code is found, maximum-likelihood decoding by means of the standard array is only feasible for codes with relatively few cosets, i.e., codes with less than, say, 15 or 20 parity checks.

For these reasons usually no attempt is made to design an optimal error-control system. Rather, a procedure such as the following one is more typically employed. On the basis of fragmentary knowledge of channel statistics, attention is directed to one of the several types of error-correcting codes that are capable of correcting the type of errors which occur on the channel and that can be simply implemented. Then systems requirements such as equipment cost, data rate, and error rate are considered, and the code which best meets these requirements is selected from the appropriate class.

Clearly, the second step in this procedure is the difficult one; it presumes a fairly detailed knowledge of the interrelationships among cost, data rate, and error rate. The following sections are devoted to illuminating these interrelationships. In Secs. 11.2 through 11.5 classes of codes useful for correcting random errors, bursts of errors, combinations of random and burst errors, and synchronization errors, respectively, are presented. Also described are various decoding procedures and methods of implementing them.

11.2 Random-error Correction

In this section cyclic codes with relatively large minimum distance (and with the less well-defined property that they can be decoded with a reasonable amount of equipment) are considered. Codes of this type are useful on channels which can be modeled reasonably well by the binary symmetric channel. That is, they perform well on channels on which bit errors occur nearly independently at random.

11.2.1 Bounds

In Sec. 10.2.3 it was pointed out that in order for a code to be t-error-correcting, no two error patterns of weight t or less could be in the same coset. Since there are 2^{n-k} cosets in an (n,k) code and $\binom{n}{i}$ n-tuples of weight i, this implies that for any t-error-correcting (n,k) code

$$2^{n-k} \geq \sum_{i=0}^{t} \binom{n}{i} \tag{11.5}$$

The above inequality is the well-known *Hamming*,[4] or *sphere-packing*, *bound*. In addition to providing an upper bound on t, it can be used to provide a lower bound on the probability of erroneous decoding P_e for the best (n,k) code used on the BSC. The resulting sphere-packing bound on P_e (see Sec. 3.2) is, roughly speaking, based on the argument that the best (n,k) code cannot perform better than a hypothetical sphere-packed (n,k) code. For this latter

$$P_e = 1 - \sum_{i=0}^{t} \binom{n}{i} p^i q^{n-i} - N_{t+1} p^{t+1} q^{n-t-1} \tag{11.6}$$

where t is the largest integer such that Eq. (11.5) holds and

$$N_{t+1} = 2^{n-k} - \sum_{i=0}^{t} \binom{n}{i} \tag{11.7}$$

Another lower bound on the number of parity checks that a linear code must have in order to have minimum distance d is the Plotkin[5] bound, which states that

$$n - k \geq 2d - 2 - \log d \tag{11.8}$$

where the logarithm is taken to the base 2.*

The derivation of this result is more involved than that of Eq. (11.5) and will be omitted. It is based on the argument that the existence of an (n,k) code with minimum distance greater than d in Eq. (11.8) would imply the existence of a shortened version of this code with the same or larger minimum distance. This implies that the minimum weight in the code exceeds the average weight of a code word, which is known to be $n2^{k-1}/(2^k - 1)$, and this is impossible, of course.

Other lower bounds on the number of parity checks required to attain a given minimum distance have been derived (e.g., see the

---

* Unless otherwise specified, all logarithms employed in this and the following chapter will be binary.

Elias bound in Fig. 11.4),[6,7] but these are much more complex and contribute relatively little insight into the problem. In most cases these bounds are usually tighter than those of Eqs. (11.5) and (11.8), but usually the improvement is small for the values of n, k, and d with which we are concerned. For this reason they are omitted.

Lower bounds on minimum distance are of interest in that they exclude the construction of codes with excessive minimum distance. A natural question is: If (n,k) codes with error-correcting ability larger than that specified by Eqs. (11.5) and (11.8) cannot be constructed, what is achievable? A partial answer can be obtained with the aid of Theorem 10-2, which states that no set of $d-1$ or fewer columns and at least one set of d columns in the parity-check matrix of a code with minimum distance d must be linearly dependent. As the first column of H for a code with n-k parity checks, choose any nonzero $(n$-$k)$-tuple; for the second, choose any other. For the third, choose any one that is not a linear combination of the first two. In general for the jth, choose an $(n$-$k)$-tuple that is not a linear combination of any $d-2$ or fewer columns. This assures that no linear combination of $d-1$ or fewer columns will be 0. Clearly, it is possible to choose n such columns provided

$$2^{n-k} > \sum_{i=0}^{d-2} \binom{n-1}{i} \tag{11.9}$$

In the worst case all linear combinations of $n-1$ columns taken $d-2$ or fewer at a time will be unique; then it is always possible to add one more column to H if Eq. (11.9) holds. This result is commonly referred to as the Varsharmov[8]-Gilbert[9]-Sacks[10] bound.

These three results are useful in selecting codes for use in error-control systems. The first two are reassuring because they guarantee that most of the easily implemented codes presented in the following sections are reasonably close to optimum. The use of these results is illustrated in the following, fairly typical, code-selection problem.

Example 11-2. It is desired to construct a double-error-correcting block code whose efficiency is 3/4. That is, it is required that $d = 5$ and $k = 3/4n$. Since the probability of erroneous decoding increases with block length with R and d fixed, we are interested in making the code as short as possible. As a guide as to how short we can expect to make such a code, we examine the bounds of Eqs. (11.5), (11.8), and (11.9).

By the Hamming bound of Eq. (11.5),

$$\frac{n}{4} \geq \log\left[1 + n + \frac{n(n-1)}{2}\right] \tag{11.10}$$

This yields

$$n \geq 40 \tag{11.11}$$

By the Plotkin bound of Eq. (11.8)

$$n - k = \frac{n}{4} \geq 6 \tag{11.12}$$

or

$$n \geq 24 \tag{11.13}$$

Based on these results, one might attempt to construct a code of length 40. The Varsharmov-Gilbert-Sacks bound of Eq. (11.9) guarantees that it is possible to construct a code of length 64 which meets the stated requirements. In fact, it will be seen that there are several (48,36) shortened cyclic codes with $d = 5$ but that no analytical techniques are known for constructing a code of length 40 or 44 with the desired properties.

It is interesting to consider the behavior of Eqs. (11.5) through (11.9) as n increases without bound and the code efficiency is kept constant. The Hamming bound gives

$$H\left(\frac{d}{2n}\right) \leq 1 - R \tag{11.14}$$

where, as in Chap. 3,

$$H(x) = -x \log x - (1 - x) \log (1 - x) \tag{11.15}$$

The Plotkin bound gives

$$\frac{2d}{n} \leq 1 - R \tag{11.17}$$

while the Varsharmov-Gilbert-Sacks bound yields

$$H\left(\frac{d}{n}\right) \geq 1 - R \tag{11.17}$$

This behavior is illustrated in Fig. 11.4 in which the ratio d/n is plotted as a function of R. A bound developed by Elias which is tighter than both the Hamming and Plotkin bounds for very large n is also shown. This bound has the limiting form

$$\frac{d}{n} < 2\lambda(1 - \lambda) \tag{11.18}$$

where λ is given by

$$H(\lambda) = 1 - R \tag{11.19}$$

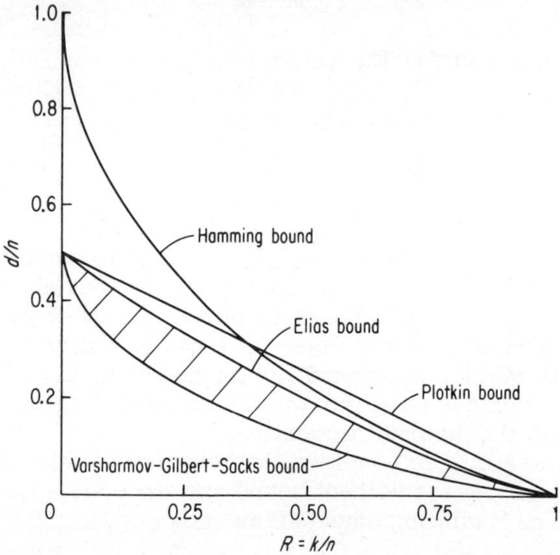

Fig. 11.4. Bounds on minimum distance d/n as a function of code efficiency ($R = k/n$).

Many of the codes described in the following sections have relatively large minimum distance for given values of code length and efficiency and as such fall reasonably close to the Hamming-Plotkin-Elias bound.

11.2.2 Bose-Chaudhuri[11,12]-Hocquenghem (BCH) Codes[13]

The codes discussed in this section are, as a class, the most powerful random-error-correcting codes known. For most useful values of code length and efficiency the error-correcting ability of these codes (and their shortened versions) is quite close to the bounds of the preceding section. Also, a decoding algorithm that can be implemented with a reasonable amount of equipment has been devised for these codes. A more fundamental and hence mathematically involved description of the BCH codes and their decoding algorithm is given in Peterson.[3]

The bound. The BCH codes are defined in terms of the BCH bound. The proof of this result is too involved to be presented here; consequently, only a statement of that bound will be given.

First, however, a brief review of the relevant terminology defined in Chap. 10 is in order.

In Sec. 10.2.4 it was shown that every binary cyclic code of length n is completely determined by its generator polynomial $g(x)$, a divisor of $x^n + 1$. It can be shown that the polynomial $x^n + 1$ with binary coefficients can be factored into n linear factors

$$(x + \alpha_1)(x + \alpha_2) \cdots (x + \alpha_n) \tag{11.20}$$

where the roots α_i are elements of some larger *field*.† (This is directly analogous to factoring $x^2 + 1$, with real coefficients, into $(x - i)(x + i)$, where $i = \sqrt{1}$ is an element of a field larger than the real numbers, i.e., the complex numbers.) These n roots can be shown to form a cyclic group under the operation multiplication. That is, for some (primitive) root α the n roots can be expressed as

$$\alpha^1, \alpha^2, \ldots, \alpha^{n-2}, \alpha^{n-1}, \alpha^n = 1 = \alpha^0 \tag{11.21}$$

The lowest-degree polynomial with binary coefficients which divides $x^n + 1$ and of which α^i is a root is referred to as the *minimum polynomial* of α^i; it is designated $m_i(x)$. If such a polynomial is considered for the moment to have integer coefficients, clearly all coefficients of the polynomial $m_i^2(x) - m_i(x^2)$ are even. Consequently in the binary case

$$m_i^2(x) = m_i(x^2) \tag{11.22}$$

and if α^i is a root of $m_i(x)$, so are α^{2i}, α^{4i}, α^{8i}, ... The number of such roots is, of course, simply the degree of $m_i(x)$. For practically all values of n of interest tables exist that enable one to determine which roots of $x^n + 1$ are roots of a given divisor of $x^n + 1$, or equivalently, which polynomial is the minimum function of a given root of $x^n + 1$. A short table of this type is given in Table 11.3; a longer version is given in Peterson.[3]

We can now state the celebrated BCH bound: The minimum distance of the code generated by $g(x)$ must be greater than the largest number of consecutive roots of $g(x)$. (The j roots α^{i+1}, α^{i+2}, . . . , α^{i+j} are *consecutive* for $0 \leq i \leq n - 1$.) Since $g(x)$ has degree n-k, exactly n-k of the roots of $x^n + 1$ are roots of $g(x)$. The bound applies to all cyclic codes and in the case of the BCH codes seems remarkably tight. The numerous ideas condensed into the preceding paragraphs can be understood more fully by considering the case of $n = 15$ in detail.

† A field is a set of elements which forms a commutative group under addition and a group under multiplication and for which the distributive laws hold. Three familiar fields are the rational, the real and the complex numbers.

Example 11-3. Table 11.3 can be used to find the factors of $x^{15}+1$. Thus

$$x^{15}+1 = (x+1)(x^4+x+1)(x^4+x^3+x^2+x+1)(x^2+x+1)(x^4+x^3+1) \tag{11.23}$$

This table also specifies α as a root of (x^4+x+1). Then by Eq. (11.22)

$$(\alpha^4+\alpha+1)^2 = \alpha^8+\alpha^2+1 = 0 \tag{11.24}$$

But this says that α^2 is a root of x^4+x+1. In like fashion we could show that α^4 and α^8 must also be roots of $m_1(x)$. Since $m_1(x)$ has degree four, this exhausts the set of roots. (Since α is a root of $x^{15}+1$ it follows that $\alpha^{16}=\alpha$.) In addition to $m_1(x)$, Table 11.3 also specifies $m_3(x)$, $m_5(x)$, ... The roots of the various minimum polynomials are tabulated in Table 11.2.

Table 11-2. Divisors of $x^{15}+1$ and Their Roots.

Minimum polynomials	Exponents of roots
$m_0(x) = x+1$	0
$m_1(x) = x^4+x+1$	1, 2, 4, 8
$m_3(x) = x^4+x^3+x^2+x+1$	3, 6, 12, 9
$m_5(x) = x^2+x+1$	5, 10
$m_7(x) = x^4+x^3+1$	7, 14, 13, 11

Here, as elsewhere, only the exponents of α have been written in order to simplify the representation. Note that the number of roots always equals the degree of the minimum polynomial, as it must.

The cyclic code generated by

$$g(x) = m_0(x)\,m_1(x)\,m_7(x) \tag{11.25}$$

has, by the BCH bound, minimum distance at least 6, because the exponents of the roots are

$$\underline{0,\ 1,\ 2,}\ \ ,\ 4,\ \ ,\ \ ,\ 7,\ 8,\ \ ,\ \ ,\ 11,\ \ ,\ \underline{13,\ 14,} \tag{11.26}$$

and thus there are five consecutive roots (underlined).

The codes. Since the roots of a polynomial completely determine the polynomial, a cyclic code can be specified by the roots of its generator polynomial. The BCH codes are cyclic codes whose generator polynomial is chosen so as to have as large a set of consecutive roots as is possible for a generator of a given degree. In other words for a given umber of parity checks the generator is selected to maximize the minimum distance guaranteed by the BCH bound. More precisely, the BCH code generated by

$$g(x) = \text{LCM}\left[m_i(x), m_{i+1}(x), \ldots, m_{i+d-2}(x)\right] \tag{11.27}$$

consists of all binary polynomials of degree less than n which are divisable by $g(x)$, and no others. (The symbol LCM means least common multiple.) It has minimum distance at least d and hence is capable of correcting all error patterns of weight not greater than $t = [(d-1)/2]$. Note that with this definition $g(x)$ is the polynomial of smallest degree such that Eq. (11.27) holds; that is, it includes no factors which do not contribute to the largest set of consecutive roots.

The most important subclass of the class of BCH codes, both in theory and in practice, is the class of codes of length $n = 2^m - 1$, that is, the *primitive codes*. Within this subclass the most interesting codes have $i = 1$ in Eq. (11.27). From Eq. (11.22) and its argument, $m_i(x) = m_{2i}(x)$ and $2t$ consecutive roots are guaranteed by t consecutive odd roots. Thus Eq. (11.27) reduces to

$$g(x) = \text{LCM}\left[m_1(x), m_3(x), \cdots m_{2t-1}(x)\right] \tag{11.28}$$

But by arguments too involved to justify including here, we can show that the degree of every divisor of $x^{2^m-1} + 1$ cannot exceed m (see Peterson[3] or any standard text on modern algebra). Since at most t minimum polynomials are required to guarantee that the generator polynomial has t consecutive odd roots, at most mt parity checks are required. This can be stated as in Theorem 11-2.

Theorem 11-2: There exists a BCH code of length $n = 2^m - 1$ and minimum distance at least $d = 2t + 1 \leq n$ which has at most mt parity checks.

In Table 11.3 is given a list of the generator polynomials and important parameters of the primitive codes of length ≤ 255. In the setting up of this table several different polynomials could have been chosen to be $m_1(x)$, but once $m_1(x)$ was selected the other $m_i(x)$ were determined. In this case, the $m_1(x)$ used in Appendix C of Peterson,[3] i.e., the $m_1(x)$ of lowest weight, was chosen. Table 11.3 contains a list of several useful nonprimitive BCH codes.

The following example is intended to clarify the concepts just developed.

Example 11-4. Consider again the primitive codes of length 15 given in Example 11-3. If $g(x)$ is chosen to be the least common multiple of $m_1(x)$, $m_2(x)$, $m_3(x)$, and $m_4(x)$, by the BCH bound the cyclic code generated by $g(x)$ has $d \geq 5$. Since $m_1(x)$ and $m_3(x)$ have no common factors, Eq. (11.28) yields

$$g(x) = m_1(x)m_3(x) = (x^4 + x + 1)(x^4 + x^3 + x^2 + x + 1) \tag{11.29}$$

This is the generator polynomial of a (15,7) double-error-correcting BCH code.

If instead $g(x)$ is chosen to be the generator polynomial of Example 11-3; namely,

$$g(x) = \text{LCM}\,[m_0(x), m_1(x), m_7(x)] \tag{11.30}$$

then, as we pointed out earlier, there are five consecutive roots and

$$g(x) = (x + 1)(x^4 + x + 1)(x^4 + x^3 + 1) \tag{11.31}$$

This polynomial generates a (15,6) BCH code that has $d \geq 6$.

Finally, suppose $g(x)$ had been chosen to be

$$g(x) = (x^4 + x + 1)(x^2 + x + 1) \tag{11.32}$$

The roots of this polynomial are α^1, α^2, α^4, α^5, α^8, and α^{10} and thus the BCH bound guarantees that $d \geq 3$. But

$$\text{LCM}[m_1(x)] = m_1(x) \tag{11.33}$$

is a polynomial of smaller degree which has the same largest set of consecutive roots. It follows that this code generated by $g(x)$ is not a BCH code.

Table 11.3 lists some BCH codes of length 15 and their generator polynomials. Other codes can be generated by taking different combinations of minimum polynomials.

Table 11-3. Some BCH Codes of Length 15.
The $m_i(x)$ are defined in Table 11-2. The number (0) denotes the fact that α^0 is a root of $g(x)$.

n	k	d	$g(x)$	Roots of $g(x)$
15	14	2	$m_0(x)$	(0)
15	11	3	$m_1(x)$	(1,2,4,8)
15	10	4	$m_0(x)\,m_1(x)$	(0) (1,2,4,8)
15	7	5	$m_1(x)\,m_3(x)$	(1,2,4,8) (3,6,12,9)
15	6	6	$m_0(x)\,m_1(x)\,m_3(x)$	(0) (1,2,4,8) (3,6,12,9)
15	5	7	$m_1(x)\,m_3(x)\,m_5(x)$	(1,2,4,8) (3,6,12,9) (5,10)
15	4	8	$m_0(x)\,m_1(x)\,m_3(x)\,m_5(x)$	(0) (1,2,4,8) (3,6,12,9) (5,10)

For codes in the preceding example and, indeed, for all primitive codes for which it is known, the actual minimum distance equals the value of the BCH bound. However, this is not true in general and for several of the nonprimitive codes listed in Table 11.4 the minimum distance actually exceeds the bound.

Decoding. The *Peterson*[14] *decoding algorithm* described in his section is applicable to all BCH codes. However, the following description assumes a primitive code.

Before proceeding here it is necessary to discuss some of the more elementary properties of feedback shift registers. Consider a circuit that divides any input polynomial by the mth-degree polynomial $m_1(x)$ defined in the preceding section. [Remember that every root of x^n+1 can be expressed as a power of α, a root of $m_1(x)$.] The period of a circuit of this type is exactly $n = 2^m - 1$; if it were $u < n$, $m_1(x)$ would have to divide x^u+1, which it does not. This implies that the remainders obtained by dividing x^i, $0 \le i \le n-1 = 2^m - 2$ by $m_1(x)$ are distinct. These m-tuples, basic to the following discussion, have several interesting properties. It should be clear that together with the all-zero m-tuple, they form a commutative group under term-by-term addition. Also true, but not so obvious, is that without the all-zero element they form a commutative group under polynomial multiplication. That is, the remainder obtained by dividing $x^i \cdot x^j$ or $x^j \cdot x^i$ by $m_1(x)$ is simply the product of the remainders associated with x^i and x^j reduced modulo-$m_1(x)$. Also the product of the remainders of x^i and x^{n-i} is the remainder of x^0, or unity. Furthermore, the operation multiplication distributes over addition. That is, if a, b, and c denote the remainders associated with various powers of x, $a(b+c) = ab+ac$. These properties are exactly the conditions necessary for the 2^m m-bit remainders to form a field. In other words any arithmetic operation, such as addition (or subtraction), multiplication, or division that one can perform on the real numbers can be performed on these remainders.

We are now in a position to describe the decoding algorithm for the BCH codes. As before, consider a code word $c(x)$ transmitted through a noisy channel; let $e(x)$ denote the error polynomial added to $c(x)$ by the channel. Decoding consists of determining $e(x)$ from $s(x)$, the syndrome calculated from the received polynomial $r(x) = c(x) + e(x)$. Then $c(x)$ is determined by adding $e(x)$ to $r(x)$.

The syndrome for a BCH code is calculated in a manner slightly different from, but completely equivalent to, that explained in Sec. 10.2. By Eq. (11.28), for a primitive BCH code with $i = 1$,

$$g(x) = \mathrm{LCM}\left[m_1(x), m_3(x), \ldots, m_{2t-1}(x)\right] \tag{11.34}$$

That is, $m_1(x)$, $m_3(x)$, ..., and $m_{2t-1}(x)$ divide $g(x)$ and hence $c(x)$. Whereas earlier the syndrome was the remainder obtained by dividing $r(x)$ by $g(x)$, for this decoding algorithm is is necessary to compute t partial syndromes by dividing $r(x)$ successively by $m_1(x)$, $m_3(x)$, ..., $m_{2t-1}(x)$. It can be shown that either representation of the syndrome can be obtained uniquely from the other. Denote the remainder obtained by dividing $r(x)$ by $m_i(x)$ by the symbol $S_i(x)$. The first partial syndrome can be obtained by dividing $r(x)$ by $m_1(x)$; this can be implemented simply with a circuit of the type shown in Fig. 11.1(a) in which the connections correspond to $m_1(x)$. After n shifts, the register contains

$$S_1(x) = \text{rem}\ \frac{r(x)}{m_1(x)} \tag{11.35}$$

However, since $r(x) = c(x) + e(x)$ and since $m_1(x)$ divides $c(x)$, this gives

$$S_1(x) = \text{rem}\ \frac{e(x)}{m_1(x)} \tag{11.36}$$

If $e(x)$ has only one nonzero term, that is, $e(x) = x^i$, $0 \le i \le n-1$, $S_1(x)$ is simply the unique remainder corresponding to that value of i. If j errors occur, $S_1(x)$ is the sum of the remainders corresponding to the various errors. That is, if β_j represents the remainder corresponding to the jth error,

$$S_1 = \beta_1 + \beta_2 + \cdots + \beta_j \tag{11.37}$$

In the sequel the β_j will be referred to as the *error-location numbers*. If it is possible to determine each of these numbers, then, since each is associated with a particular position in the word, the errors can be corrected.

If two 0s are inserted between successive terms of

$$r(x) = r_{n-1}x^{n-1} + r_{n-2}x^{n-2} + \cdots + r_1 x + r_0 \tag{11.38}$$

the result is

$$r_{n-1}x^{3(n-1)} + 0x^{3n-4} + 0x^{3n-5} + r_{n-2}{}^{3(n-2)} + \cdots$$
$$+ 0r^4 + r_1 x^3 + 0x^2 + 0x + r_0 = r(x^3) = c(x^3) + e(x^3) \tag{11.39}$$

Since $m_3(x)$ divides $c(x)$, it follows that $c(\alpha^3) = 0$. But this says that α is a root of $c(x^3)$, or equivalently, that $c(x^3)$ is divisible by $m_1(x)$.
Thus

$$\text{rem}\ \frac{r(x^3)}{m_1(x)} = \text{rem}\ \frac{e(x^3)}{m_1(x)}$$

and if the polynomial of Eq. (11.39) is shifted into another register identical to the S_1 *register*, the result is

$$S_3 = \beta_1{}^3 + \beta_2{}^3 + \cdots + \beta_j{}^3 \tag{11.40}$$

For obvious reasons this register will be referred to as the S_3 register.

In general, substituting $i-1$ 0s between terms of $r(x)$ and shifting the resulting polynomial $r(x^i)$ into a register idential to S_1 yields the result

$$S_i = \beta_1{}^i + \beta_2{}^i + \cdots + \beta_j{}^i \tag{11.41}$$

provided that $m_i(x)$ divides $g(x)$. As will be seen shortly, these functions S_i, known as the *power-sum symmetric functions* of the error-location numbers, are sufficient to enable the decoder to correct up to as many errors as there are consecutive, odd-subscripted S_i's. These S_i's can be found from Eq. (11.41); the even ones can be found by noting that in the binary case

$$S_{2i} = S_i^2 \tag{11.42}$$

If exactly t errors occurred, the *elementary symmetric functions* of the error-location numbers would be

$$\begin{aligned}
\sigma_1 &= \beta_1 + \beta_2 + \cdots + \beta_t \\
\sigma_2 &= \beta_1\beta_2 + \beta_1\beta_3 + \cdots + \beta_1\beta_t + \beta_2\beta_3 + \cdots + \beta_2\beta_t + \cdots + \beta_{t-1}\beta_t \\
\sigma_3 &= \beta_1\beta_2\beta_3 + \beta_1\beta_2\beta_4 + \cdots \\
&\vdots \\
\sigma_t &= \beta_1\beta_2\beta_3 \cdots \beta_t
\end{aligned} \tag{11.43}$$

The S_i, where $i = 1, 2, \ldots, 2t-1$, are related to these functions by Newton's identities:

$$\begin{aligned}
S_1 + \sigma_1 &= 0 \\
S_3 + S_2\sigma_1 + S_1\sigma_2 + \sigma_3 &= 0 \\
S_5 + S_4\sigma_1 + S_3\sigma_2 + S_2\sigma_3 + S_1\sigma_4 + \sigma_5 &= 0 \\
\vdots \qquad\qquad\qquad & \vdots \\
S_{2t-1} + S_{2t-2}\sigma_1 + \cdots \qquad + S_{t-1}\sigma_t &= 0
\end{aligned} \tag{11.44}$$

Recall that the S's and σ's represent m-place polynomials for which the operations addition, multiplication, and division are defined. These operations can be implemented in a remarkably simple manner. Addition of two m-tuples requires only a single two-input EXCLUSIVE-OR CIRCUIT and m shifts of the registers containing the addends. Multiplication of two arbitrary m-place polynomials can be implemented with a simple adaptation of the division circuit of Fig. 11.1. Division, while more difficult than either addition or multiplication, can be implemented with not more than two m-stage feedback shift registers (see Peterson[3] for details).

In matrix notation, Eq. (11.44) states that

$$\mathrm{S} = \mathbf{M}_t \mathbf{\Sigma} \tag{11.45}$$

where

$$\mathrm{S} = \begin{bmatrix} S_1 \\ S_3 \\ \vdots \\ S_{2t-1} \end{bmatrix} \qquad \mathbf{\Sigma} = \begin{bmatrix} \sigma_1 \\ \sigma_2 \\ \vdots \\ \sigma_t \end{bmatrix} \tag{11.46}$$

and

$$\mathbf{M}_t = \begin{bmatrix} 1 & 0 & 0 & 0 & 0 & 0 & \cdots & 0 \\ S_2 & S_1 & 1 & 0 & 0 & 0 & \cdots & 0 \\ S_4 & S_3 & S_2 & S_1 & 1 & 0 & \cdots & 0 \\ & & & \cdots & & & & \\ S_{2t-2} & S_{2t-3} & S_{2t-4} & S_{2t-5} & S_{2t-6} & S_{2t-7} & \cdots & S_{t-1} \end{bmatrix} \tag{11.47}$$

It can be shown that this matrix is nonsingular provided exactly t errors occurred. Thus in this case the σ_i can be found by inverting $\mathbf{M}_t$ and multiplying the result by S. This gives

$$\mathbf{\Sigma} = [\mathbf{M}_t]^{-1}\,\mathrm{S} \tag{11.48}$$

Given the σ_i, it is possible to solve for the error location numbers by means of the following decoding equation:

$$\begin{aligned} \Sigma(x) &= (x+\beta_1)(x+\beta_2)\cdots(x+\beta_t) \\ &= x^t + \sigma_1 x^{t-1} + \sigma_2 x^{t-2} + \cdots + \sigma_{t-1}x + \sigma_t \end{aligned} \tag{11.49}$$

If a particular location number, say γ, is substituted into the right side of this equation, the sum will clearly be 0 if γ equals one of the β_i and nonzero otherwise. Thus each of the n possible error-location numbers can be tested, and the bits corresponding to the condition $\Sigma(x) = 0$ can be changed. This step of the decoding procedure can be very simply implemented (see Chien[15] for details).

This completes the description of the decoding algorithm for $j = t$ errors. This procedure also works without alteration for $j = t - 1$. The only difference is that now $\sigma_t = 0$. For $j < t - 1$, several procedures have been proposed; all of these make use of the fact that for mmmm the matrix M_t is not invertible. The oldest and simplest procedure is simply to reduce the size of M_t by 2 when it is found that M_t is not invertible and to try again. Eventually one obtains either M_j or M_{j-1}, which can be inverted. The rest of the correction process then proceeds as discussed earlier.

In summation, then, the decoding algorithm consists of three major steps:

1. Compute the S_i.
2. Calculate the σ_i from the S_i.
3. Form the equation $\Sigma(x)$, test each location for error, and correct the erroneous bits.

Step 1 employs t identical m-stage registers; conveniently these registers are essentially all that are required to implement step 3 also. Step 2 can be accomplished in any one of several different ways.[16-19] The most obvious and direct is to construct the matrix M_t and then invert it. This requires roughly $3t^2/4$ m-stage registers, an arithmetic unit capable of performing multiplication, division, and addition in a finite field, and the necessary control circuitry. For small t, say less than 6 or so, this seems to be the best procedure. It is possible, because of the systematic form of M_t, to break the order-t matrix into two order-$(t/2)$ matrices, one of which is triangular.[16] Although a fair amount of computation is required to do this, for moderately large t, say in the neighborhood of 10 or so, this is more than compensated for by the reduction in the complexity of the matrix inversion. For large t the best procedure is that of Berlekamp[17,18] in which the σ_i are determined from the S_i without actually constructing and inverting the matrix. The complexity of this procedure has been shown to approach linear growth with n for fixed values of code efficiency as n increases without bound. However, it appears to be unattractive for small values of t.

Nothing has been said above about the time required for decoding. Generally speaking, it increases as n^2 for small t and then if Berlekamp's procedure[17] is employed, linearly as n gets large. However, a particular decoder can be built in several fundamentally different ways, e.g., serial or parallel operation, and decoding time varies widely among these different implementations.

Nonbinary codes. The BCH codes and, in fact, cyclic codes in general, are defined in the nonbinary as well as the binary case. That is, cyclic codes whose symbols are chosen from a set of size q can be constructed provided q is a power of a prime.[3] However, because of the prevalence of binary data and the relative ease with which binary codes can be implemented, nonbinary codes are primarily of theoretical interest. One exception to this statement is the class of *Reed-Solomon*[20] *codes*. The most interesting codes in this class have symbols chosen from the field of 2^m elements and have length $2^m - 1$. Thus each symbol can be represented as a binary m-tuple, and the code can be considered as a type of binary code. These codes are capable of correcting up to t-symbol errors provided the code has at least $2t$ parity-check symbols. If the modulation system employs 2^m signals and if all the off-diagonal entries in the transition matrix of the channel are roughly equal, this type of coding is extremely efficient. Also, generalizations both of the (binary) decoding procedure just described and of Berlekamp's improved procedure have been devised for these codes.[17,21,22] However, the nonbinary nature of the codes makes decoding inherently more involved than in the binary case.

Interpreted as a binary code, a t-symbol-error-correcting Reed-Solomon code is capable of correcting, in addition to all error patterns containing t or fewer bit errors, all single bursts of length up to $m(t-1) + 1$, several shorter bursts in a single word, and numerous combinations of burst and random errors. However, compared with the strictly burst-correcting codes of Sec. 11.3, this burst-correcting ability employs a great deal of equipment.

With respect to the Reed-Solomon codes, it should be pointed out that if the transition matrix is not "flat," as above, but instead peaked about the main diagonal (as is so with pulse-amplitude modulation and additive gaussian noise, for example), binary coding is superior. For with a suitable mapping of levels to binary m-tuples, e.g., Gray coding,[23,24] the average number of bit errors per symbol error is quite close to unity.

11.2.3 Majority-logic Decodable Codes

The codes discussed in this section form a much smaller subclass of the class of cyclic codes than do the BCH codes. Also they are slightly inferior to the latter codes in terms of error-correcting ability for most interesting values of code length and efficiency. However, majority-logic decoding is very simply implemented—often much more simply than either of the decoding algorithms for the BCH codes. Thus, the two types of codes are often competitive and provide the designer of an error-control system with a certain amount of freedom in choosing a code.

Before describing the construction of the codes, which is rather involved, we shall explain the notion of majority-logic decoding,[25-27] which is not.

Basic majority-logic decoding. The concept of a set of orthogonal *parity-check sums* will be used extensively throughout this section. A brief discussion of these terms is necessary.

In the description of the parity-check matrix presented in Sec. 10.2.1, it was pointed out that for any code word the dot product of the code word and any row or linear combination (sum) of rows of the parity-check matrix $\mathbf{H}$ must be 0. That is, if $c = (c_{n-1}, c_{n-2}, \ldots, c_1, c_0)$ and $h = (h_{n-1}, h_{n-2}, \ldots, h_1, h_0)$ represent, respectively, a code word and some linear combination of rows of $\mathbf{H}$, the modulo-2 sum

$$c_{n-1}h_{n-1} + c_{n-2}h_{n-2} + \cdots + c_1h_1 + c_0h_0 = 0 \qquad (11.50)$$

For an n-tuple which is not a code word, some sums of this form must be nonzero. In general this type of sum is referred to as a *parity-check sum*. Since h_i is either 1 or 0 in the binary case, such a sum is simply the modulo-2 sum of the symbols of the n-tuple in positions which correspond to 1s in the vector h. A set of check sums is said to be *orthogonal* on a particular bit position if that bit is involved in every sum in the set and no other bit is involved in more than one sum. Thus the set of sums in Eq. (11.51) is orthogonal on the bit c_0.

$$\begin{array}{llll} c_6 & + c_3 & & + c_0 \\ c_5 & & + c_2 & + c_0 \\ c_4 & & c_1 & + c_0 \end{array} \qquad (11.51)$$

If it is possible to construct a set of $d-1$ parity-check sums orthogonal on a particular bit in a cyclic code, then, because of the cyclic symmetry of the code, such a set can be constructed on every bit. Now consider an n-tuple of weight $d-1$ or less and the set of check sums orthogonal on one of the nonzero bits. There are $d-1$ sums and at most $d-2$ 1s in these sums in addition to the 1 common to every sum in the set. Thus at least one sum will be nonzero. But this says that no n-tuple of weight $d-1$ or less can be a code word.

Theorem 11-3: A cyclic code in which it is possible to form a set of $d-1$ parity-check sums orthogonal on any bit has minimum distance at least d.

Besides stating that certain codes are capable of correcting up to $t \leq [(d-1)/2]$ random errors, this result suggests a method of constructing the necessary decoders. The first step of the decoding procedure is simply to calculate the $d-1$ parity-check sums orthogonal on the first (highest-order) bit in the code. If $t \leq [(d-1)/2]$ errors occurred and one of them occurred in the first bit, at least

$$(d-1) - \left(\left[\frac{d-1}{2}\right] - 1\right) = \left[\frac{d+1}{2}\right] \tag{11.52}$$

sums will be nonzero. However, if the first bit was not in error, at most $[(d-1)/2] < [(d+1)/2]$ sums will be 1.

Two different implementations of the process, roughly corresponding to the two different methods of encoding cyclic codes, are known.[26] The first of these, designated Type I, is shown in Fig. 11.5.

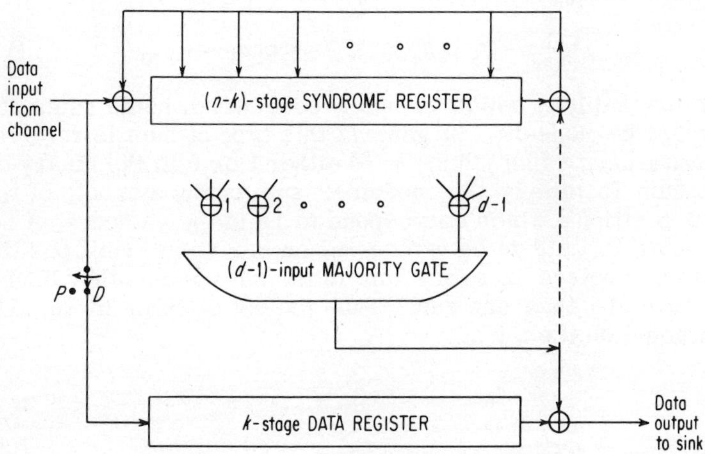

Fig. 11.5. Type I majority-logic decoder.

It corrects errors by performing the following steps.

1. The n-bit received word is shifted into the $(n\text{-}k)$-stage SYNDROME REGISTER which, as explained earlier, divides it by $g(x)$ to produce the syndrome. The k-bit data section of the word is simultaneously shifted into the k-stage DATA REGISTER.
2. Each of the $d-1$ EXCLUSIVE-OR gates forms one of the $d-1$ check sums orthogonal on the highest-order data bit.
3. Each of these check sums is fed into a $(d-1)$-input MAJORITY GATE. This gate has the property that its output is a 1 if and only if more than half its inputs are 1s and is 0 otherwise.

Thus the first output of this gate is a 0 if the first information bit is correct and 1 if it is not, provided $t = [(d-1)/2)$ or fewer errors occurred.

4. This output is added to the first information bit, correcting it if it was erroneous or leaving it unaltered if it was correct.
5. The DATA REGISTER and SYNDROME REGISTER are shifted once. Thus the uncorrected bit is shifted out to the data sink and replaced by the second data bit. Also, by Theorem 11-1, the new syndrome is related to this bit in exactly the same manner that the previous syndrome was related to the first bit. Thus the second bit can be corrected in exactly the same manner as the first.
6. This process is repeated a total of k times, thereby correcting all data-bit errors, provided, of course, that the total number of errors was t or fewer. If for some reason it is desired to correct parity errors, this can be done by simply storing the received parity bits in an extension of the DATA REGISTER and repeating the correction process n-k more times.

The second method of implementing majority-logic decoding employs the Type II decoder illustrated in Fig. 11.6.

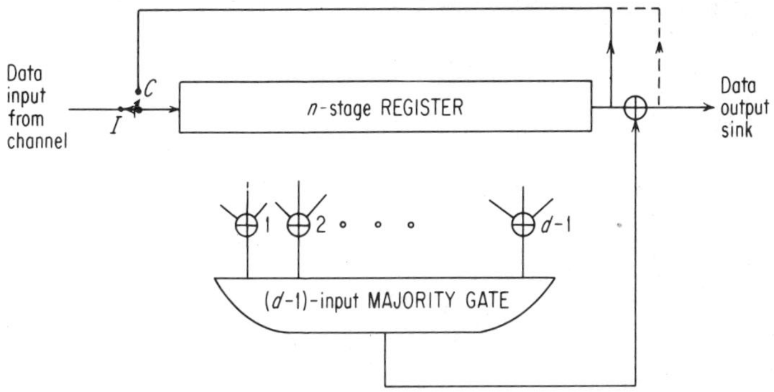

Fig. 11.6. Type II majority-logic decoder. The dashed line indicates that information is to be fed back *after* correction.

It operates as follows.

1. With the switch in position I (input), the n-bit received word is shifted into the register, and the switch is thrown to position C (correct).
2. The check sums are computed by the $d-1$ EXCLUSIVE-OR CIRCUITS.
3. As before, provided t or fewer errors occurred, the output of the MAJORITY GATE is a 1 if and only if the first data bit is in error. Thus correction can be performed by adding the MAJORITY GATE output to this bit.
4. Then the register is shifted cyclically once. Because the code is cyclic, the second bit can be corrected exactly as the first bit was. The process is repeated a total of k times until all errors in the data bits have been corrected. Again parity errors can be corrected by simply performing n-k more shift-correct operations.

The implementation of the decoding procedures described above and illustrated in Figs. 11.5 and 11.6 can often be simplified slightly. However, because of the explanatory difficulties involved these simplifications are ignored here.

Since both preceding methods employ exactly n stages of shift register, this criterion cannot be used to decide which can be implemented more economically. However, for the most important codes to which the majority-logic decoding applies, the total number of inputs to the EXCLUSIVE-OR is determined by the length of the register on which they operate. That is, for a Type I decoder a total of n-k inputs are required, while a Type II decoder employs roughly n inputs. For low-rate codes this advantage is offset somewhat by the EXCLUSIVE-OR necessary to calculate the syndrome in a Type I decoder, but for high-rate codes it seems clear that the Type I decoder is the more simply implemented.

Example 11-5 is intended to clarify the concepts just presented.

Example 11-5. Consider the (7,3) BCH code for which

$$g(x) = \mathrm{LCM}[m_0(x), m_1(x)] \tag{11.53}$$

In other words, from Table 11.3,

$$g(x) = (x+1)(x^3+x+1) = x^4+x^3+x^2+1 \tag{11.54}$$

From Sec. 10.2.4 the generator matrix of the code is

$$\mathbf{G} = [\mathbf{I}_3\mathbf{P}] = \begin{bmatrix} 1&0&0&1&1&1&0\\ 0&1&0&0&1&1&1\\ 0&0&1&1&1&0&1 \end{bmatrix} \tag{11.55}$$

while its parity-check matrix is

$$\mathbf{H} = [\mathbf{P}^T\mathbf{I}_4] = \begin{bmatrix} 1&0&1&1&0&0&0\\ 1&1&1&0&1&0&0\\ 1&1&0&0&0&1&0\\ 0&1&1&0&0&0&1 \end{bmatrix} \tag{11.56}$$

From the definition of $\mathbf{H}$ it follows that the dot product of a code vector $\mathbf{c}$ and any linear combination of rows of $\mathbf{H}$ is 0. Consider the three combinations, row 1, row 3, and row 2 plus row 4. If $\mathbf{c} = (c_6\, c_5\, c_4\, c_3\, c_2\, c_1\, c_0)$, these combinations represent, respectively, the sums

$$\begin{aligned} &c_6 + c_4 + c_3 \\ &c_6 + c_5 + c_1 \\ &c_6 + c_2 + c_0 \end{aligned} \tag{11.57}$$

Since the sums are orthogonal on c_6, by Theorem 11-3 the code has minimum distance at least 4 (in fact, it is exactly 4). The Type I decoder for this code in shown in Fig. 11.7(a); the equivalent Type II decoder is shown in Fig. 11.7(b).

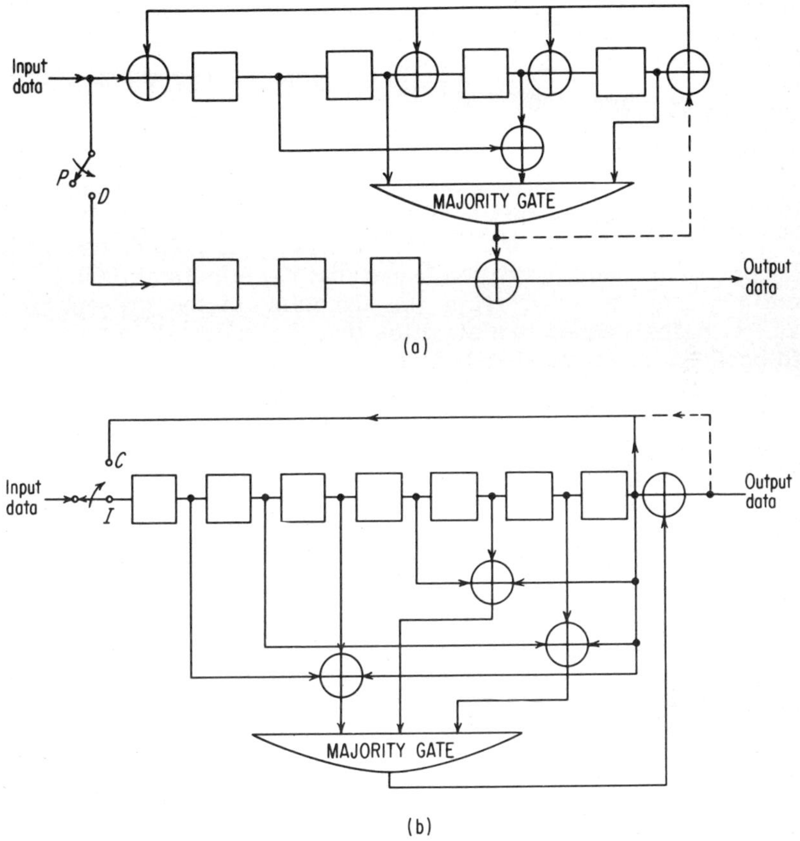

Fig. 11.7. (a) Type I and (b) Type II majority-logic decoders for (7,3) code.

To simplify the required explanation, we have assumed that t or fewer errors occurred. While there certainly exist patterns of weight greater than t that will cause an erroneous decoding, it should be clear that there are also error patterns of weight greater than t that are correctable. Thus majority-logic decoding differs in this respect from the bounded-distance decoding procedures for the BCH codes.

Although some patterns of weight greater than t can be corrected with majority-logic decoding, this decoding procedure is not optimal. That is, many patterns of weight slightly greater than t which are within the power of the code to correct result in decoding errors. The desire to improve the algorithm and apply it to a larger class of codes has led to the following procedures.

Modifications of the basic decoding procedure. 1) Use of feedback. Since the first bit to be decoded is presumably known before the decoding of successive bits, its effect can be removed before these latter are decoded. Thus if $t+1$ errors occurred and an error in the first bit was successfully corrected, the remaining t errors can also be corrected. Similar arguments hold for error patterns of higher weight. This modification has the effect of reducing slightly the conditional probability of decoding erroneously given that slightly more than t errors occurred. It has the additional virtue that it can be implemented quite simply. With reference to Figs. 11.5 and 11.6, it is only necessary to include one dashed connection in each figure to ensure that the effects of earlier corrections are removed from the decoding of the present bit. In Fig. 11.6 the dashed line signifies that the information bit is to be fed back after, rather than before, correction.

2) Variable-threshold decoding.[28,29] This is another modification that results in substantial reductions in the above-mentioned conditional probability when long codes are employed. It can be applied to any majority-logic-decodable codes. However, it requires substantially more time to decode a word than does the procedure described above. On the other hand, the only change required in the decoders of Figs. 11.5 and 11.6 is that the MAJORITY GATE threshold must be capable of being varied in time and the threshold-setting circuitry necessary to control this gate must be added (see below).

Variable threshold decoding operates as follows. With the threshold set at its highest value, $d-1$, the decoder attempts to decode each bit of the received word. A given bit is changed if and only if all $d-1$ inputs to the majority gate are 1 at the appropriate time. When the decoder has attempted to decode each of the n bits of the word unsuccessfully, the threshold is lowered by one. If another cycle is completed with no changes, the threshold is again lowered by one. If a change is made, the threshold is immediately increased by one. At the completion of every cycle it is lowered by one, except initially, when it is set at $d-1$. Thus the threshold is gradually lowered until it reaches its minimal value $[(d+1)/2]$, where decoding ceases. Although no analytic results have been derived, computer simulations have shown variable threshold decoding to be substantially superior to fixed-threshold decoding when long codes are used.[28]

3) L-step decoding.[25,26] The concept of a set of check sums orthogonal on a single bit generalizes without undue complication. A set of check sums is said to be orthogonal on a sum of bits if every bit in the sum is involved in every check sum in the set and no other bit is involved in more than one check sum.

For example, if we let c_0 in Eq. (11.51) represent the sum $c_0' + c_0''$, the three check sums are orthogonal on this sum. By the argument employed in the proof of Theorem 11-3, the value of any sum on which it is possible to construct $d-1$ orthogonal check sums can be determined correctly, provided $t = [(d-1)/2]$ or fewer errors occurred.

Once determined in the above manner, a sum of bits can be regarded as a check sum and thus can be used in decoding. For example, it can be added to other known check sums and possibly used to solve algebraically for the values of various bits. Further, a sum of bits can also be used as a check sum in a set of sums orthogonal on a bit or another sum. This latter process of employing a succession of L majority decisions to decode a bit has been called *L-step decoding*,[26] or the *Reed algorithm*,[25] after its inventor. Although its implementation is obviously more complex than that of one-step majority decoding, it has the advantage that it applies to a class of rather attractive cyclic codes. These codes as well as codes decodable in one step will now be described.

The codes.[25–34] The construction of the majority-logic decodable cyclic codes discussed below relies heavily on the properties of finite geometries.[35–37] Although the definition of a finite geometry is given and a few of its properties employed in the development of the codes, this brief introduction can do little more than acquaint the reader unfamiliar with the notion of a finite geometry with the necessary terminology. While this will be sufficient for the reader interested only in determining the macroscopic properties of the codes (in order to estimate their performance in a data-communications system, say), a detailed understanding of the codes requires more background than can be included here (see Veblen and Young[36] and Carmichael[37] for a treatment of finite geometries and Massey,[26] Kasami, Lin, and Peterson,[31] and Weldon[33] for a more detailed discussion of the codes based on them).

Codes can be constructed by considering either Euclidean or projective geometries, but since the latter are simpler to explain, they will be considered in more detail.

A z-dimensional projective geometry is defined as a set of points which are subject to the following set of postulates:[37]

1. It contains a finite number of points and one or more subsets called *lines*, each of which contains at least three points.
2. Two distinct points lie on one and only one line.
3. If A, B, and C are points not all on the same line and if a line l contains a point D of the line AB and a point E of the line BC but does not contain A or B or C, then l contains a point F on the line CA.
4. Not all the points considered are in the same L space, L an integer less than z.

5. If postulate 4 is satisfied, there exists in the set of points considered no $(z+1)$ space.

In such a geometry a line consists of the set of $q+1$ points linearly dependent on two linearly dependent points. More generally an L-space consists of the set of $q^L + q^{L-1} + \cdots + q + 1$ points linearly dependent on $L+1$ linearly independent points. Because of our interest in the binary codes, q will always be taken to be of the form 2^s.

The bit positions (numbered $0, 1, \ldots, n-1$) in a cyclic code of length

$$n = 2^{sz} + 2^{s(z-1)} + \cdots + 2^s + 1 \tag{11.58}$$

can be put into 1-to-1 correspondence with the n points in a z-dimensional finite projective geometry over $GF(2^s)$, that is, $PG(z,2^s)$. Denote the bit positions corresponding to the points of an L-dimensional subspace of this geometry, $L < z$, by the numbers e_1, $e_2, \ldots, e_p$ where $p = 2^{sL} + 2^{s(L-1)} + \cdots + 2^s + 1$. Then we associate with this L-*flat*, as it is called, the polynomial

$$p(x) = x^{e_1} + x^{e_2} + \cdots + x^{e_p} \tag{11.59}$$

Now suppose that every such polynomial is in the null space of a cyclic (n,k)-code. That is, the sum of the bits corresponding to the points of each L-flat is zero for every code word in the code. Then from the following statement stems the usefulness of geometry in constructing majority-logic decodable codes: the check sums corresponding to the set of L-flats which intersect only on a single flat of smaller dimension are orthogonal on the check sum of that flat. Thus if the number of L-flats orthogonal on an $(L-1)$-flat is $d'-1$ and $t = [(d'-1)/2]$ or fewer errors occur, the decoder can determine the check sums corresponding to all $(L-1)$-flats. Because the number of "orthogonal" flats increases as their dimensionality decreases, the decoder can then determine the check sums associated with the $(L-2)$-flats, $(L-3)$-flats, and finally the 0-flats. But a 0-flat contains only one point, so correction can then be performed. Thus we have proved that any code whose null-space contains all L-flats can be L-step orthogonalized to correct up to $[(d'-1)/2]$ errors, where now d' represents the number of L-flats which intersect on a single $(L-1)$-flat.

Given a projective geometry, i.e., given s and z, it is relatively simple to determine the number of L-flats orthogonal on an $(L-1)$-flat. Since every point in the geometry will appear in one of the flats and since $2^{s(L-1)} + \cdots + 2^s + 1$ points appear in every L-flat, it follows that

$$d' - 1 = \frac{(2^{sz} + \cdots + 2^s + 1) - (2^{s(L-1)} + \cdots + 2^s + 1)}{(2^{sL} + \cdots + 2^s + 1) - (2^{s(L-1)} + \cdots + 2^s + 1)} \tag{11.60}$$

Then since d is clearly not less than d', we have the following (usually tight) lower bound on d:

$$d \geq 2^{s(z-L)} + \cdots + 2^s + 2 \tag{11.61}$$

The problem of determining k, the number of information symbols in the code, is not so simple, however. It turns out that the generator polynomial contains as roots all $\alpha^{i(2^s-1)}$ such that the number $i(2^s - 1)$, when written in binary form, has weight $s(z - L)$. The symbol α here denotes a primitive element $GF(2^{s(z+1)})$. The proof of this assertion is too involved to present here; the interested reader is referred to Kasami, Lin, and Peterson,[31] Weldon,[32] and Graham and MacWilliams.[34]

The properties of projective geometry codes can be summarized thus:

$$n = \frac{2^{s(z+1)} - 1}{2^s - 1} = 2^{sz} + 2^{s(z-1)} + \cdots + 2^s + 1$$

k = number of integers less than $2^{s(z+1)} - 1$ divisible by $2^s - 1$, whose weight, when written in binary form, exceeds $s(z - L)$ (11.62)

$$d \geq 2^{s(z-L)} + \cdots + 2^s + 2$$

When $s = 1$, the class of projective geometry codes reduces to the even-minimum-distance Reed-Muller codes. For these codes and, indeed, for many other projective geometry codes, the bound on d is known to hold with equality. In addition, a simple combinatorial expression for k has been found for these codes.

There is another class of majority-logic-decodable cyclic codes, based on Euclidean rather than projective geometries, that are similar in structure to the codes described above. Euclidean geometry codes have the following properties:[33]

$$n = 2^{sz} - 1$$

k = number of integers less than $2^{sz} - 1$ whose s weight is not greater than $L - 1$ (11.63)

$$d \geq 2^{s(z-L)} + \cdots + 2^s + 1$$

The parity-check polynomial $h(x)$ for these codes contains all roots α^i such that the s weight of i does not exceed $L - 1$, where α

is a primitive element of $GF(2^{sz})$. The *s-weight* of a number is defined as the maximum number of disjoint multiples of $2^s - 1$ in binary form contained in the binary representation of the number. (For details see Kasami, Lin, and Peterson[31] and Weldon.[32]) Again as with the projective geometry codes, when $s = 1$, these codes reduce to Reed-Muller codes—this time to the odd-minimum-distance codes. For these latter the lower bound on d is known to hold with equality, and a simple expression for k is known. These last statements hold for several other subclasses of Euclidean geometry codes as well.

Although the algebraic properties of the codes in these two classes are rather different, they differ very little in the way they are encoded or decoded. Hence in the remainder of this section no distinction will be drawn between them.

From the brief description of majority-logic decoding given earlier it should be clear that the complexity, and hence cost, of a majority-logic decoder increases with L, the number of decoding steps. (In fact, it increases exponentially with L.) For this reason one-step decoding is much more attractive for use in real systems. Fortunately there are many useful codes for which $L = 1$.[30,34] A number of these are tabulated in Table 11.5. Besides the generator polynomial of each code, the decoding rules derived from its geometric structure are also tabulated. In addition an example illustrating the use of Table 11.5 is given. With it and sufficient diligence the reader should be able to design a decoder for any of the codes listed.

Where $L > 1$ it is still possible to construct a one-step decoder, provided one is willing to sacrifice some of the error-correcting ability of the code.[28] A one-step decoder is constructed by finding the largest possible set of L-flats which are orthogonal on a single point, rather than on an $(L-1)$-flat, as earlier. This obviously reduces the number of orthogonal flats that can be constructed. Sometimes this reduction in error-correcting ability is small, however, and compensated for by a rather substantial reduction in decoder complexity.

The reader may notice, upon examining the BCH codes listed in Tables 11.3 and 11.4 and the majority-logic-decodable codes of Table 11.5, that several BCH codes are majority-logic-decodable. In fact, some of the geometrically constructed codes of this section are known to be BCH codes, while others have been shown to be related to BCH codes in an interesting way.[31-34]

Some majority-logic-decodable noncyclic codes. Up to this point we have considered cyclic codes and the closely related shortened cyclic codes exclusively. We now turn our attention briefly to an interesting class of codes which, though not cyclic, seem to share the implementation advantages of cyclic codes. These *quasi-cyclic codes*[28] are best defined in terms of their parity-check matrices.

The $(n\text{-}k)$-by-n parity-check matrix of an (n,k) quasi-cyclic code has the form

$$\mathbf{H} = \begin{bmatrix} \mathbf{p}_{m-1}\,\mathbf{0}\,\mathbf{p}_{m-2}\,\mathbf{0} \cdots & \mathbf{p}_0\,\mathbf{I} & & \\ & \mathbf{p}_{m-1}\,\mathbf{0} \cdots & \mathbf{p}_2\,\mathbf{0}\,\mathbf{p}_0\,\mathbf{I} & \\ & \ddots & \ddots & \\ & & \mathbf{p}_{m-1}\,\mathbf{0}\,\mathbf{p}_{m-2}\,\mathbf{0} \cdots & \mathbf{p}_0\,\mathbf{I} \end{bmatrix} \tag{11.64}$$

where $\mathbf{0}$ and $\mathbf{I}$ denote, respectively, the all-zero and identity matrices of order n_0-k_0, $\mathbf{p}_i$ is an arbitrary (n_0-k_0)-by-k_0 matrix, and the blanks denote 0s. There are $m(n_0-k_0) = n\text{-}k$ rows in $\mathbf{H}$ and $mn_0 = n$ columns. Of course, the matrix of Eq. (11.64) can be put into systematic form without affecting the minimum distance of the code. In fact, the matrix $\mathbf{H}$ can be rearranged so that it is composed entirely of circulant submatrices.

In a quasi-cyclic code any cyclic shift of a code word by n_0 bits results in another code word. Hence the codes lend themselves readily to shift register implementation since each n_0-bit section can be processed identically.

The class of quasi-cyclic codes is considerably larger than the class of cyclic codes and in a nontrivial way contains numerous cyclic codes as members. (Trivially all parity-check codes are quasi-cyclic codes for which $m = 1$.) Another subclass of the class of quasi-cyclic codes which have been investigated are the *self-orthogonal codes*.[28] These codes derive their name from the property that in such a code various parity-check digits are orthogonal on each information digit. This permits a very simple form of majority-logic decoding to be employed. Unfortunately the restriction that parity-check digits, rather than sums of such digits, be orthogonal on information digits forces the codes to be relatively long. Specifically

$$n \geq \frac{R(d-1)(d-2)}{(1-R)^2} + \frac{1}{1-R} \tag{11.65}$$

therefore, n increases roughly as the square of the minimum distance for a given value of code efficiency R. Consequently, given values of R and d, self-orthogonal quasi-cyclic codes are longer than the BCH or majority-logic-decodable cyclic codes. However, the codes are quite easy to implement[28] and therefore may be attractive in some situations.

11.2.4 Other Decoding Techniques

Presented in the preceding two sections were a class of random-error-correcting cyclic codes and a decoding algorithm that used specialized algebraic properties of the codes. Although these codes are defined over practically all values of code length and efficiency, together they constitute only a very small subclass of the class of cyclic codes. Unfortunately, however, relatively little is known at present about constructing or decoding efficient random-error-correcting cyclic codes not in this subclass.

In this section several decoding techniques applicable to cyclic codes in general are presented. As will be seen, the implementation of these techniques seems to be simpler than that of the methods of the preceding sections for short codes but rapidly becomes impractically complex as the code length increases.

Systematic-search decoding. It has been pointed out that decoding consists of associating an error pattern with a syndrome. The simplest method of doing this, conceptually at least, is to generate the syndrome associated with each of the error patterns one wishes to correct and then compare it with the actual syndrome. When a match is achieved, the error pattern is known. Then, provided t or fewer errors occurred, correction can be effected by adding the pattern to the received word. Required to implement this technique are a syndrome generator programmed to generate successively the syndromes of all correctable error patterns and a multiple input gate to recognize the occurrence of a match between the locally generated syndrome and the syndrome of the received word.

Since there are $\binom{n}{i}$ error patterns of weight i, it is necessary to search through

$$\sum_{i=0}^{t} \binom{n}{i} \tag{11.66}$$

error patterns to correct all patterns of weight t or less. Obviously this procedure is impractical if t and n are not small. For example, if $n = 100$ and $t = 5$, approximately 10^8 error patterns must be examined in 10^2 data-bit times. With the data rates currently in use and with presently available logic circuitry, not more than a few tens of thousands of operations can be done conveniently by the decoder in one bit time. As a result this procedure is typically limited in applicability to the cases of $t = 2$ or 3 with moderate n.

In principle it is also possible to decode by systematically generating the 2^k code words and comparing them with the received n-tuple. This sort of decoding is limited to codes with relatively few code words since the number $2^k/n$ is limited by the practical considerations mentioned above. For such low-rate codes the decoding technique described below is usually more simply implemented.

Error-trapping decoding. Suppose a code word $c(x)$ in a t-error-correcting cyclic code is transmitted through a channel and t or fewer errors occur in the parity section of the word. The syndrome

$$s(x) = \text{rem}\,\frac{r(x)}{g(x)} = \text{rem}\,\frac{c(x)}{g(x)} + \text{rem}\,\frac{e(x)}{g(x)}$$

$$= \text{rem}\,\frac{e(x)}{g(x)} = \text{rem}\,\frac{e_d(x)}{g(x)} + e_p(x) \tag{11.67}$$

where, as earlier, the received polynomial $r(x)$ is the sum of $c(x)$ and the error pattern $e(x)$. The polynomials $e_d(x)$ and $e_p(x)$ refer to the data and parity sections, respectively, of $e(x)$. But by the hypothesis, $e_d(x) = 0$, and therefore $s(x) = e_p(x)$. In other words, the syndrome preceded by k zeros *is* the error pattern and correction can be accomplished simply by adding it to the received n-tuple.

Now suppose that the error pattern which occurred was not $e(x)$ but $e'(x) = x^{-1}e(x)$, which is $e(x)$ cyclically shifted one step toward the low-order end of the word. By Theorem 11-1 the syndrome of this error pattern is

$$s'(x) = \text{rem}\,\frac{x^{-1}e(x)}{g(x)} = \text{rem}\,\frac{x^{-1}s(x)}{g(x)}$$

or equivalently

$$s(x) = \text{rem}\,\frac{xs'(x)}{g(x)} \tag{11.68}$$

Thus if $s'(x)$ is stored in a circuit which divides by $g(x)$, such as the syndrome generator of Fig. 11.1(a), a single shift of the register yields $s(x)$. As a result, any error pattern whose t or fewer errors are confined to n-k consecutive positions can be corrected with the circuit shown in Fig. 11.8. This decoder operates as follows.

The received word is shifted from the channel into the DATA REGISTER; simultaneously the syndrome is calculated so that when the entire word has entered, $s(x)$ is in the SYNDROME REGISTER. If the weight of the syndrome is t or less, the WEIGHT SENSOR causes the contents of the SYNDROME REGISTER to be added to the last n-k bits in the DATA REGISTER. (This operation, while shown as being performed in parallel, can just as easily be done serially.) If the weight is greater than t, both registers are

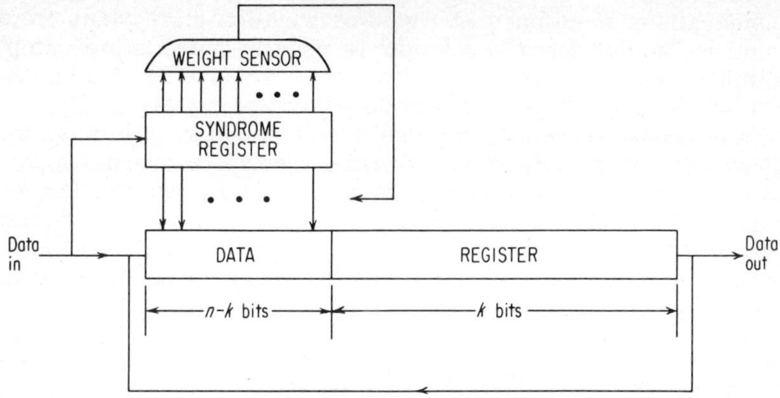

Fig. 11.8. Error-trapping decoder.

shifted cyclically once. If the weight of the new "syndrome" is t or less, the error pattern which occurred was confined to the nth through $(n-k+1)$th bits. But these bits are now in the last n-k stages of the DATA REGISTER; thus if the weight of the syndrome is $\leq t$, correction can be accomplished as above by adding the syndrome to the DATA REGISTER. Similarly any error pattern of weight t or less can be corrected provided all errors are confined to some span of n-k or fewer bits.

This technique is extremely easy to implement; in fact it is perhaps the most simply implemented technique in situations where it can be used at all. Unfortunately, it is limited to decoding relatively short inefficient codes, single-error-correcting codes, and some double-error-correcting codes. In such situations it seems to be almost uniformly superior to the second of the two systematic-search algorithms.

In general this technique can always be used to correct $t' \leq t$ errors provided

$$t' < \frac{n}{k} = \frac{1}{R} \tag{11.69}$$

This result follows directly from the fact that the number of parity checks n-k must exceed the maximum-minimum span between errors in an n-tuple of weight t'. Since the error-correcting ability of this algorithm is determined only by the code efficiency, it cannot be used to decode long, high-rate codes unless one is willing to sacrifice much of the error-correcting ability of the code.

Various generalizations of this technique have been proposed in an effort to extend its range of usefulness. One of these, *permutation decoding*,[41] employs other code-preserving permutations of the n-bit positions in addition to the cyclic permutation. These allow one to "collect" numerous errors into some $(n$-$k)$-bit section of the received word. The cyclic permutation described above is then employed to correct the errors. While permutation decoding is applicable to some codes that cannot be corrected by simple error trapping, at present it, too, seems limited to relatively short, inefficient codes. However, this may be a consequence of the present limited knowledge of code-preserving permutations and thus could change as more is learned about them.

Another generalization of error trapping relies on the fact that if one error occurs outside some $(n-k)$-bit span, its position can be determined with little added equipment.[40,42] That is, if it is always possible to shift all but one error into the $(n-k)$-bit parity section of the word and at the same time shift the other error into one of a small number of predesignated positions in the block, the error pattern can be corrected. For corresponding to an error in a given position in the data section of the word is a particular syndrome; if $t'-1$ other errors lie in the parity section of the word, the presence of an isolated error can be detected by a gate which senses a weight difference of less than t' between the register contents and the aforementioned syndrome.

This technique is applicable to a few codes which cannot be decoded with simple error trapping. Perhaps its most important application is to double-error-correcting codes for which Eq. (11.64) does not hold; that is, for which $R > 1/2$. Any such code can be decoded in this manner provided a sufficient number of syndrome-recognition gates are employed.

In theory this modified error-trapping technique can also be used to locate two or more errors in the data section. Since one multiple-input gate is required for each pattern to be recognized, however, other decoding techniques are usually simpler to implement.

Hybrid techniques. By combining systematic-search and error-trapping decoding, it is possible to design for some codes decoders which are simpler than if either technique were used separately. Also, combinations of these techniques and those of Secs. 11.2.2 and 11.2.3 may sometimes be effective. One obviously attractive combination is that of systematic-search and majority-logic decoding. This approach would be useful for decoding codes in which the maximum error-correcting ability cannot be realized by majority-logic decoding alone.

11.3 Burst-error Correction

We shall now treat cyclic and shortened cyclic codes which are able to correct a single burst of errors and we present a simply implemented decoding algorithm for such codes. Codes for correcting combinations of random and burst errors are discussed in Sec. 11.4.

11.3.1 Bounds

In Sec. 10.2.3 we saw that if a burst of length l is a code word in a linear code, the burst-correcting ability of the code could not exceed $[(l-1)/2)]$. We now show that in an (n,k) code there is at least one code word in which all the 1s in the word appear in a burst of length $(n-k+1)$ or less,[44] which begins on a given bit. There are only 2^{n-k} cosets in such a code, and two of the $(2^{n-k}+1)$ n-tuples (including the all-zero n-tuple) that have all their 1s in such a burst must be in the same coset. Thus their sum, which by Theorem 10-3 is a code word, will also have all of its 1s confined to a burst of length not greater than $n-k+1$. Equation (10.17) therefore gives

$$b \leq \frac{n-k}{2} \tag{11.70}$$

Note that we have not specified the position of the burst of length $n-k+1$ in the code word in the above development; clearly such a code word must exist for all positions of the burst within the word. This fact will be useful in applying the above result to convolutional codes in Sec. 12.3.

Another upper bound on the burst-correcting ability of a linear code can be obtained as follows. There are 2^{b-1} bursts of length b or less which start on a given bit. These correspond to the set of $(b-1)$-tuples preceded by a single 1. In a cyclic (n,k) code, counting bursts that overlap both ends of an n-tuple, there are n positions where a burst can start. Thus since every correctable error patterns must have a unique syndrome, of which there are 2^{n-k}, for a cyclic code to be burst-b-correcting it is necessary that

$$2^{n-k} \geq n2^{b-1} \tag{11.71}$$

or

$$n-k > b-1+\log n \tag{11.72}$$

In a shortened cyclic code, error patterns that consist of a burst at each end of the n-tuple are not regarded as bursts; thus there are only $n-b+1$ positions where a burst of length b can start. Straightforward algebraic manipulations yield the bound:[45]

$$n-k \geq b-1+\log(n-b+2) \tag{11.73}$$

This result and Eq. (11.70) provide, as did the Hamming and Plotkin bounds with random-error correction, a lower bound on the number of parity checks necessary for a code to have a desired error-correcting ability.

A bound analogous to the Varsharmov[8]-Gilbert[9]-Sacks[10] bound can be derived for burst-correcting codes. This bound states that it is possible to construct an (n,k) burst-b-correcting code for all b such that

$$n - k > 2(b - 1) + \log(n - 2b + 2) \tag{11.74}$$

The proof of this result, omitted here, is given in Peterson.[3]

Examination of these results shows that Eq. (11.70) provides a better bound than Eqs. (11.72) and (11.73) for large n and fixed code efficient R. In the limit Eq. (11.74) states that all values of b less than $(n\text{-}k)/2$ are attainable while Eq. (11.70) states that no higher values can ever be attained. In brief, for large n, $b = (n\text{-}k)/2$ is an (almost) attainable optimum for linear codes.

11.3.2 Decoding

A burst-correcting cyclic code can be decoded with a slight variation on the error-trapping decoder described in Sec. 11.2.4.[3] Suppose a correctable burst of length $b' < n\text{-}k$ occurs in the parity section of the word. Since no errors occurred in the data section, $e_d(x) = 0$ and

$$e(x) = e_p(x) \tag{11.75}$$

As before $e(x)$ denotes the polynomial representation of the error pattern and $e_d(x)$ and $e_p(x)$ denote, respectively, the data and parity sections of the polynomial. Then since the syndrome

$$s(x) = \text{rem}\,\frac{e(x)}{g(x)} = \text{rem}\,\frac{e_p(x)}{g(x)} = e_p(x) \tag{11.76}$$

the burst can be corrected simply by adding the syndrome to the received word.

If the error pattern was not $e(x)$ but $e'(x) = x^{-1}e(x)$ [in other words, $e(x)$ shifted cyclically 1 bit in the low-order direction], by Theorem 11-1 the syndrome of this error pattern $s'(x)$ would be related to $s(x)$ as follows:

$$s(x) = \text{rem}\,\frac{xs'(x)}{g(x)} \tag{11.77}$$

From Eq. (11.77) we observe that this burst can be corrected simply by cyclically shifting the syndrome $s'(x)$ once in its syndrome generator and adding the result $s(x)$ to a shifted version of the received word. The above argument holds with x replaced by x^i; thus a correctable burst in any position in the word can be corrected with this technique.

The above decoding scheme can be implemented as shown in Fig. 11.8 with the modification that the WEIGHT-SENSING GATE is replaced by the BURST-RECOGNIZING CIRCUIT (see Sec. 11.2.4 for a description of the operation of this decoder). This circuit takes on various forms depending on how the burst-correcting code is to be used. If only bursts of length not greater than the guaranteed burst-correcting ability b are to be corrected, the BURST-RECOGNIZING CIRCUIT consists of simply of a single large AND-GATE whose $(n\text{-}k\text{-}b)$ inputs are the high-order bits of the syndrome or a $1 + \log(n\text{-}k\text{-}b)$-stage counter connected to the output of the register. This circuit recognizes the presence of $n\text{-}k\text{-}b$ zeros in the high-order positions of the syndrome (or, in other words, the presence of a correctable burst of length b or less in the b low-order positions of the syndrome). As in the case of error-trapping the feedback connection to the syndrome generator is broken when a burst is detected, and then the burst is added to the appropriately shifted received word.

In many codes, especially long ones, the number of bursts of length not greater than $b \leq (n\text{-}k)/2$ is only a small fraction of the number of cosets. For example, there is a (150,90) cyclic code which is capable of correcting all bursts of length less than or equal to 30. That is, it meets the bound of Eq. (11.73), and it is not possible to find another (150,90) code with a higher b. But the code has 2^{60} cosets while only

$$n2^{b-1} \approx 2^{36} \tag{11.78}$$

of these are employed to correct all bursts of length less than 31. Stated differently, this says that only one out of every 16 million or so cosets are used. In Sec. 11.4 it will be seen that the unused cosets can often be employed to improve the random-error-correcting ability of the code. However, if one is interested only in correcting bursts, the decoding algorithm just described can be modified to use the cosets as follows.[46]

After the syndrome is calculated, it is cyclically shifted through a complete n-bit cycle. During this cycle the BURST-RECOGNIZING CIRCUIT records the length l of the shortest burst appearing in the register. (In this application this circuit can be implemented with a slightly more complex counter than used earlier.) The correction process is then performed as above, but now correction takes place when the first l-bit burst appears in the register.

Clearly this decoding procedure decodes every error pattern as if it were a burst of length not greater than $n\text{-}k$; consequently it uses all 2^{n-k} cosets. By Eq. (11.72) there are enough cosets to decode all bursts of length b where

$$b < n - k + 1 - \log n \tag{11.79}$$

As indicated by Eq. (11.79), for large $n\text{-}k$ and n, the overwhelming majority of bursts of length slightly less than $n\text{-}k$ are correctable. For instance, with reference to the (150,90) code mentioned earlier, there are enough cosets to correct all bursts of length 53 or less, although it is known that some bursts as short as 31 bits in length are uncorrectable.

The above decoding algorithm applies to shortened cyclic codes as well as to cyclic codes. The only modification necessary is to shift the syndrome generator through the states corresponding to untransmitted bits before correction begins.

It is interesting to note that if the probability of occurrence of every error pattern whose errors lie in a burst of length i is less than that of all patterns whose errors lie in a burst of length $i-1$ for all i, this algorithm yields maximum-likelihood decoding.

We now turn our attention to the codes to which the decoding algorithm just discussed applies.

11.3.3 Burst-correcting Codes

Little work has been done on the problem of designing codes that efficiently corrects bursts of length greater than b, the guaranteed burst-correcting ability of the code. Indeed the preceding section suggests that little is to be gained by trying to find such codes—almost any long cyclic code will perform nearly as well as the best code of that length, efficiency, and guaranteed burst-correcting ability.

The effort expended on the design of burst-correcting codes has been directed toward the problem of constructing codes for which b is relatively close to the bound of Eq. (11.73). Unlike random-error-correcting codes in which the most useful codes have been constructed by analytic techniques, the best burst-correcting codes have been found by computer-aided search procedures. Before discussing these codes and the few good burst-correcting codes which have generated analytically, it is worthwhile to consider the following theorem.

Theorem 11-4:[47] *Let* $g'(x)$ *denote the generator polynomial of an* (n',k') *cyclic code with burst-correcting ability* b'. *Then* $g(x) = g'(x^i)$ *generates an* $(n'i, k'i)$ *cyclic code with burst-correcting ability* $b = b'i$.

Proof: Since $g'(x)$ divides $x^{n'}+1$, then clearly $g'(x^i)$ divides $x^{n'i}+1$ and hence generates a cyclic code of length $n = n'i$. But the degree of $g'(x^i)$ is i times the degree of $g'(x)$ so the code has $k'i$ information symbols. It remains to be shown that the code corrects all bursts of length $b'i$.

Consider an i-by-n array in which each row is an n-bit code word generated by $g'(x)$. Now is the array is shifted so that the jth bit in the kth row replaces the jth bit in the $(k+1)$th row if $k < i$ and the $(j+1)$th bit mod-n in the first row if $k = i$, clearly every row is still a code word in the cyclic code generated by $g'(x)$. But the array is simply a code word in some linear $(n'i, k'i)$ code and it has just been shown that after a cyclic shift all parity checks are satisfied; ie., another code word results. Consequently the $(n'i, k'i)$ code is cyclic. It is a relatively straightforward task to verify that the code word whose only nonzero information bit occurs in the lowest-order (leftmost) information bit in the top row is in fact $g(x) = g'(x^i)$. But a burst of length $b'i$ bits will be correctable since at most b' bits can appear in any row of the array constructed above. Since the code generated by $g'(x)$ can correct all bursts of length $b = b'i$, the theorem is proved.

A more elegant proof of this theorem is available, but this one has the advantage that it introduces the concept of "interleaving," which we shall discuss in the following section. Thus a shortened cyclic (n', k') code interleaved as above to degree i yields a shortened cyclic $(n'i, k'i)$ code capable of correcting bursts i times as long as the original code.

If the only concern is with correcting bursts of length $b'i$ or less, a burst-correcting code constructed by interleaving can be decoded by decoding the i subwords independently. Although this does not reduce the number of bits that must be processed by (and thus stored in) the decoder, for certain types of limited-access storage media this modification considerably simplifies the decoder. This point is discussed more fully in Sec. 11.4.

Theorem 11-4 reduces the problem of finding long, powerful burst-correcting codes to that of finding short codes of the same efficiencies. This problem is ideally suited to trial-and-error solution with the aid of a digital computer. Several investigators have tabulated such codes;[48–53] Table 11.6 is based on their results. In it are listed, for various values of efficiency, the shortest burst-correcting codes for which the ratio b/n is the largest known. That is, if two codes with a given efficiency have the same ratio b/n and no code is known for which b/n is larger, the shorter of the two codes is given. For other codes that are nearly as good as those listed the interested reader is referred to the references given above. Codes with values of R not listed can be constructed by shortening the listed code with the next-highest value of efficiency.

A few of the codes tabulated in Table 11.6 have been constructed analytically. Numerous other codes, some only slightly inferior to those in the table, have also been constructed in this way. Several such analytic procedures are described below. The first of these yields a class of codes that contains the well-known Fire[45] codes as a subclass. In situations that dictate the use of a truly cyclic, as opposed to a shortened cyclic code, these codes are usually the best available. Therefore the result is stated as a theorem, and a proof is provided. As we shall see in Sec. 11.4, this results also provides a method of constructing codes capable of correcting either burst or random errors.

Analytic techniques for constructing burst-correcting cyclic codes.
1. *Theorem 11–5:*[54] *Let* $g'(x)$ *denote the generator polynomial of an* (n',k') *cyclic code with burst-correcting ability* b, *and let* $p(x)$ *denote a polynomial whose smallest irreducible factor has degree* b *or greater. The code generated by*

$$g(x) = p(x)g'(x) \tag{11.80}$$

is capable of correcting all bursts of length b *or less and has length*

$$n = \mathrm{LCM}(e, n') \tag{11.81}$$

where e *denotes the exponent of* $p(x)$. *That is,* e *is the smallest integer such that* $p(x)$ *divides* $x^e + 1$.

Proof: In order that the code generated by $g(x)$ be capable of correcting all bursts of length b or less, it is necessary and sufficient that no two such bursts appear in the same coset and thus add to a code word.

Let $B_1(x)$ and $B_2(x)$ denote two bursts of length not greater than b, and assume that they add to a code word; that is,

$$x^r B_1(x) + x^s B_2(x) = q(x)g(x) = q_1(x)p(x)g'(x) \tag{11.82}$$

It follows that

$$x^i B_1(x) + B_2(x) = q_2(x)g'(x) \tag{11.83}$$

but since $g'(x)$ generates a cyclic code of length n', it divides $x^{n'}+1$. Then because $x^{n'}+1$ divides $x^{vn'}+1$, $v = 1, 2, 3, \ldots$, it follows that

$$x^i B_1(x) = x^{i-vn'} B_1(x)(x^{vn'}) = x^{i-vn'} B_1(x) + q_3(x)g'(x) \tag{11.84}$$

Thus Eq. (11.84) gives

$$x^{i-vn'} B_1(x) + B_2(x) = q_4(x)g'(x) \tag{11.85}$$

where we choose $0 \le v \le i-1$ to make the degree of $x^{i-vn'}B_1(x)$ less than n'. But this says that two bursts of length b or less add to a code word in the original (n',k') code generated by $g'(x)$. Thus $B_1(x) = B_2(x)$ and $i - vn' = 0$.

The subscripts on the B's can now be dropped and Eq. (11.84) becomes

$$B(x)(x^{vn'} + 1) = q_2(x)g'(x) = q_5(x)p(x) \tag{11.86}$$

But by hypothesis $p(x)$ has no factors of degree less than b, which is at least one greater than the degree of $B(x)$. This implies that $p(x)$ divides $(x^{vn'}+1)$. But this is impossible since it says that both $p(x)$ and $g'(x)$ divide $x^{vn'}+1$. For vn' is necessarily less than n and x^n+1 is the smallest polynomial of this form divisible by $g(x) = p(x)\,g'(x)$. Q.E.D.

By noting that the trivial $(2b-1, 0)$ code generated by $x^{2b-1}+1$ is capable of correcting all bursts of length not greater than b, one obtains the well-known class of Fire[45] codes.

Corollary: The code generated by

$$g(x) = p(x)(x^{2b-1} + 1) \tag{11.87}$$

and exponent $2^m - 1$ (that is, $p(x)$ is primitive) where $p(x)$ has degree $m \geq b$, has length

$$n = \mathrm{LCM}[2^m - 1, 2b - 1] \tag{11.88}$$

and is capable of correcting all bursts of length b or less.* Note that the number n-k-$2b$, which is in some sense a measure of the "badness" of burst-correcting codes, is smaller for the Fire codes with $m = b$ than for any other codes constructed with Theorem 11-5 and for large n is roughly equal to b. That is, for these codes

$$b = \frac{n - k + 1}{3} \tag{11.89}$$

2. The cyclic $(n, 1)$ code generated by $x^{n-1} + x^{n-2} + \cdots + x + 1$ has only two code words, $g(x)$ and the all-zero word, and hence is capable of correcting all error patterns of weight less than $n/2$. Consequently if n is odd, it is capable of correcting all bursts of length not greater than $(n-1)/2$ and thus is an optimal burst corrector. This result and Theorem 11-4 provide a means of constructing optimal burst-correcting codes for all values of efficiency equal to or less than 1/3.

3. The following result guarantees that codes with large minimum distance can be used to correct either bursts or random errors.

Theorem 11-6:[55] *The guaranteed burst-correcting-ability of a cyclic code with minimum distance d is bounded by*

$$b \geq \frac{3d - 8}{4} \tag{11.90}$$

Proof. Assume that there exists a code word all of whose 1s are confined to two bursts $B_1(x)$ and $B_2(x)$ each of length $b+1$ or less. That is.

---

*For a proof of this corollary see Peterson.[49]

$$c(x) = x^r B_1(x) + x^s B_2(x) \tag{11.91}$$

Let z_1 and z_2 denote the number of 0 terms in the polynomials $B_1(x)$ and $B_2(x)$, respectively. Clearly since $c(x)$ is a code word

$$2(b + 1) - (z_1 + z_2) \geq d \tag{11.92}$$

Since the code is cyclic, $(1+x)(c(x))$ is also a code word. Its weight w is bounded by

$$4 + 2(z_1 + z_2) \geq w \geq d \tag{11.93}$$

Equations (11.92) and (11.93) can be manipulated into the form of Eq. (11.90). Since no code word can consist of two bursts each of length $(3d-8)/4$ or less, this provides a lower bound on the burst-correcting ability of the code. Q.E.D.

4. If $g'(x)$ generates an (n, k) cyclic code with minimum distance $d = 2t+1$ and if one of the factors of $g'(x)$ has exponent n, the $(n, k-1)$ cyclic code generated by $g(x) = (x+1)g'(x)$ has burst-correcting ability at least $t+1$.[56] When specialized to the single-error-correcting BCH codes, i.e., Hamming codes, this says that there exists a $(2^m-1,\ 2^m-m-2)$ burst-2-correcting code for all $m > 1$.
5. Polynomials of the form

$$g(x) = (x^2 + x + 1)p(x) \tag{11.94}$$

often generate burst-3-correcting codes[57] where $p(x)$ is an irreducible polynomial of degree $2m$ and order $2^{2m}-1$. The most important code generated in this manner is the previously mentioned (15,9) code (see Example 11-6).

Example 11-6 is intended to illustrate the use of Theorems 11-4 and 11-5 and Table 11.6 in the construction of burst-correcting codes.

Example 11-6. Table 11.6 shows that the (15,9) cyclic code generated by $x^6 + x^5 + x^4 + x^3 + 1 = g'(x)$ is capable of correcting all bursts of length 3 or less. Theorem 11-4 guarantees that the $(15i, 9i)$ codes generated by $g'(x^i)$ are capable of correcting all bursts of length not greater than 3i. Also each such code can be shortened so that there exist $(15i-j,\ 9i-j)$ codes for all $i > 0$, $9i \geq j \geq 0$ which are burst-$3i$-correcting. Note that all of these codes have $b = (n-k)/2$ and thus are optimal burst correctors.

Theorem 11-5 states, in effect, that adding b more parity checks to any of the cyclic codes produced by Theorem 11-4 yields a considerably longer, more efficient burst-b-correcting code. Thus if $g_1(x)$ is taken as

$$g_1(x) = g'(x)(x^3 + x + 1) \tag{11.95}$$

since $p(x) = x^3 + x + 1$ is an irreducible polynomial and $x^7 + 1$ is the polynomial of least degree of this form divisible by $p(x)$, the code generated by $g_1(x)$ has length

$$n = \text{LCM}[7, 15] = 105 \tag{11.96}$$

This code has nine parity checks, and is capable of correcting all bursts of length 3, or less.

By taking $p(x)$ to be an irreducible polynomial of higher degree and order, it is possible to construct much longer, more efficient burst-3-correcting codes. Then, if it is desired, Theorem 11-4 can be applied again to increase b while keeping R fixed. On the other hand, it is possible to apply Theorem 11-4 first and then Theorem 11-5. Thus the code generated by

$$g_2(x) = g'(x^2)\,p(x) \tag{11.97}$$

where $p(x)$ is an irreducible polynomial of degree 6 and order 63 has natural length 630, has 18 parity checks and is capable of correcting all bursts of length 6 or less. This compares favorably with the code generated by $g_1(x^2)$, which has the same burst-correcting ability and number of parity checks but only has length 210. In general, applying Theorem 11-4 before 11-5 yields longer, more efficient codes.

The present position of burst-correcting codes can be summarized briefly. Mathematically the codes are poorly understood, in fact, even less well understood than random-error-correcting codes. However, from the all-important engineering point of view, the problem of constructing easily implemented, near-optimal, burst-correcting codes is solved. That is, codes have been found (by computer search, primarily) for nearly all useful values of code efficiency and length not excluded by known bounds. Also a very simply instrumented decoding algorithm is available for these codes. This situation stands in contrast to the wide-open problem of random-error-correcting code construction discussed in Sec. 11.2.

11.4 Codes for Correcting Bursts and Random Errors

In Secs. 11.2 and 11.3 we described codes designed to correct specific, well-defined classes of error patterns. These codes perform well on channels which error patterns outside these classes occur with low probability. That is, a strictly burst-correcting code will perform well on a channel on which random errors and bursts longer than the burst-correcting ability of the code occur infrequently. Similarly, a random-error-correcting code performs well when bursts and higher-weight patterns occur with low probability. In this section codes for combatting both types of errors are discussed.

Many, perhaps most, real communications channels corrupt signals with both gaussian and impulsive noise, producing random and burst errors, respectively. In other situations the properties of a given channel may vary with time, or, as with the switched telephone network, the channel may be selected at random from an ensemble of channels with widely differing characteristics. Unless one wishes to construct a very sophisticated error-control system which adapts to various types of channels, the most practical error-control technique may be to use a code capable of correcting both bursts and random errors.

In the three preceding sections, meaningful bounds that specified conditions under which codes could and could not be constructed were presented. No such bounds have been derived for codes capable of correcting both random errors and bursts. Of course, it is possible to modify the coset-counting bounds of Eqs. (11.5) and (11.72), but except for very short codes, the result is hardly superior to the random-error-correction and burst-correction bounds applied separately. It may be that such bounds cannot be derived and that simultaneous random- and burst-error correction places no serious constraints on codes in general.

The most effective method of correcting burst and random errors is to interleave a relatively short code to produce a much longer, more powerful code. This technique is also effective against multiple bursts and other error patterns that are neither typical bursts nor random errors. Before discussing interleaving, however, we shall consider cyclic codes which are not produced by interleaving and yet which are capable of correcting both bursts and random errors. These codes will be seen to be especially well suited for interleaving.

11.4.1 Basic Codes

Consider an (n, k) cyclic code which has minimum distance $d = 2t + 1$ and which has burst-correcting ability b. Such a code can be used to correct *either* all error patterns of weight t or less *or* all bursts of length not greater than b. There are

$$N_R = \sum_{i=0}^{t} \binom{n}{i}$$

correctable patterns of weight t or less and $N_B = n2^{b-1}$ correctable bursts of length b or less. Of these exactly

$$N_{RB} = n\,2^{t-1} + \sum_{i=t+1}^{b} \sum_{j=2}^{t} \binom{i-2}{j-2} \tag{11.98}$$

patterns are bursts of length b or less which have weight t or less. These patterns are correctable regardless of whether the code is used as a random- or burst-error corrector.

In general, the cosets corresponding to the $N_R - N_{RB}$ random-error patterns which are not also bursts of length b or less may be disjoint from the $N_B - N_{RB}$ cosets corresponding to bursts which have weight t or less. Since only one correctable pattern can be in a coset, minimum distance $2t+1$ and burst-correcting ability b are not sufficient to guarantee that a code can simultaneously correct *both* burst and random errors.

No general analytic method of constructing simultaneous burst- and random-error-correcting codes is known. A handful of results of limited practical importance have been found, however.

1. Cyclic product codes, a small subclass of the class of cyclic codes, can sometimes correct both types of errors efficiently, but unfortunately such codes must either have low rates or be very long.[58]
2. The general form of Theorem 11-6 shows that random-error-correcting codes have some simultaneous burst-correcting ability, provided the code is not used to its fullest random-error-correcting capability. However, such codes are typically not efficient correctors of both types of errors.
3. As pointed out in Sec. 11.3.3, multiplying a generator polynomial of a cyclic code with $d = 2t+1$ by $x+1$ produces a code which is capable of simultaneously correcting all bursts of length $t+1$. This result is obviously important only for small values of t.

Specific codes capable of efficient simultaneous correction can be found by examining the coset structure of codes which are capable of correcting either bursts or random errors. An extensive list of computer-generated codes of the latter type is given by Elspas; others, not quite so good, can be constructed using results given in this reference.* Still others can be constructed by Theorem 11-5 and its corollary. In this case $p(x)$ and $g'(x)$ are taken to be, respectively, the generator polynomials of a random-error-correcting and a burst-correcting cyclic code of length n.

11.4.2 Interleaved Codes

In Sec. 11.3 we pointed out that if $g'(x)$ generates a cyclic code (n', k') code with burst-correcting ability b', $g(x) = g'(x^i)$ generates a cyclic (ni, ki) code capable of correcting all bursts of length $b = b_i$ or less. The latter code is formed by interleaving (interlacing) the former.

---

*After this was written an extensive list of computer-generated codes for efficiently correcting bursts and random errors was tabulated.[97]

An n-bit code word in such a code can be visualized as an i-by-n' rectangular array in which the coefficient x^e occupies position (y, z), $0 \le y \le i - 1$, $0 \le z \le n' - 1$, where

$$y = e \bmod i \tag{11.99a}$$

and

$$z = \left[\frac{e}{i}\right] \tag{11.99b}$$

In other words, the array can be thought of as having been filled by shifting the code word in, high-order bits first, starting at the upper-left corner. Every position in the array shifts into the one below it except for those in the bottom row, which shift into the position at the top of the next column.

In this array every row is a code word in the (n', k') code (subcode) generated by $g'(x)$. Consequently, the minimum distance of the code and subcode are identical. Adjacent bits in subcode words are generated by i bits in the n-bit code word and hence are separated by i bits on the channel. Because of this separation, bursts of errors of moderate length can effect relatively few bits in any subcode word. The parameter i is commonly referred to as the *interleaving degree*.

Since every subcode word is a code word in a cyclic code, it can be decoded without regard to the i–1 other subcode words in the code word. Because each of these subcode words can be decoded in an identical fashion and the subcode is ordinarily fairly short, the decoding of an n-bit code word can be accomplished with relatively little equipment. This is the primary advantage of interleaved codes. The decoder consists of a large store and the decoder required for a much shorter code. Surprisingly perhaps, in most cases no reduction in the burst-correcting ability attainable with general systematic-search decoding results when this type of decoding is employed.

The performance of the interleaved code against purely random errors is identical to that of the subcode since, as has been mentioned, the codes have the same minimum distance. However, on channels with memory, i.e., those afflicted by dependent errors, the performance of the interleaved code is superior and often considerably so. This point is considered further in Sec. 11.5, where the performance of error-correcting block codes is analyzed.

The following example is intended to illuminate the concept of interleaving.

Example 11-7. In Example 11-4, in Sec. 11.2, it was pointed out that the (15,7) BCH code generated by

$$g'(x) = (x^4 + x^3 + 1)(x^4 + x^3 + x^2 + x + 1) = x^8 + x^4 + x^2 + x + 1 \tag{11.100}$$

has minimum distance 5 and hence is double-error-correcting. Trivially, its burst-correcting ability b is at least 2.

It is desired to interleave this code to degree 5 to produce a code for which b is at least 10. This can be done as explained above by taking

$$g(x) = g'(x^5) = x^{40} + x^{20} + x^{10} + x^5 + 1 \tag{11.101}$$

The array associated with this code is

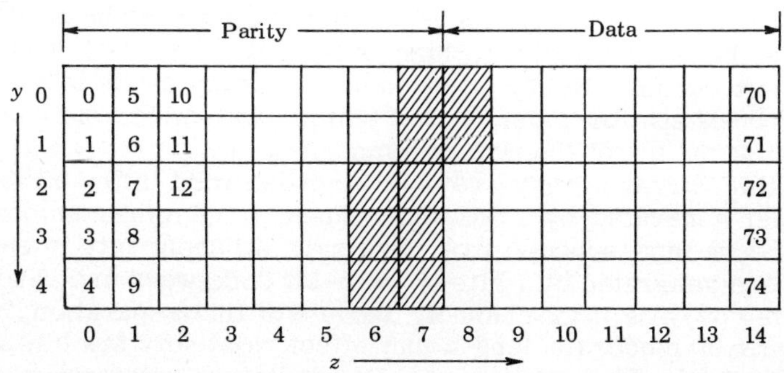

The exponents of the terms of the code words are shown in the positions in the array to which they correspond. The reader may wish to verify that Eq. (11.100) holds; i.e., that

$$y = e \bmod 5 \tag{11.102}$$

and
$$z = \left[\frac{e}{5}\right] \tag{11.103}$$

where e is an integer in the array.

The shaded region in the figure depicts a burst of length 10. The error polynomial of this burst is

$$B(x) = x^{32}(1 + b_1 x + b_2 x^2 + \cdots + b_8 x^8 + x^9) \tag{11.104}$$

where the b_i represents arbitrary binary digits. Note that no subcode word is affected by more than two bits of the burst.

In conclusion, it seems fair to state that while the codes discussed in this section possess the error-correcting powers of burst- and random-error-correcting codes, their construction is fraught with the difficulties attendant to the construction of both these types of codes. Thus while the known simultaneous burst- and random-error-correcting codes are attractive for use in real systems, much more could profitably be learned about their construction.

11.5 Performance

In this section the performance of the various types of forward-acting error-correcting codes and of codes used for error detection is considered. Specifically we shall attempt to evaluate the probability that erroneous data will be delivered under different channel conditions.

Before discussing the performance of error-correcting codes it seems appropriate to consider, at least briefly, the knotty problem of performance measures. For a given data rate the ideal measure of the effectiveness of an error-control system would be a single parameter that would, among other things, (1) be easily measured, (2) be simply defined, (3) be mathematically tractable, (4) characterize a system so well that competing systems could be compared readily and thus ordered as to their utility to a data user. Unfortunately, no single parameter with all the above properties is known. However, several performance measures that possess a number of these qualities have been used.

The oldest measure, also used in evaluating modulation systems, is the average bit-error rate; that is, the average rate at which bit errors are delivered to the sink. This parameter has the advantages that it is easily measured, is mathematically simple, and, in some cases, does portray the reliability of the channel. Unfortunately, it suffers from the very serious disadvantage that while it does give the mean of the output error distribution, it says nothing at all about the variance or other properties of the distribution. Thus in many cases of practical importance this parameter is incapable of ordering different types of error control systems.

A second performance measure, word error rate, possesses almost all of the advantages of the first, and in addition, assesses the utility of error-control systems more realistically in many situations.

The third and perhaps the most meaningful measure of the usefulness of a particular system to a data user is the mean time between disturbances. This parameter treats all types of errors in the same manner and simply measures the average duration of error-free intervals of useful length. Unfortunately, a detailed knowledge of channel statistics is necessary to compute this average. Consequently, in the following performance evaluation we shall employ the second-mentioned measure, word error rate, almost exclusively. In the interests of consistency this will be done even in situations where it is possible to compute the third performance measure or where the first would suffice.

Let us denote the 2^k code words in an (n, k) code by $\{c_i\}$, $1 \le i \le 2^k$, and the 2^n possible received n tuples by $\{r_j\}$, $1 \le j \le 2^n$. Let $p(j|i)$ denote the probability of receiving r_j, when c_i is transmitted. Then under the assumption that the c_i are transmitted with equal probability, the probability that a decoding error will be made is

$$P_e = 1 - \frac{1}{2^k}\left[\sum_{i=1}^{2^k} \sum_{j \in S_i} p(j|i)\right] \tag{11.105}$$

where S_i denotes the set of subscripts of n-tuples which are decoded into c_i. If the channel is time-varying, then $p(j|i)$ is a function of time and the average word-error rate can be found by time-averaging P_e in Eq. (11.105).

Unless some simplifying assumptions are made, the evaluation of this simple-looking result is immensely difficult even for a relatively short code used on a time-invariant channel. Thus (as earlier) we shall assume that the errors introduced by the channel are independent of the transmitted code word, that is,

$$p[r_j|c_i] = p[r_j + c_k|c_i + c_k] \tag{11.106}$$

for any code word c_k. This reduces Eq. (11.105) to

$$P_e = 1 - \sum_{j=1}^{S \le 2^{n-k}} p(j) \tag{11.107}$$

where now $p(j)$ denotes the probability of receiving the jth coset leader given that the all-zero word was transmitted. The number S is the number of coset leaders actually decoded into the all-zero word. For many of the decoding algorithms discussed in the preceding sections S is less than 2^{n-k}, often considerably so.

At this point is is necessary to consider the performance of the various types of error-correcting codes separately.

11.5.1 Performance of Random-error-correcting Codes

On channels perturbed by errors which occur independently at random all error patterns of a given weight occur with the same probability. Because it simplifies the analysis considerably, it is customary to assume that such patterns are also equiprobable on channels on which the noise can only be approximately modeled by a sequence of independent random events. As pointed out in Chap. 2, this assumption enables us to describe the channel statistically by the distribution of $P(m,n)$, the probability of m errors occurring in an n-bit block. Letting β_m denote the probability of decoding an error pattern of weight m correctly reduces Eq. (11.107) to

$$P_e = 1 - \sum_{m=0}^{n} \beta_m P(m,n) \tag{11.108}$$

or

$$P_e = \sum_{m=0}^{n} \alpha_m P(m,n) \tag{11.109}$$

where $\alpha_m = 1 - \beta_m$ denotes the probability of erroneously decoding an error pattern of weight m.

The parameter α_m is strictly a function of the code and decoding algorithm employed. If a t-error-correcting BCH code is employed and it is decoded with the Peterson (bounded-distance) decoding algorithm,[3]

$$\alpha_m \begin{cases} = 0 & 0 \leq m \leq t \\ = 1 & t < m \leq n \end{cases} \tag{11.110}$$

and Eq. (11.109) reduces to

$$P_e = \sum_{m=t+1}^{n} P(m,n) \tag{11.111}$$

Things are not quite so simple when, say, a cyclic code is decoded with majority-logic decoding. For while α_m is 0 for $0 \leq m \leq t$, it is less than 1 for m slightly greater than t, and no analytic method of determining α_m is known in general. It is certainly true that

$$\alpha_m \begin{cases} = 0 & 0 \leq m \leq t \\ \leq 1 & t < m \leq n \end{cases} \tag{11.112}$$

and consequently

$$P_e \leq \sum_{m=t+1}^{n} P(m,n) \tag{11.113}$$

but in some cases, particularly if n is large and variable threshold decoding is employed, α_m may be very much less than 1 for m slightly greater than t. Thus Eq. (11.113) may yield only a very crude upper bound to P_e. In some cases such as this, brute-force techniques such as Monte Carlo estimation of the α_m may be dictated if an accurate estimate of P_e is desired. See Townsend and Weldon,[28] for example.

If bit errors occur independently at random with probability p, the channel is a binary symmetric channel and

$$P(m,n) = \binom{n}{m} p^m (1-p)^{n-m} \tag{11.114}$$

That is, $P(m,n)$ is simply the well-tabulated binomial distribution and P_e in Eq. (11.111) is simply the "tail" of this distribution. Application of Stirling's approximation for $n!$ to this result shows that if the ratio t/n is kept fixed, P_e can be made to decrease exponentially with increasing code length. In practice this sort of behavior can be approached quite closely with the BCH codes (sec. 11.2.2) provided n is not very large (say, less than 5,000 or so).[3] Thus extremely low error rates can be obtained with known codes on truly memoryless channels.

11.5.2 Performance of Burst-correcting Codes

In order to evaluate the performance of a burst-correcting code it is obviously necessary to know something about the frequency with which bursts of various lengths occur. In a real channel this knowledge can be gained by direct measurement or by assuming that the channel can be modeled using one of the several mathematical burst-error channels presented in Chap. 2. If it can be assumed that all bursts of a given length occur with equal probability, the channel can be characterized by the distribution of $P(l,n)$, the probability of a burst of length l in a block of length n. In this case Eq. (11.107) yields

$$P_e = \sum_{l=0}^{n} \alpha_l P(l,n) \tag{11.115}$$

where α_l denotes the probability of decoding a burst of length l erroneously.

If the maximum-likelihood decoding algorithm described in Sec. 11.3.2 is employed, Eq. (11.115) becomes

$$P_e = \sum_{l=b+1}^{n-k} \alpha_l P(l,n) + \sum_{l=n-k+1}^{n} P(l,n) \tag{11.116}$$

since

$$\alpha_l \begin{cases} = 0 & 0 \le l \le b \\ \le 1 & b < l \le n-k \\ = 1 & n-k < l \le n \end{cases} \tag{11.117}$$

The correction of bursts which overlap the ends of a code word in a cyclic code is ignored here.

Although the α_l, $b < l \le n-k$, are not known precisely, the following general bounds have been derived:[46]

$$\begin{aligned} 2^{-b-1} &\le \alpha_l \le n2^{-b} & b < l < n-k-b \\ 2^{l-(n-k)-1} &\le \alpha_l \le n2^{l-(n-k)} & n-k-b \le l \le n-k \end{aligned} \tag{11.118}$$

For b (and hence $n-k$) large, these bounds are reasonably accurate; for smaller values of b one is reduced to either a detailed analysis of specific codes or to Monte Carlo estimation.

If only bursts of length not greater than b are corrected, then $\alpha_l = 1$, $b < l \leq n-k$, and Eq. (11.117) yields

$$P = \sum_{l=b+1}^{n} P(l,n) \tag{11.119}$$

11.5.3 Performance of Codes Used for Error Detection

In systems that employ error detection there are three probabilities of interest. In addition to P_e, defined earlier, P_d, the probability of detecting an error, as well as P_c, the probability of correct decoding, are of interest. These three are, of course, related by the expression

$$1 = P_c + P_d + P_e \tag{11.120}$$

If no error correction is employed

$$P_c = P(0,n) \tag{11.121}$$

The probability of erroneous decoding is just the probability that the error pattern which occurs is one of the $2^k - 1$ nonzero code words. Let $W(m)$ denote the *weight spectrum*, or *weight distribution*, of the code, that is, $W(m)$ is the number of code words of weight m. If $W(m)$ is known and if it can be assumed, as earlier, that all error patterns of a given weight occur with equal probability,

$$P_e \approx \sum_{m=d}^{n} \frac{W(m)}{\binom{n}{m}} P(m,n) \tag{11.122}$$

For many random-error-correcting codes such as the BCH codes the spectrum has been observed to have the shape of the distribution of binomial coefficients.[59] That is,

$$W(m) \cong \frac{\binom{n}{m}}{2^{n-k}} \tag{11.123}$$

Note that with this approximation

$$\sum_{m=0}^{n} W(m) = 2^k \tag{11.124}$$

as it must. Subject to the typical inaccuracies of a few percent inherent in Eq. (11.123), Eq. (11.122) yields

$$P_e \cong \frac{1}{2^{n-k}} \sum_{m=d}^{n} P(m,n) \tag{11.125}$$

This result, which is rather easy to evaluate, seems to provide a reasonably accurate approximation to P_e even where the spectrum of the code is known to be substantially different from the binomial distribution. This, and the fact that other methods of determining P_e are often impractically difficult, make Eq. (11.125) a generally useful result.

Finally from Eqs. (11.120), (11.121), and (11.125) we have

$$P_d \cong \sum_{m=1}^{n} P(m,n) - \frac{1}{2^{n-k}} \sum_{m=d}^{n} P(m,n) \tag{11.126}$$

In most cases the second term in this expression is considerably smaller than the first (that is, $P_e \ll P_d$) and hence can be ignored.

In some situations error detection is performed after a limited amount of error correction. For example, if a random-error-correcting code is used to correct all error patterns of weight t or less and none of higher weight, and, simultaneously to detect all uncorrectable error patterns, the results of the preceding paragraphs can be modified to give

$$P_c = \sum_{m=0}^{t} P(m,n) \tag{11.127}$$

and

$$P_e \approx \frac{\sum_{i=0}^{t} \binom{n}{i}}{2^{n-k}} \left[\sum_{m=d-t}^{n} P(m,n) \right] \tag{11.128}$$

with P_d again given by Eq. (11.120).

Similar but more exact results can be exhibited if the $P(l,n)$ rather than $P(m,n)$ distribution is used. It can be shown that for every cyclic code the fraction of undetectable bursts is 0 for bursts of length not greater than $n-k$, $2^{-(n-k-1)}$ for bursts of length $n-k+1$, and $2^{-(n-k)}$ for all longer bursts.[3]

Equations (11.111) and (11.125) indicate that, as expected, a considerably lower error rate can be attained with a given code by using error detection than by using forward-acting error correction. This is one of the two important advantages of error detection over correction. (The other, ease of implementation, was discussed earlier.) Explicit comparisons of error correction with detection and retransmission are presented in Peterson[3] and Hamming.[4] These should prove useful to the systems designer faced with the problem of deciding between these two alternatives.

11.6 Synchronization-error Correction

with the problem of deciding between these two alternatives.

The discussion of block codes contained in Chap. 10 and in Secs. 11.2 through 11.5 has been based on an assumption heretofore unstated. We have assumed that *word framing* is known at the receiver. In other words we have taken for granted that the decoder knows where each word starts and ends. Under ordinary circumstances this is so; given that the preceding word ended on a particular bit, the decoder only need count n bits to locate the last bit of the present word. However, if one or more bits are deleted from or inserted into the bit stream, the decoder will not be able to determine correctly the ends of the present word or subsequent words. In such situations ordinary additive-error correction as discussed earlier is completely ineffective. In this section coding techniques capable of correcting synchronization errors of this type are presented.

Data-communications systems divide naturally into two types: synchronous sytems, in which data are transmitted continuously, and asynchronous systems, in which messages are transmitted at irregular intervals. Whether or not error coding is used in an asynchronous system, it is ordinarily necessary to resynchronize the receiver on each message. This is usually done either by preceding each message by a prefix specifically designed to enable the receiver to determine the location of the start of the message or by choosing as messages bit sequences that enable the receiver to determine this information automatically.[60,61] This type of synchronization, which incidentally is identical to that encountered at the start of synchronous transmission, will not be discussed here. Rather we shall consider the somewhat more difficult problem of recapturing synchronization in a synchronous system.

The model of the communications channel we shall consider is the following. Additive-error-correcting code words are transmitted through a noisy channel that, besides adding errors to them, occasionally inserts or deletes bits or sequences of bits. The word "occasionally," while not defined precisely here, will be taken to mean that synchronization errors occur much less frequently than additive errors. (The problem of coding for channels in which synchronization errors predominate can also be handled. However, because this situation occurs in so few real channels, this problem will not be considered here.) The decoder examines the received words, and on the basis of their syndromes attempts to determine the proper word framing and the positions of any errors which may have occurred.

At this point it is necessary to distinguish between synchronization-error correction and synchronization recovery. Assume a code word is transmitted and a single bit is deleted by the channel. *Synchronization recovery*, normally performed on the word following the corrupted word, consists of determining whether this word has been shifted left or right relative to word framing, and if so, how far. Several methods of designing additive-error-correcting codes capable of recovering synchronization are presented below. One drawback to this type of recovery is that since nothing is done about the word in which bits were deleted or inserted, errors are usually delivered to the data sink whenever a synchronization error occurs. Correcting these errors is the problem of *synchronization-error correction*. By adding considerably more redundancy to each code word it is possible, in certain cases, to correct these synchronization errors.[62] However, in most real data-transmission systems, errors of this type occur infrequently enough to be, if not negligible, at least not worth the relatively large reduction in data required to correct them. Consequently we shall not consider this aspect of the synchronization problem further. (The reader may have reason to believe that synchronization-error correction guarantees synchronization recovery. This is not so, however, as can be readily verified by example.)

For additive errors, correction is performed by associating an error pattern with the syndrome of the received word. Syncronization recovery is performed in exactly the same way: the code is constructed so that the syndromes corresponding to the most probable losses of synchronization will be disjoint from those of correctable additive-error patterns. Of course, uncorrectable additive-error patterns can cause the decoder to shift word framing when it is, in fact, already in synchronization. However, since this erroneous shift in word framing is by definition within the synchronization recovery capability of the code, it will be corrected during the succeeding word, provided only that a correctable additive-error pattern occurs. Thus false resynchronization is usually a relatively unimportant consideration.

We define r_d, the *synchronization-loss detection capability*, to be one less than the smallest undetectable synchronization slippage. With this definition the set of syndromes resulting from synchronization losses of r_d or fewer bits are disjoint from the set of syndromes associated with correctable additive-error patterns. Similarly we define r_c, the *synchronization-loss correction capability*, to be one less than the smallest uncorrectable synchronization slippage. Since any pattern must be detectable to be correctable,

$$r_c \leq r_d \tag{11.129}$$

In addition to being disjoint from correctable additive-error syndromes, it is necessary that the syndromes of correctable synchronization losses be distinct; i.e., no two slippages of r_c or fewer bits can yield the same syndrome. It is also true that

$$r_c \geq \left[\frac{r_d}{2}\right] \tag{11.130}$$

for if every slippage of r_d bits is detectable, every slippage of $r_d/2$ bits can be corrected by simply trying all synchronization locations within $r_d/2$ bits of the organization position. The only location not to be detected as a synchronization loss will be the correct position.

If one were transmitting data over a channel on which synchronization errors occurred much more frequently than additive errors or if the data were such that additive errors could be ignored, synchronization could be recovered by inserting a specially chosen synchronization sequence between n-bit code words. It is possible to design such sequences, e.g., Barker sequences[63] of length r_d+1 for many values of r_d with the property that a left or right shift of the sequence by r_d or fewer bits can be detected provided the sequence is not too badly corrupted by errors. Essentially the same result can be achieved by coding the messages with the techniques described in References [60] and [61].

These techniques are not very effective on channels perturbed primarily by additive errors. Even if the messages are encoded in an additive-error-correcting code and a Barker sequence is used, errors in this sequence will, with relatively high probability, result in decoding errors. The technique described in Sec. 11.6.1 is equivalent in a general sense to the use of a Barker sequence protected by coding.

11.6.1 Extended Cyclic Codes[64,65]

The synchronization-recovery technique described in this section can be used with any cyclic code. Its disadvantage is that it requires that additional redundancy be added to each code word. A less general technique which can sometimes circumvent this difficulty is discussed in the following section.

Consider the additive-error-correcting cyclic (n, k') code generated by $g_a(x)$. Every code word in this code has the form

$$c(x) = i(x) g_a(x) \tag{11.131}$$

where $i(x)$ is ordinarily an arbitrary polynomial of degree less than k'. To impart synchronization recovery ability to the code, we restrict $i(x)$ to have the form

$$i(x) = i(x) g_s(x) + 1 \tag{11.132}$$

That is, we employ only a subset of the set of code words in the code generated by $g_a(x)$. The polynomial $g_s(x)$ has degree j and exponent $m \leq n$; that is, $g_a(x)$ divides x^m+1 but not x^u+1, $u<m$. The polynomial $i(x)$ is an arbitrary polynomial of degree less than $k'-j$.

After $i(x)$ is encoded into the cyclic code word $c(x)$ by means of Eqs. (11.131) and (11.132), an "extended cyclic code word" is formed for transmission as follows.

Let

$$c_{n-1}, \ldots, c_{n-[(m-1)/2]}, \ldots, c_{[(m-1)/2]}, \ldots, c_1, c_0 \qquad (11.133)$$

denote the coefficients of $c(x)$. The corresponding coefficients of the extended word are

$$\overbrace{c_{[(m-1)/2]-1}, \ldots, c_1, c_0}^{\text{Prefix}}, c_{n-1}, \ldots, c_{n-[(m-1)/2]}, \ldots, c_{[(m-1)/2]}, \ldots,$$

$$c_1, c_0, \underbrace{c_{n-1}, \ldots, c_{n-[(m-1)/2]}}_{\text{Suffix}} \qquad (11.134)$$

That is, the lowest order $(m-1)/2$ bits of $c(x)$ are used as a prefix and the highest-order $(m-1)/2$ bits as a suffix to form a code word in the extended cyclic code. This is illustrated in Fig. 11.9. These bits, which are really additional parity bits, are appended to the word because a synchronization slippage of s bits merely transforms the center of such a code word [that is, $c(x)$] into a cyclically shifted version of itself [that is, $x^s c(x)$]. Therefore, even if such a slippage occurs, the additive-error-correcting ability of the code generated by $g_a(x)$ is not impaired.

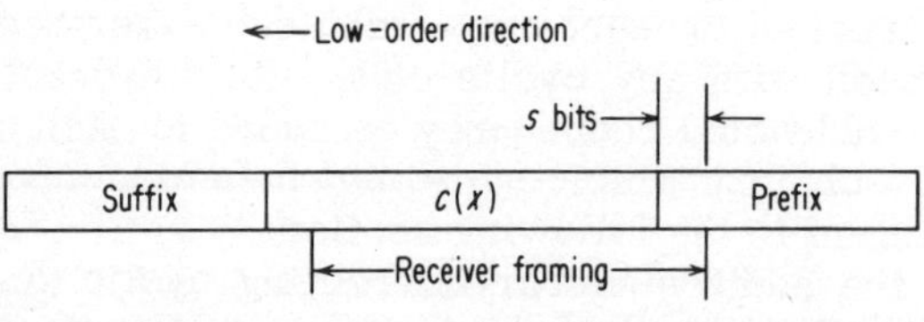

Fig. 11.9. An extended cyclic code word with a framing slippage of s bits in the low-order direction.

More precisely, let us assume that the code word of Eq. (11.134) is transmitted through a channel which shifts word framing by up to s bits in either direction. This would result if some bits were inserted into, or deleted from, the preceding word by the channel. The decoder strips off what it considers to be the prefix and suffix and proceeds to operate on the center n-tuple. Let us assume that an error pattern $e(x)$ correctable with the cyclic code generated by $g_a(x)$ was added to this n-tuple by the channel. Thus the received n-tuple is

$$r(x) = x^s c(x) + e(x) \qquad (11.135)$$

Note that a framing shift in the high-order direction of the bit stream as in Fig. 11.9 corresponds to a positive value of s. But since $x^s c(x)$ is just a code word in the cyclic code generated by $g_a(x)$ and since $e(x)$ is, by hypothesis, correctable, the decoder can determine the code word $x^s c(x)$. But by Eqs. (11.131) and (11.132)

$$x^s c(x) = g_a(x)[x^s i(x) g_s(x) + x^s] \tag{11.136}$$

Dividing this result by the known polynomial $g_a(x)$ yields the bracketed term as a quotient. Then dividing this sum by $g_s(x)$ yields the remainder

$$\text{rem}\ \frac{x^s}{g_s(x)} \tag{11.137}$$

For $-[(m-1)/2] \leq s \leq [(m-1)/2]$ no two of these remainders are identical since this would imply that $g_s(x)$ divides $x^i + 1$, $i < m$, and, by hypothesis, it does not. This says that the decoder can associate uniquely with each such remainder a particular value of s corresponding to a framing shift of $[(m-1)/2]$ or fewer bits in either direction. With s known it is a straightforward matter to determine $i(x)$ since $x^s i(x)$ has already been determined. This completes the decoding process. Despite its apparent complexity this procedure can be implemented in a rather straightforward manner.

If one is interested in a synchronization-loss detection, the prefix and suffix can each be made twice as long for a given polynomial $g_s(x)$. This guarantees that every slippage of fewer than m bits will be detected.

The restriction made earlier that m not exceed n is not really necessary; it was made solely for notational convenience. Thus if one is willing to pay the price in redundancy, very large synchronization slippages can be corrected by employing a very long prefix and suffix. The same is true of the restriction that the high- and low-direction synchronization-error-correcting ability be the same. If slippage in one direction is more likely to occur than in the other, greater correcting ability can be provided where it is needed. Observe that the only restriction necessary is that the sum of the maximum correctable slippages in both directions must be less than m.

In the technique just described no reduction in the additive-error-correcting ability of the code results when synchronization errors occur. For this luxury a fairly substantial reduction in overall code efficiency is necessary. Where synchronization detection or recovery ability without added redundancy is required, the techniques described in Sec. 11.6.2 are applicable.

11.6.2 Coset Codes

One drawback to the use of truly cyclic codes in error-control systems is that these systems are extremely sensitive to synchronization loss. Unless a subset of the code is used as in the preceding section, every cyclic shift of a code word yields another code word; thus a synchronization slippage of one bit, for example, will produce with equal probability, either a legitimate code word or a code word which differs from a code word by distance one. In either event if any additive-error-correction is used, a decoding error will result. A similar situation results when a slippage of a few bits occurs.

Coset codes provides a simple and effective method of eliminating this very undesirable sensitivity to synchronization loss while retaining the desirable features of cyclic codes. We have seen that every coset of a parity-check code is composed of the 2^k n-tuples formed by adding a fixed n-tuple to the 2^k code words. Every coset contains one and only one n-tuple all of whose 1s are confined to the n-k low-order (parity-check) positions. There are 2^{n-k} such n-tuples, counting the all-zero (n-k)-tuple, and 2^{n-k} cosets, counting the code. Thus we shall always consider a coset word as the sum of a code word and an (n-k)-tuple added to its parity-check section. It should be apparent that adding a fixed (n-k)-tuple to every code word does not in any way affect the additive-error-correcting ability of the code. The pattern can always be subtracted out at the decoder if synchronization has been maintained, and the resulting n-tuple will be the same as if the original code word were transmitted.

By choosing the added (n-k)-tuple, denoted $p(x)$ and referred to as the *synchronizing polynomial*, in an appropriate manner, coset codes can be designed to prevent synchronization loss. Several techniques of this type are described briefly below. All of these results are taken from Tong[66] to which the interested reader is referred for details.

If a cyclic random-error-correcting code able to correct all error patterns of weight $t \geq 2r_c + 1$ or less is shortened by at least $2r_c + 1$ bits, it is possible to find a synchronization polynomial $p(x)$, which will enable the coset code specified by $p(x)$ to correct all synchronization slippages of r_c or fewer bits. Although a derivation of this result is not given here, the basic idea behind tho technique is as follows.

If synchronization is maintained, adding $p(x)$ to a code word $c(x)$ and then subtracting (adding) it out at the receiver does not affect any of the properties of the code. If a synchronization slippage of $s \leq r_c = [(t-1)/2]$ bits occurs however, the received word differs from a code word in two ways. First, s bits of the adjacent words are added to a cyclically shifted version of the transmitted word and, second, the polynomial $p(x)(1+x^s)$ is also added to this word.

Note that for this technique to be workable at least a triple-error-correcting code must be employed.

In Tong[66] a method is described for choosing the synchronizing polynomial $p(x)$. The polynomial $p(x)$ is chosen so as to guarantee that the syndrome of every synchronization slippage of not more than $[(t-1)/2]$ bits is different from that associated with any correctable coset. In addition, $p(x)$ is chosen to guarantee that the syndrome of every high-order-direction slippage is different from that assodiated with any low-order slippage, and vice versa. Briefly, this is done by choosing $p(x)$ so as to make its syndrome that associated with any low-order slippage, and vice versa. untransmitted bits in the shortened cyclic code. Since such a bit can never be in error, the syndrome can then be interpreted as having resulted from a synchronization slippage. In addition to their synchronization-recovery capabilities, codes constructed in this manner can also correct a moderate number of additive errors in the presence of a synchronization slippage. Generally speaking, in the presence of a large slippage relatively few errors can be corrected, and vice versa.

A similar technique has been derived for double-error-correcting shortened cyclic codes. In this technique a 2-bit prefix must be added to each word to guarantee that $r_c = 1$.

Methods for recovering synchronization and detecting synchronization losses with cyclic, as opposed to shortened cyclic, codes have also been devised. Again because of the number and length of the proofs, only a brief statement of the more important results will be given here.

First of all, it is shown that for any (n, k) coset code derived from a cyclic code

$$r_c \leq \frac{n - k - 1}{2} \tag{11.138}$$

If a cyclic code with minimum distance d is used to correct t errors, every error pattern of weight up to $d - t - 1$ is detectable. Thus if every slippage of r_d or fewer bits results in an error pattern of weight greater than t and less than $d - t$, it can be detected. In general, it is possible to construct synchronizing polynomials $p(x)$ for which the general bound

$$r_d \leq d - 2t - 2 \tag{11.139}$$

is satisfied with equality. The synchronization-recovery ability of these codes is given by

$$r_c = \min\left[r_d, \frac{n - k - t - 2}{2 + [(t/2)]}\right] \tag{11.140}$$

Although these codes do not employ a prefix and suffix as did the extended cyclic codes of the preceding section, they do require that the code not be used to its fullest additive-error-correcting capabilities; thus in a sense they also require the insertion of additional redundancy.

It has been shown that a primitive t-error-correcting BCH code has at least one coset of weight $2t-1$.[67] In general there are a large number of cosets of weight greater than t that are, of course, available for synchronization recovery if bounded-distance decoding is used. In Tong[66] it is shown that any such code for which $t \geq 3$ can be used to recover synchronization after a slippage of a single bit without the insertion of any additional redundancy.

The implementation of the techniques described in this section is remarkably simple. This is because (1) the codes are cyclic and (2) there are very few different synchronization slippages to be corrected in any case.

11.7 Summary

In this chapter we have considered the problem of correcting errors with a cyclic block code. In so doing we have relied heavily on the material of Sec. 10.2. It was pointed out that because of their cyclic properties, the implementation of these codes is usually straightforward and does not require an excessive amount of equipment. The three most important uses of cyclic codes, i.e., random-error correction, burst-correction, and correction of combinations of bursts and random errors were examined. It was seen that classes of cyclic codes are known for each of these three applications and that while there is some room for improvement in terms of implementation, the performance of known codes is quite good in most cases of interest. Nearly optimum performance can often be attained.

Finally, codes for correcting losses in synchronization were considered. We say that although less is known about these codes than about additive-error-correcting codes, nevertheless a few workable techniques have been devised.

In Chap. 12 convolutional codes are described in a similar manner. Since these codes are not as well understood as block codes, an attempt is made to relate them to the latter where possible.

Appendix

Some Useful Block Codes

This Appendix contains tables (summarized in the list below) of the important parameters of cyclic codes of various types.

Table No.	Codes tabulated
11.3	Primitive BCH
11.4	Nonprimitive BCH
11.5	Majority-logic decodable
11.6	Burst-correcting

The codes tabulated in Table 11.3 are primitive codes of length n having k information symbols whose generator polynomials have the form

$$g(x) = \text{LCM}\left(m_1(x)\, m_3(x) \cdots m_{2t-1}(x)\right)$$

Specifically,

$$g_i(x) = m_1(x)\, m_3(x) \cdots m_i(x)$$

The factors $m_i(x)$ are represented by the exponents of their nonzero terms [e.g., (3,1,0) represents $x^3 + x + 1$] and are taken from Peterson.[3] The value of error-correcting ability t given in Table 11.3 is the value guaranteed by the BCH bound. Values of t not marked with an asterisk (*) are known to be exact;[31] thus the actual error-correcting ability of an asterisked code may exceed the value listed, but no cases of this are known and it seems unlikely that any exist.

If $g(x)$ is taken as the product of $g_i(x)$ and $g_0(x) = x + 1$, the code has minimum distance at least $2t + 2$. Inclusion of this factor, equivalent to a parity check on all code symbols, thus increases d by one but does not affect t.

The generator polynomial $g_{(n-1)/2}(x)$ contains as factors all the divisors of $x^n + 1$ except $x + 1$ for all values of n. Thus (1,0) and the polynomials represented in Table 11.3 for a given n are all of the divisors of $x^n + 1$. For example, the three divisors of $x^7 + 1$ are (1,0)(3,1,0)(3,2,0).

Table 11.3. Generator Polynomials and Important Parameters of Selected Primitive BCH Codes.

n	k	t	Generator polynomial
7	4	1	$g_1(x) = (3, 1, 0)$
	1	3	$g_3(x) = g_1(x) \cdot (3, 2, 0)$
15	11	1	$g_1(x) = (4, 1, 0)$
	7	2	$g_3(x) = g_1(x) \cdot (4, 3, 2, 1, 0)$
	5	3	$g_5(x) = g_3(x) \cdot (2, 1, 0)$
	1	7	$g_7(x) = g_5(x) \cdot (4, 3, 0)$
31	26	1	$g_1(x) = (5, 2, 0)$
	21	2	$g_3(x) = g_1(x) \cdot (5, 4, 3, 2, 0)$
	16	3	$g_5(x) = g_3(x) \cdot (5, 4, 2, 1, 0)$
	11	5	$g_7(x) = g_5(x) \cdot (5, 3, 2, 1, 0)$
	6	7	$g_{11}(x) = g_7(x) \cdot (5, 4, 3, 1, 0)$
	1	15	$g_{15}(x) = g_{11}(x) \cdot (5, 3, 0)$

Table 11.3. Generator Polynomials and Important Parameters of Selected Primitive BCH Codes *(Continued)*.

n	k	t	Generator polynomial
63	57	1	$g_1(x) =$ (6, 1, 0)
	51	2	$g_3(x) = g_1(x)$ · (6, 4, 2, 1, 0)
	45	3	$g_5(x) = g_3(x)$ · (6, 5, 2, 1, 0)
	39	4	$g_7(x) = g_5(x)$ · (6, 3, 0)
	36	5	$g_9(x) = g_7(x)$ · (3, 2, 0)
	30	6	$g_{11}(x) = g_9(x)$ · (6, 5, 3, 2, 0)
	24	7	$g_{13}(x) = g_{11}(x)$ · (6, 4, 3, 1, 0)
	18	10	$g_{15}(x) = g_{13}(x)$ · (6, 5, 4, 2, 0)
	16	11	$g_{21}(x) = g_{15}(x)$ · (2, 1, 0)
	10	13	$g_{23}(x) = g_{21}(x)$ · (6, 5, 4, 1, 0)
	7	15	$g_{27}(x) = g_{23}(x)$ · (3, 1, 0)
	1	31	$g_{31}(x) = g_{27}(x)$ · (6, 5, 0)
127	120	1	$g_1(x) =$ (7, 3, 0)
	113	2	$g_3(x) = g_1(x)$ · (7, 3, 2, 1, 0)
	106	3	$g_5(x) = g_3(x)$ · (7, 4, 3, 2, 0)
	99	4	$g_7(x) = g_5(x)$ · (7, 6, 5, 4, 2, 1, 0)
	92	5	$g_9(x) = g_7(x)$ · (7, 5, 4, 3, 2, 1, 0)
	85	6	$g_{11}(x) = g_9(x)$ · (7, 6, 4, 2, 0)
	78	7	$g_{13}(x) = g_{11}(x)$ · (7, 1, 0)
	71	9	$g_{15}(x) = g_{13}(x)$ · (7, 6, 5, 3, 2, 1, 0)
	64	10	$g_{19}(x) = g_{15}(x)$ · (7, 6, 3, 1, 0)
	57	11	$g_{21}(x) = g_{19}(x)$ · (7, 6, 5, 2, 0)
	50	13*	$g_{23}(x) = g_{21}(x)$ · (7, 6, 0)
	43	14*	$g_{27}(x) = g_{23}(x)$ · (7, 6, 4, 1, 0)
	36	15	$g_{29}(x) = g_{27}(x)$ · (7, 5, 3, 1, 0)
	29	21*	$g_{31}(x) = g_{29}(x)$ · (7, 6, 5, 3, 2, 1, 0)
	22	23	$g_{43}(x) = g_{31}(x)$ · (7, 5, 2, 1, 0)
	15	27	$g_{47}(x) = g_{43}(x)$ · (7, 5, 4, 3, 0)
	8	31	$g_{55}(x) = g_{47}(x)$ · (7, 6, 5, 4, 3, 2, 0)
	1	63	$g_{63}(x) = g_{55}(x)$ · (7, 4, 0)
255	247	1	$g_1(x)$ (8, 4, 3, 2, 0)
	239	2	$g_3(x) = g_1(x)$ · (8, 6, 5, 4, 2, 1, 0)
	231	3	$g_5(x) = g_3(x)$ · (8, 7, 6, 5, 4, 1, 0)
	223	4	$g_7(x) = g_5(x)$ · (8, 6, 5, 3, 0)
	215	5	$g_9(x) = g_7(x)$ · (8, 7, 5, 4, 3, 2, 0)
	207	6	$g_{11}(x) = g_9(x)$ · (8, 7, 6, 5, 2, 1, 0)
	199	7	$g_{13}(x) = g_{11}(x)$ · (8, 5, 3, 1, 0)
	191	8	$g_{15}(x) = g_{13}(x)$ · (8, 7, 6, 4, 2, 1, 0)
	187	9	$g_{17}(x) = g_{15}(x)$ · (4, 1, 0)
	179	10*	$g_{19}(x) = g_{17}(x)$ · (8, 6, 5, 2, 0)
	171	11	$g_{21}(x) = g_{19}(x)$ · (8, 7, 3, 1, 0)
	163	12*	$g_{23}(x) = g_{21}(x)$ · (8, 6, 5, 1, 0)
	155	13*	$g_{25}(x) = g_{23}(x)$ · (8, 4, 3, 1, 0)
	147	14*	$g_{27}(x) = g_{25}(x)$ · (8, 5, 4, 3, 2, 1, 0)

Table 11.3. Generator Polynomials and Important Parameters of Selected Primitive BCH Codes *(Continued)*.

n	k	t	Generator polynomial
255	139	15	$g_{29}(x) = g_{27}(x) \cdot (8, 7, 3, 2, 0)$
	131	18*	$g_{31}(x) = g_{29}(x) \cdot (8, 5, 3, 2, 0)$
	123	19*	$g_{37}(x) = g_{31}(x) \cdot (8, 6, 4, 3, 2, 1, 0)$
	115	21*	$g_{39}(x) = g_{37}(x) \cdot (8, 7, 6, 5, 4, 3, 0)$
	107	22*	$g_{43}(x) = g_{39}(x) \cdot (8, 7, 6, 1, 0)$
	99	23	$g_{45}(x) = g_{43}(x) \cdot (8, 5, 4, 3, 0)$
	91	25	$g_{47}(x) = g_{45}(x) \cdot (8, 7, 5, 3, 0)$
	87	26*	$g_{51}(x) = g_{47}(x) \cdot (4, 3, 2, 1, 0)$
	79	27*	$g_{53}(x) = g_{51}(x) \cdot (8, 7, 2, 1, 0)$
	71	29*	$g_{55}(x) = g_{53}(x) \cdot (8, 7, 5, 4, 0)$
	63	30*	$g_{59}(x) = g_{55}(x) \cdot (8, 6, 3, 2, 0)$
	55	31	$g_{61}(x) = g_{59}(x) \cdot (8, 7, 6, 3, 2, 1, 0)$
	47	42	$g_{63}(x) = g_{61}(x) \cdot (8, 7, 6, 4, 3, 2, 0)$
	45	43*	$g_{85}(x) = g_{63}(x) \cdot (2, 1, 0)$
	37	45*	$g_{87}(x) = g_{85}(x) \cdot (8, 7, 5, 2, 0)$
	29	47*	$g_{91}(x) = g_{87}(x) \cdot (8, 7, 6, 5, 4, 2, 0)$
	21	55	$g_{95}(x) = g_{91}(x) \cdot (8, 7, 4, 3, 2, 1, 0)$
	13	59	$g_{111}(x) = g_{95}(x) \cdot (8, 6, 5, 4, 3, 1, 0)$
	9	63	$g_{119}(x) = g_{111}(x) \cdot (4, 3, 0)$
	1	127	$g_{127}(x) = g_{119}(x) \cdot (8, 6, 5, 4, 0)$

Table 11.4 contains a list of nonprimitive BCH (n, k) codes for which shortening a primitive BCH code is not known to produce an (n, k) code with the same or larger minimum distance.

Table 11.4. Selected Nonprimitive BCH Codes.

n	k	t	Generator polynomial
17	9	2	(8, 7, 6, 4, 2, 1, 0)
21	12	2	(6, 4, 2, 1, 0) (3, 2, 0)
23	12	3	(11, 9, 7, 6, 5, 1, 0)
33	22	2*	(10, 9, 5, 1, 0) (1,0)
33	12	4*	(10, 9, 8, 7, 6, 5, 4, 3, 2, 1, 0)
41	21	4	(20, 19, 17, 16, 14, 11, 10, 9, 6, 4, 3, 1, 0)
47	24	5	(23, 19, 18, 14, 13, 12, 10, 9, 7, 6, 5, 3, 2, 1, 0)
65	53	2	(12, 8, 7, 6, 5, 4, 0)
65	40	4*	(12, 8, 7, 6, 5, 4, 0) (1, 0) (12, 10, 9, 8, 6, 4, 3, 2, 0)
71	35	5*	
73	46	4	(9, 7, 4, 3, 0) (9, 4, 2, 1, 0) (9, 8, 0)

The codes tabulated in Table 11.5 can be decoded with one-step majority-logic decoding as explained in Sec. 11.2.3. In Table 11.5 are listed, in addition to the usual code parameters, the orthogonal check polynomials necessary for decoding and the geometry in which these polynomials represent subspaces.

Table 11.5. Majority-logic-decodable Cyclic Codes.

Associated geometry	n	k	d	Generator polynomial	Orthogonal check polynomials
$PG(2,2)$	7	3	4	(7, 0)/(3, 2, 0)	(3, 2, 0)
$EG(2,2^2)$	15	7	5	(15, 0)/(7, 3, 1, 0)	(7, 3, 1, 0)
$PG(3,2)$	15	4	8	(15, 0)/(4, 1, 0)	(4, 1, 0) (10, 5, 0)
$PG(2,2^2)$	21	11	6	(21, 0)/(11, 8, 7, 2, 0)	(11, 8, 7, 2, 0)
$PG(4,2)$	31	5	16	(31, 0)/(5, 2, 0)	(5, 2, 0)
$EG(2,2^3)$	63	37	9	(26, 24, 20, 16, 14, 13, 12, 11, 10, 2, 0)	(39, 33, 31, 26, 14, 11, 10, 0)
$EG(3,2^2)$	63	13	21	(50, 49, 47, 38, 37, 33, 32, 31, 26, 25, 23, 20, 19, 18, 15, 14, 13, 9, 6, 5, 3, 1, 0)	(16, 4, 1, 0) (26, 19, 9, 0)
$PG(5,2)$	63	6	32	(63, 0)/(6, 1, 0)	(6, 1, 0) (26, 7, 0) (42, 21, 0) (45, 9, 0)
$PG(2,2^3)$	73	45	10	(73, 0)/(45, 42, 36, 29, 25, 24, 10, 2, 0)	(45, 42, 36, 29, 25, 24, 10, 2, 0)
$PG(3,2^2)$	85	24	22	(61, 60, 58, 50, 49, 47, 41, 40, 39, 36, 35, 33, 32, 31, 28, 27, 25, 21, 20, 19, 17, 14, 13, 11, 9, 8, 6, 5, 4, 3, 1, 0)	(27, 16, 7, 1, 0) (68, 51, 34, 17, 0)
$PG(6,2)$	127	7	64	(127, 0)/(7, 1, 0)	(7, 1, 0) (54, 5, 0) (55, 13, 0)
$EG(2,2^4)$	255	175	17	(80, 74, 72, 70, 68, 62, 60, 58, 50, 48, 46, 41, 40, 38, 36, 35, 34, 33, 32, 31, 30, 29, 27, 24, 22, 20, 18, 16, 14, 8, 6, 4, 2, 0)	(253, 243, 240, 208, 197, 181, 115, 101, 97, 77, 76, 71, 48, 40, 7, 0)
$EG(4,2^2)$	255	21	85	(234, 231, 229, 227, 226, 225, 224, 223, 219, 218, 215, 214, 212, 211, 210, 208, 207, 204, 203, 195, 193, 191, 190, 185, 183, 182, 181, 177, 164, 163, 162, 159, 158, 156, 155, 153, 150, 147, 142, 139, 137, 135, 132, 129, 124, 121, 116, 114, 107, 104, 102, 101, 100, 99, 94, 92, 91, 90, 88, 87, 84, 83, 82, 81, 80, 76, 75, 72, 69, 67, 66, 63, 60, 58, 57, 54, 51, 46, 44, 43, 35, 32, 30, 27, 26, 24, 20, 18, 15, 10, 8, 7, 4, 1, 0)	(119, 51, 17, 0) (213, 141, 8, 0) (196, 91, 15, 0) (197, 101, 7, 0)
$PG(7,2)$	255	8	128	(255, 0)/(8, 6, 5, 1, 0)	(0, 1, 25) (0, 5, 138) (0, 7, 112) (0, 9, 120) (0, 13, 99) (0, 19, 92) (0, 17, 68) (0, 85, 170)

The $d-1$ polynomials orthogonal on the high-order bit in a code can be obtained from the listed polynomials by either the squaring or cyclic-shifting operation. For example, the seven orthogonal polynomials associated with the (15,7) code are (4,1,0), (8,2,0), (10,5,0), (14,3,0), (13,6,0), (12,11,0), and (9,7,0). Note that these polynomials are orthogonal on the digit associated with the number 0.

Table 11.6. Shortest Burst-correcting Codes of a Given Efficiency for Which Ratio b/n is Largest Known.[18,34,36]

Code efficiency $R = k/n$	(n, k)	Burst-correcting ability b	$n-k-2b$	b/n	Generator polynomial $g(x)^a$
$1/(2b+1)$	$(2b+1, 1)$	b	0	$b/(2b+1)$	$(x^{2b+1}+1)/(x+1)$
0.42	(7, 3)	2	0	0.286	35
0.60	(15, 9)	3	0	0.200	171
0.63	(27, 17)	5	0	0.105	2671
0.65	(34, 22)	6	6	0.176	15173
0.68	(50, 34)	8	0	0.160	224531
0.81	(67, 54)	6	1	0.090	36365
0.85	(103, 88)	7	1	0.068	114361
0.87	(63, 55)	3	2	0.048	711
0.88	(85, 75)	4	2	0.047	2651
0.91	(131, 119)	5	2	0.038	15163
0.92	(169, 155)	6	2	0.035	55725
0.93	(121, 112)	3	3	0.025	1411
0.96	(290, 277)	5	3	0.017	24711
0.98	(511, 499)	4	4	0.008	10451
0.99	(1023, 1010)	4	5	0.004	22365

[a]The polynomials $g(x)$ are related to the listed numbers as in the following example: $35 = 011\ 101 = x^4 + x^3 + x^2 + 1$.

Notes

The procedures for encoding cyclic block codes are from Peterson.[14] The best known class of random-error-correcting codes are the BCH codes.[11-13] The implementation of Peterson's decoding algorithm for these codes was simplified by Chien.[15] Recently Berlekamp devised a simply implemented method of obtaining the elementary symmetric functions from the power-sum-symmetric functions without inverting a matrix.[17,18] The first majority-logic-decodable codes were described by Reed[25] and Muller,[29] and Rudolph[96] was the first to employ the properties of finite geometries to construct majority-logic-decodable cyclic codes. Most of the burst-correcting codes of Table 11.6 were found by computer search by Kasami.[51] The material of Secs. 11.3.1 and 11.3.2 is taken from Peterson.[3] Kasami, Hsu, and Chien[97] contains a large list of good codes for correcting both burst and random errors.

12

Error Control Using Convolutional Codes

In the preceding chapter we presented techniques for constructing block codes capable of correcting various types of errors. Economically attractive methods of implementing these codes were also described. Classes of easily implemented codes capable of correcting random errors, bursts, and combinations of random and burst errors are presented. The significant difference between the state-of-the-art of block and convolutional codes will undoubtedly be noticed. Although the latter are developing rapidly, this difference—caused by the fact that convolutional codes are inherently more difficult to analyze than block codes—does not appear to be decreasing.

Like block codes, convolutional codes can be used either to detect or correct errors. However, because data are usually retransmitted (and deleted) in blocks, block codes seem better suited to situations requiring only error detection. Thus in the sequel only the forward-acting error-correcting capabilities of convolutional codes will be considered.

In Sec. 11.5 we considered the performance of block codes. It would be desirable to do the same for convolutional codes in this chapter. However, because successive decodings are highly dependent (rather than independent as with block codes), determining the probability of erroneously decoding of a convolutional code is a very difficult problem. Some results pertinent to this problem have been obtained,[1-3] but these are of limited applicability. For this reason the performance of convolutional codes will not be considered further. As we shall see, the known procedures for decoding convolutional codes are relatively simple to implement and hence to simulate on a digital computer. In the absence of analytical results, simulation, perhaps employing the channel models of Chap. 2, seems to be the best way of estimating the performance of these codes.

In Chap. 10 we mentioned that for nearly all practical convolutional codes, either $n_0 - k_0$ or $k_0 = 1$. for practical values of code efficiency the block length of codes of this type is quite low; e.g., for $R = 3/4$, the block length is 4. Thus the problem of synchronization slippage is not nearly so severe for convolutional codes as for block codes with their considerably larger block lengths. A trial-and-error search for correct framing is usually satisfactory. For this reason convolutional codes capable of synchronization recovery are not discussed in the sequel.

Because successive decodings are independent, decoding errors do not tend to propagate when block codes are used. This is not true of convolutional codes; a decoding error may cause subsequent blocks to be decoded in error if the results of previous decodings are fed back to modify the syndrome (see Sec. 10.3). Since most known, easily implemented, convolutional codes employ this type of feedback, the error-propagation effect must be considered. However, as we shall see, decoding errors do not tend to propagate for most of the practical convolutional codes known at present.

12.1 Encoding, Syndrome Calculation, and Error Correction

In convolutional codes information digits other than those in the present block are involved; otherwise the encoding, syndrome-calculation, and error-correction operations for convolutional codes are identical to the corresponding operations for block codes.

12.1.1 Encoding

This operation is best understood by examining Fig. 12.1. The convolutional encoder shown in the figure operates as follows. With the *D-P* switch in position *D* (data), the k_0 data bits of the block to be encoded (i.e., the 0th block, are shifted into the DATA REGISTER and directly out to the channel. The switch is then thrown to position *P* and the $n_0 - k_0$ parity checks which complete the block are read out to the channel. At this point the encoder is ready to encode the next block.

The *i*th of the $n_0 - k_0$ EXCLUSIVE-OR circuits corresponds to the *i*th row of the basic parity-check matrix of the code. The binary digit $p_l^T(i,j)$ specifies whether or not a particular stage is connected to the *i*th EXCLUSIVE-OR. It is the entry in the *i*th row, *j*th column of the submatrix P_l^T in the basic parity-check matrix

$$\mathrm{h} = \left[\mathrm{P}_{m-1}{}^T \, \mathrm{O} \mathrm{P}_{m-2}{}^T \ldots \mathrm{P}_0{}^T \mathrm{I}\right] \tag{12.1}$$

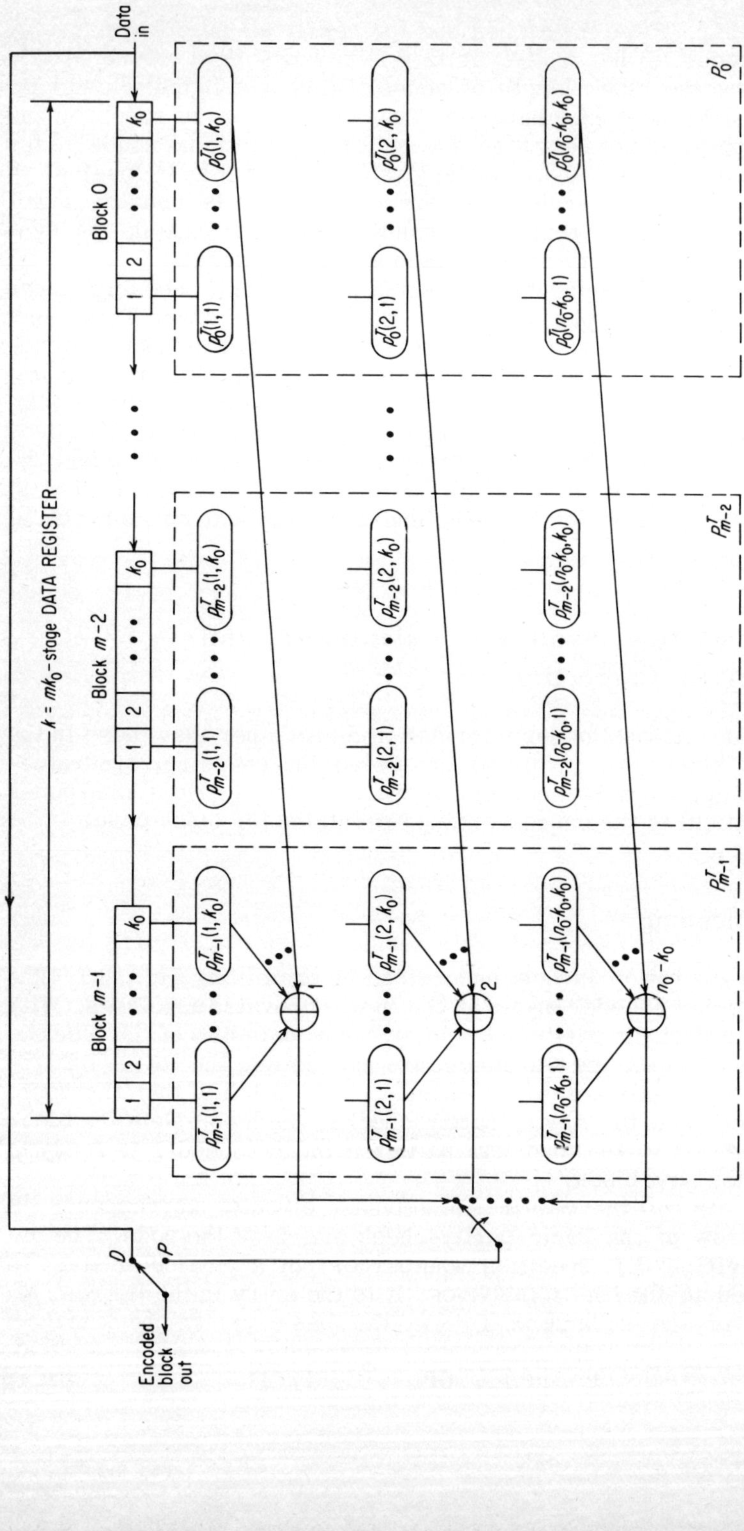

Fig. 12.1. A $k = mk_0$-stage encoder for an (mn_0, mk_0) convolutional code.

and is, of course, equal to $p_l(j,i)$. To clarify this point and to demonstrate the inherent simplicity of this type of circuit (belied by Fig. 12.1), let us consider the encoder for the (12,9) convolutional code of Examples 10.8 and 10.9.

Example 12-1. The pertinent parameters of the (12,9) code are $n_0 = 4$, $k_0 = 3$, and $m = 3$. Its basic parity-check matrix is

$$\mathbf{h} = [1100 \ 1010 \ 1111] \tag{12.2}$$

The corresponding encoder is shown in Fig. 12.2.

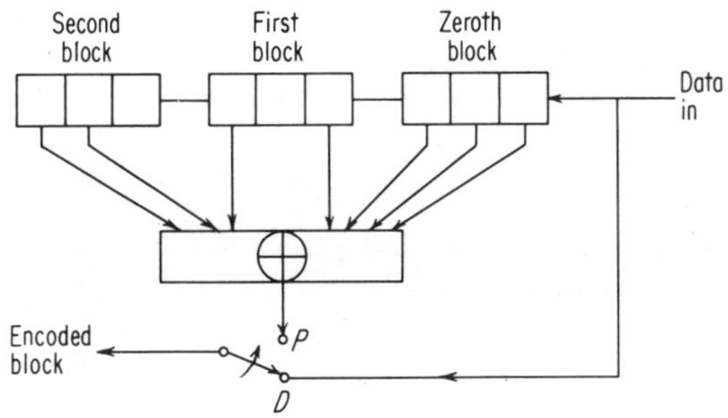

Fig. 12.2. A k-stage encoder for the (12,9) code for which $P_2^T = 110$, $P_1^T = 101$ and $P_0^T = 111$.

The k-stage convolutional-code encoder corresponds to the k-stage encoder for cyclic codes shown in Fig. 11.3. The reader, noticing the similarities between the two circuits and recalling that cyclic codes can be encoded with an $(n$-$k)$-stage register, may ask whether or not a corresponding encoder exists for convolutional codes. There is, and it is depicted in Fig. 12.3.[2] Whereas the k-stage encoder accepts data in serial form, this $(n$-$k)$-stage encoder accepts the k_0 data bits in the zeroth block on k_0 parallel leads. Thus, if the source emits data serially, a series-to-parallel converter is required. Encoding is performed by shifting the k_0-bit data block to be encoded into the encoder and simultaneously out to the channel. As this block is shifted in, the entire encoder is shifted once, as shown, emitting the $n_0 - k_0$ parity bits on parallel leads. Therefore, unless data are to be transmitted in parallel, a parallel-to-serial converter is required. At this point the encoder is prepared to encode the next block. Note that only $(n-k) - (n_0 - k_0)$ stages of shift register are actually employed in this nominal $(n-k)$-stage encoder.

The rather complicated appearance of Fig. 12.3 masks the inherent simplicity of this type of encoder for most practical values of the parameters m, n_0, and k_0. For example, in many of the codes for which simple decoders have been devised, most of the $p_l^T(i,j)$ are zero. This simplicity and the operation of the decoder are illustrated in Example 12-2.

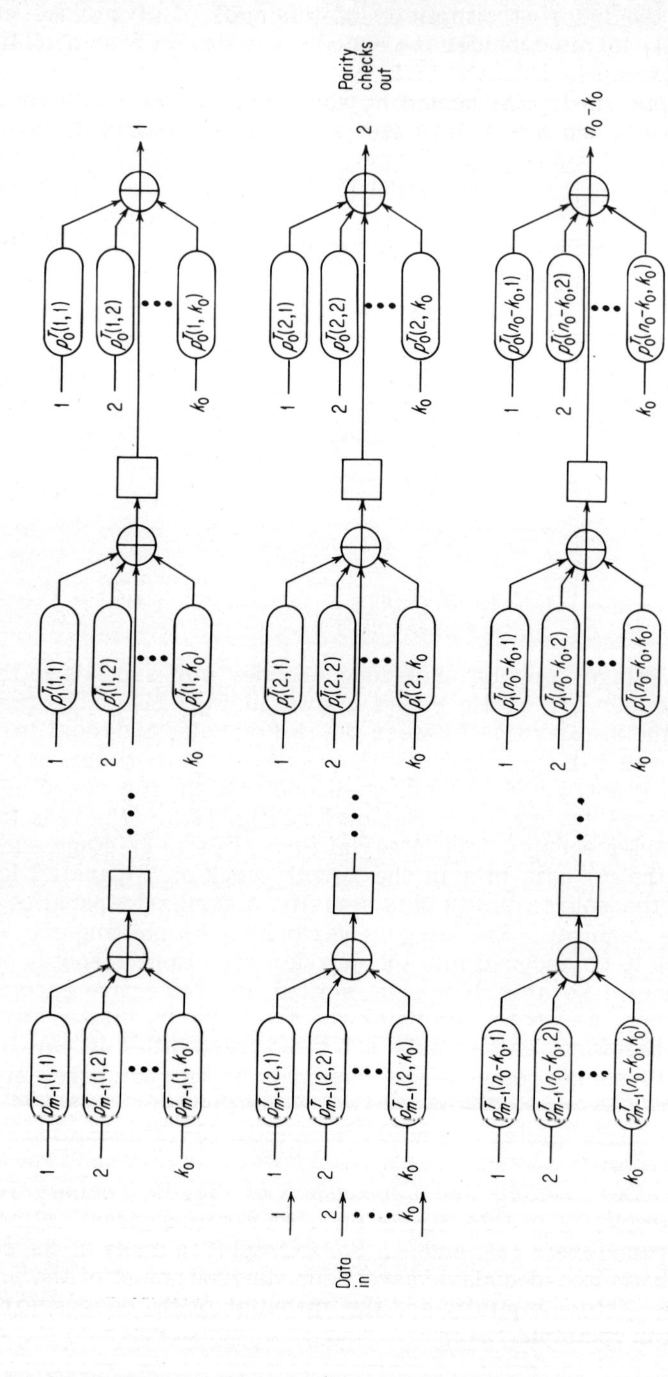

Fig. 12.3. An $(n\text{-}k) = m(n_0 - k_0)$-stage encoder for an (mn_0, mk_0) convolutional code. The $p_l^T(i,j)$ are elements of the basic parity-check matrix of Eq. (12.1).

Example 12-2. The $(n\text{-}k)$-stage encoder for the (12,9) code is shown in Fig. 12.4. That this encoder actually computes the parity bit as specified by the matrix $\mathbf{h}$ of Eq. (12.2) can be seen as follows. The check bit in the 0th block is the sum of the three information bits in this block (performed by the third EXCLUSIVE-OR), the first and third information bits of the previous (first) block (performed by the second EXCLUSIVE-OR during the encoding of the first block and stored in the second shift-register stage), and the first and second information bits of the second block (performed by the first EXCLUSIVE-OR during the encoding of the second block and stored in the first shift-register stage).

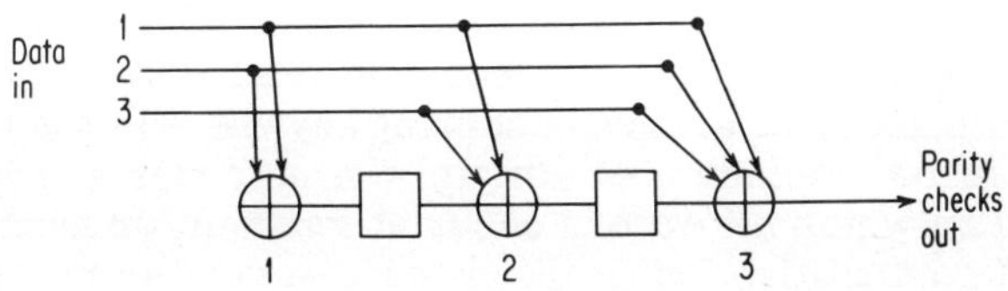

Fig. 12.4. An $(n\text{-}k)$-stage encoder for the (12,9) code.

12.1.2 Syndrome Calculation

As pointed out in Chap. 10 where it was defined, the syndrome is calculated at the decoder by encoding the received data bits and adding the resulting check bits to the received checks. Hence both encoders described in Sec. 12.1.1 can be modified to calculate the syndrome bit associated with a given block by adding the EXCLUSIVE-OR circuit(s) necessary to add the encoder output to the received check bits. Of course, since the resulting syndrome is (presumably) to be used for error correction, an $(n\text{-}k)$-bit register must be provided to store the section of the syndrome necessary to decode the oldest block in the decoder, i.e., the $(m-1)$th block.

12.1.3 Error Correction

Error correction is performed by associating with each syndrome a particular error pattern. Since correctable error patterns that are identical in the $(m-1)$th block result in the same action by the decoder, no differentiation between their syndromes is necessary. The procedures for decoding random and burst errors discussed in Secs. 12.2 and 12.3 can be implemented economically because they treat all such error patterns identically.

In Chap. 10 we noted specifically that unless otherwise stated, the syndrome associated with the block being decoded (block $m-1$)

would be modified to eliminate the effects of past, presumably corrected, errors. This is done by adding to (subtracting from) the syndrome the pattern of parity-check failures associated with the errors corrected in the decoded n_0-bit block. That is, when a bit error is corrected, a 1 is added to every syndrome bit which checked that bit. Thus, for example, the syndrome is set to 0 if no other errors occurred within a constraint length of the block being decoded.

12.2 Random-error Correction

In this section we deal with convolutional codes that can correct a fixed number of errors occurring anywhere in the bit stream and that can be implemented simply. These codes are well suited to channels on which bit errors occur at random, or nearly so, and hence are competitive with the random-error-correcting block codes of Sec. 11.2.

In that section, as a preliminary to presenting techniques for constructing random-error-correcting block codes, upper and lower bounds on the minimum distance attainable with a code of a given length and efficiency were derived. Results roughly corresponding to each of these three bounds, the Hamming bound, the Plotkin bound, and the Varsharmov-Gilbert-Sacks bound, have been derived for convolutional codes. These results, more difficult to derive than the similar results for block codes, are not presented here. The interested reader is referred to Massey,[2] Bussgang,[4] Robinson,[5] Wyner and Ash,[6] and Lin and Lyne[7] for statements and derivations of these bounds. Generally speaking, the convolutional code bounds indicate that for given values of R and n, the minimum distance of the best convolutional code is slightly larger than that of the best block code when n is not large. This is a generally useful rule of thumb for persons more familiar with block codes. For large n (Fig. 11.4) the bounds on the two types of codes coincide in all cases where they have been compared explicitly.

The bounds on the minimum distance attainable with a given (n,k) block code are useful because there are several ways of constructing good, easily implemented (cyclic), block codes, at least for short codes. The convolutional-code bounds are less useful at present because of the limitations on our ability to construct good codes that can be implemented as simply as the competitive block codes. This point will be more meaningful to the reader who compares the convolutional codes discussed in the following sections with the block codes of Sec. 11.2. He will see, for example, that useful codes are known for much wider ranges of code length and efficiency.

12.2.1 Single-error-correcting Codes

We have seen that for a linear (n,k) convolutional code to have minimum distance d, no set of $d-1$ or fewer columns of the parity-check matrix [Eq. (10.44)] can add to zero provided at least one of the columns is chosen from among the first n_0 columns of the matrix. A code capable of correcting all single errors must have minimum distance 3 or more; therefore no two columns of the H matrix can add to zero (be identical) if one or more of the columns is chosen from the first n_0.

It is possible to construct such a code with parameters $n_0 = 2^{m-1}$ and $k_0 = n_0 - 1$ as follows. Choose the columns of the matrix B, i.e., the first n_0 columns of H, to be the 2^{m-1} m-tuples formed by preceding each of the binary $(m-1)$-tuples by a single 1. The assignment of $(m-1)$-tuples to columns is irrelevant, except that, if the code is to be used in systematic form, it is necessary to assign the all-zero $(m-1)$-tuple to the n_0th column of H (which corresponds to a check bit). The $(m2^{m-1}, m(2^{m-1}-1))$ code specified by this matrix is capable of correcting all single errors since no two of the first n_0 columns are identical and no other column (headed by a 0) is identical to one of the first n_0 columns (headed by a 1). Since all $(m-1)$-tuples are employed in the B matrix, it does not seem possible to construct a better single-error-correcting code. In fact, this is true. In Wyner and Ash[6] it is proved that the codes are optimal, i.e., meet the coset-counting (Hamming-type) bound for convolutional codes.

Example 12-3. The (12,9) code considered in Examples 10-8, 10-9, 12-1, and 12-2 was constructed as above. Each of the first $n_0 = 2^{m-1} = 4$ columns of its parity-check matrix

$$\mathbf{H} = \begin{bmatrix} 1111 & & \\ 1010 & 1111 & \\ 1100 & 1010 & 1111 \end{bmatrix} \tag{12.3}$$

consists of a 1 followed by one of the four binary 2-tuples.

These single-error-correcting convolutional codes correspond to the single-error-correcting BCH block codes, i.e., the Hamming codes. A comparison of these two classes of codes serves to illustrate the point made in the discussion of bounds; namely, that for a given value of code efficiency and minimum distance the best convolutional codes tend to be slightly shorter than the best block codes.

The efficiency of the codes present above is

$$R = \frac{2^{m-1} - 1}{2^{m-1}} \tag{12.4}$$

Like the block codes of the preceding chapter, these codes can be shortened to provide other values of efficiency. This is done by the encoder and decoder agreeing that the first i information symbols in each block will be zero; hence there is no need to transmit them. The assignment of $(m-1)$-tuples to columns in the B matrix has been arbitrary up till this point. Now we assign these $(m-1)$-tuples so as to simplify the decoding procedure. One approach, suggested originally for the Hamming codes, is to associate with each column the $(m-1)$-tuple which in binary notation corresponds to the number of the column within the block. The all-zero $(m-1)$-tuple is assigned to the 2^{m-1}th (parity-check) column. Then the first bit of the syndrome is a 1 if an error occurred in the first block and a 0 otherwise; the remaining $m-1$ bits of the syndrome give the number of the bit which is in error if an error has occurred. The implementation of the decoding process is then quite straightforward and will not be discussed further.

After a single bit error has been corrected, the syndrome to be used in decoding the following block must be modified. Here this simply means setting the syndrome to 0 after correction has been performed. As a result, a decoding error cannot generate additional errors in subsequent blocks, and error propagation is not a problem.

The performance of these codes on the BSC is considered in Freiman and Robinson.[1] The authors show that there is little difference between the performance of these codes and that of single-error-correcting (Hamming) block codes.

12.2.2 Self-orthogonal Codes[2,8,9]

The codes discussed in this section have either one parity-check bit or one information bit per block, i.e., either $n_0 - k_0$ or $k_0 = 1$. We shall concentrate initially on the case of $n_0 - k_0 = 1$ and then show that for every self-orthogonal $(mn_0, m(n_0-1))$ code, there exists a corresponding self- orthogonal (mn_0, m) code which is its dual. These codes employ majority-logic decoding similar to that explained in Sec. 11.2.3. Also they are defined for practically all useful values of code efficiency and minimum distance. Unfortunately for large values of code length and number of information symbols, the codes usually have relatively low minimum distances. As a result, the performance of long, self-orthogonal codes is inferior to that of the corresponding block codes. However, for small values of n the two types of codes are competitive.

The codes. Let us consider the basic parity-check matrix of an $(mn_0, m(n_0-1))$ convolutional code. Since each n_0-bit block contains only one parity bit, this matrix is an $n = mn_0$-place row matrix of the form

$$\mathbf{h} = \left[\mathrm{P}_{m-1}{}^T\, 0\, \mathrm{P}_{m-2}{}^T\, 0 \ldots \mathrm{P}_0{}^T\, 1\right] \tag{12.5}$$

where the $\mathrm{P}_l{}^T$ represent $k_0 = (n_0 - 1)$-place row matrices; 0 and 1 are, of course, binary digits. This matrix can be specified by k_0 sets of integers chosen from the set $[1, 2, \ldots, m]$; the jth of these sets specifies which of the matrices P_l have 1 in their jth positions, $1 \leq j \leq k_0$. Equivalently this set specifies in which preceding blocks the jth information bit is checked by the parity bit in the present block.

At this point it is necessary to define what has been referred to as the *difference triangle*[8] associated with an ordered set of integers $[0, n_1, n_2, \ldots, n_{d-2}]$, where $0 < n_1 < n_2 < \cdots < n_{d-2}$. This difference triangle consists of the set of $(d-1)(d-2)/2$ differences $n_j - n_i$ for which $j > i$. If no two of these differences are identical, the triangle is said to be *full*.[9]

In Robinson and Bernstein[8] it is shown that if the k_0 sets of integers are of the same size and have full difference triangles, and if these triangles are disjoint (i.e., no element appears in more than one), it is possible to form $d - 1$ orthogonal check sums on each information bit in an n_0-bit code block. Then by an identical argument to that used in the proof of Theorem 11-4, the code has minimum distance at least d. Furthermore, as with block codes, a very simple majority-logic decoder can be constructed for the code as a result. The length of the code is n_0 times the number of blocks involved in the matrix $\mathbf{h}$, which, as it happens, is one more than the largest element appearing in any difference triangle.

This result, unproved here, provides two related methods of constructing self-orthogonal codes. First, the geometric techniques used in Sec. 11.2.3 can be used to construct disjoint, full, difference triangles in a few cases. Second and more important, because the code-construction problem has been shown to be identical to the problem of constructing sets of integers with simple, well-defined properties, a computer-aided trial-and-error solution can be obtained conveniently in most codes of interest.

Relating self-orthogonal codes to difference triangles also enables us to compute a lower bound on the length of a self-orthogonal code having efficiency $R = (n_0 - 1)/n_0$ and minimum distance d. Since the difference triangles must be disjoint and since each contains $(d-1)(d-2)/2$ elements, the largest element in the collection of integers in the union of the triangles must be at least $(n_0 - 1)(d-1)(d-2)/2$. Then since the code length is n_0 times one more than this number, we have the bound

$$n \geq n_0 \left[\frac{(n_0 - 1)(d - 1)(d - 2)}{2} + 1\right] \tag{12.6}$$

Table 12.2 contains a list of the specifying sets of integers of self-orthogonal convolutional codes for various values of code efficiency and minimum distance. This list is taken from Robinson and Bernstein,[8] which contains a more extensive tabulation. The reader may wish to verify that the length of the codes in the table is usually only slightly greater than the bound of Eq. (12.6). The following example is intended to illuminate the foregoing discussion and the use of Table 12.2.

Example 12-4. Let us construct a self-orthogonal code with $d = 5$ and $R = k_0/n_0 = 2/3$. Such a code is specified by $k_0 = 2$ sets of $d-1 = 4$ integers with the properties that no integer appears more than once among the $(d-1)(d-2)/2 = 6$ differences in either triangle and that the triangles are disjoint. From Table 12.2 we have that the sets of integers (0,8,9,12) and (0,6,11,13) yield full disjoint difference triangles. These triangles can be formed as shown below, and it can readily be checked that they are full and disjoint.

```
      4                 7
   1     3           5     2
8     9     12    6     11    13
```

The basic parity-check matrix specified by these sets is shown in Eq. (12.7). Also shown are the block numbers l associated

$$\begin{matrix} & 13 & 12 & 11 & 10 & 9 & 8 & 7 & 6 & 5 & 4 & 3 & 2 & 1 & 0 \\ \mathfrak{h} = [& 100 & 010 & 100 & 000 & 010 & 010 & 000 & 100 & 000 & 000 & 000 & 000 & 000 & 111] \end{matrix} \tag{12.7}$$

with each 3-tuple, that is, the subscript of P_l^T in the matrix $\mathfrak{h}$ of Eq. (12.5). The parity-check matrix of the code can be constructed from shifted, truncated versions of $\mathfrak{h}$ as in Eq. (10.44). It is suggested that the interested reader construct this matrix and verify that $d-1 = 4$ orthogonal check sums can be formed on each information bit. The length of this code is, of course, $3 \cdot 14 = 42$ while the bound of Eq. (12.6) indicates that it is not possible to construct a code shorter than 39 bits.

Exactly $n_0 - 1$ sets of integers of order $d-1$ specify a self-orthogonal $(mn_0, m(n_0-1))$ code in which a set of $d-1$ orthogonal check sums can be constructed on each bit. These sets can also be used to specify a (mn_0, m) code in which a set of $(n_0-1)(d-1)$ orthogonal check sums can be constructed on the single information bit in each block. The minimum distance of such a code is then at least

$$d' = (n_0 - 1)(d - 1) + 1 \tag{12.8}$$

In a code of this type the ith set of integers, $1 \le i \le n_0 - 1$, contains the block numbers of the blocks in which the information bit is check by the parity bit. The code constructed in this manner is the *dual* of the original code.

Since this code has the same length as the $(mn_0, m(n_0-1))$ code, solving for d in Eq. (12.8) and substituting the result into Eq. (12.6) gives the bound

$$n \geq n_0 \left[\frac{(d-1)}{2} \left[\frac{d-1}{n_0-1} - 1 \right] + 1 \right] \tag{12.9}$$

where for simplicity the prime has been dropped.

Example 12-5. The sets of integers used in Example 12-4 to construct a (42,28) self-orthogonal convolutional code with minimum distance 5 can also be used to construct its dual, a (42,14) self-orthogonal code. The basic parity-check matrix for this code, which has $n_0 - k_0 = 2$ rows and $n = 42$ columns, is

$$\begin{array}{cccccccccccccc} 13 & 12 & 11 & 10 & 9 & 8 & 7 & 6 & 5 & 4 & 3 & 2 & 1 & 0 \end{array}$$
$$\mathbf{h} = \begin{bmatrix} 000 & 100 & 000 & 000 & 100 & 100 & 000 & 000 & 000 & 000 & 000 & 000 & 000 & 110 \\ 100 & 000 & 100 & 000 & 000 & 000 & 000 & 100 & 000 & 000 & 000 & 000 & 000 & 101 \end{bmatrix} \tag{12.10}$$

Each row of this matrix is determined by one set of integers. For example, since the first set is [0,8,9,12], the first parity bit in the 0th block checks the information bits in blocks number 0,8,9, and 12, as shown.

The minimum distance of this code, that is, 9, is given by Eq. (12.8). It is recommended that the interested reader verify this result by actually constructing the $\mathbf{H}$ matrix of the code, and by observing that there are eight orthogonal check sums on each information bit in the twelfth block.

Decoding.[12] Let us assume that a $[(d-1)/2] = t$-error-correcting self-orthogonal code is to be decoded and that t or fewer errors occurred within a constraint length of the $(m-1)$th block. We have see that it is possible to construct a set of $d-1$ orthogonal check sums on each information bit in the $(m-1)$th block of such a code. By reasoning similar to that employed in the proof of Theorem 11-3, the value assumed by the majority of these check sums will always be the correct value of the noise digit which was added to the information bit in question. Thus correction can be effected by adding this digit to the information bit. A decoder capable of performing this operation with $n_0 - k_0 = 1$ is shown in Fig. 12.5

As do most decoders considered in this book, this decoder corrects errors by reencoding the received sequence, forming the syndrome, and then adding the error pattern associated with the syndrome to the $(m-1)$th block of the received sequence. Rather than redraw the k-stage encoder of Fig. 12.1, we represent it by the block labeled ENCODER. This ENCODER calculates a parity bit for each received n_0-bit block. This bit is added to the corresponding received parity bit to form one bit of the syndrome. The $m = (n\text{-}k)$-bit section of the (semi-infinite) syndrome which is within a constraint length of the block being decoded is stored in the SYNDROME REGISTER. We have seen that $d-1$ of the check sums represented by the syndrome bits are orthogonal on each information bit in block $m-1$. The function of each of the k_0 majority gates, labeled M_i, is to perform the "voting" for one of these sets of orthogonal check sums.

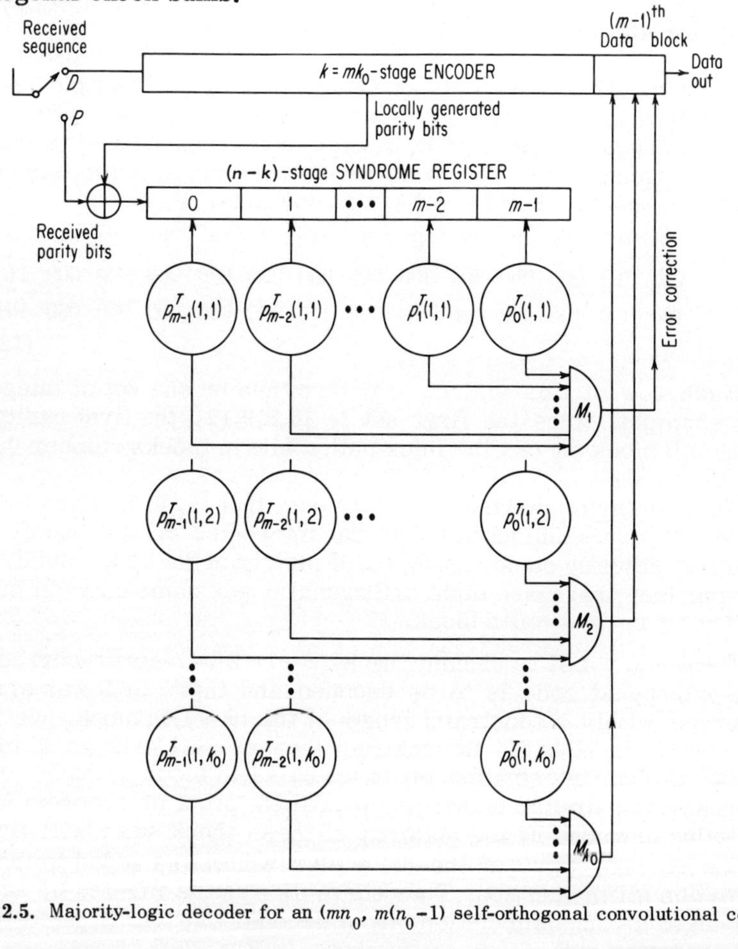

Fig. 12.5. Majority-logic decoder for an $(mn_0, m(n_0-1))$ self-orthogonal convolutional code.

The output of the ith gate, which is a 1 when more than $[d/2]$ of its inputs are 1s and 0 otherwise, is added to the ith information bit in the $(m-1)$th block in the ENCODER. Also, as represented by the two-way arrows between the SYNDROME REGISTER and majority gates, this correction signal is used to invert (add 1 to) each bit of the syndrome which is connected to M_i, that is, to each syndrome bit which checks a particular information bit. After all k_0 data bits of this block have been corrected, the block is shifted out to the sink. The next block is then shifted into the decoder, its syndrome is calculated, and decoding begins again.

A decoder for an (mn_0, m) self-orthogonal code is shown in Fig. 12.6. Its operation is identical to that described above, except that only a single majority gate is employed.

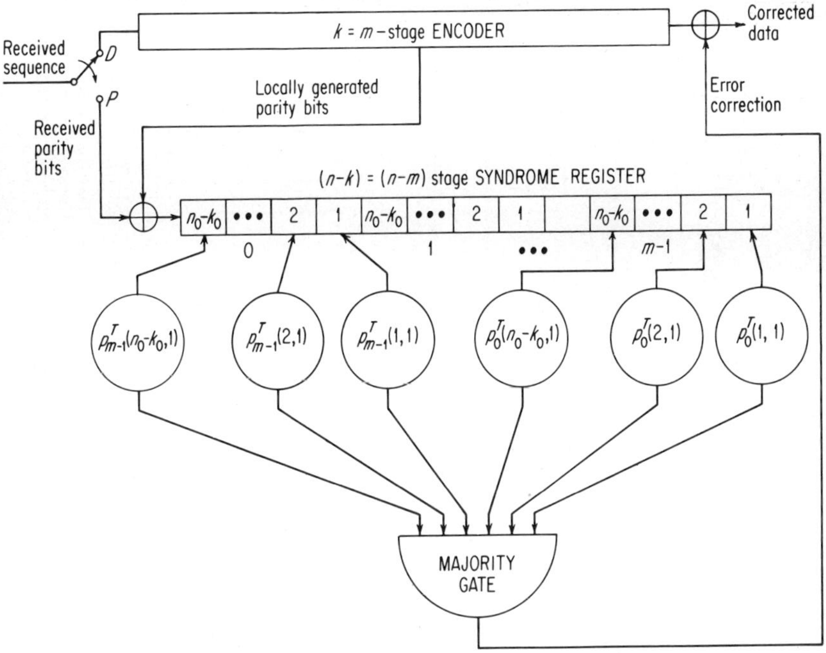

Fig. 12.6. Majority-logic decoder for an (mn_0, m) self-orthogonal convolutional code.

Example 12-6. The decoder for the (42,28) double-error-correcting, self-orthogonal convolutional code discussed in Example 12-4 is depicted in Fig. 12.7. To see that error correction is actually performed as claimed, let us assume that a single error has occurred so that the data bit in the twenty-eighth (rightmost) stage of the ENCODER is erroneous. Since this bit is in error, the 0th block syndrome bit, stored in the first (leftmost) stage of the SYNDROME REGISTER, will be a 1. During the calculation of the second block syndrome bit, the error was in the twenty-fourth stage of the ENCODER; thus this syndrome bit is also a 1. By similar arguments we can show that the only other syndrome bits which are nonzero are the eighth and fourteenth.

Since three or more (i.e., four) inputs to the majority gate M_1 are 1s, its output is a 1; adding this correction bit to the data bit in the twenty-eighth stage completes the correction process.

Had a second error occurred, at most one of the four inputs to M could have changed to a 0 and correction would have been performed as above. The second error would then have been corrected in a similar manner.

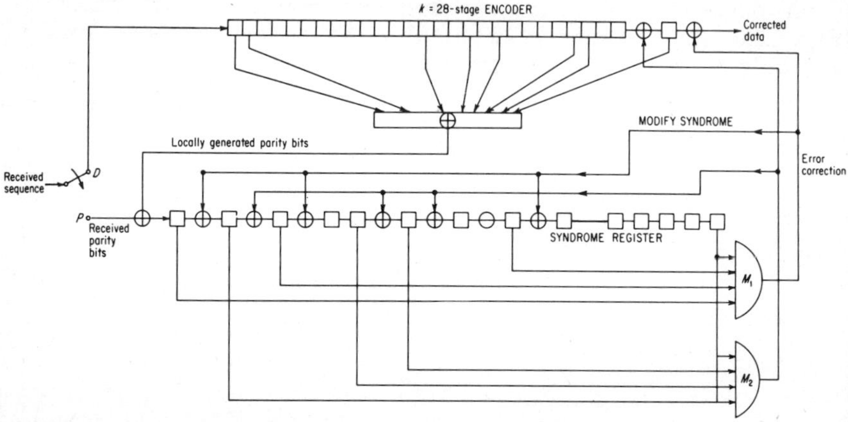

Fig. 12.7. Majority logic decoder for the (42,28) self-orthogonal code of illustration 12.6.

The leads labeled Modify Syndrome in Fig. 12.7 feed back the results of correction to the SYNDROME REGISTER. Every syndrome bit which feeds a particular majority gate checks the information bit corresponding to that gate. Thus each of these syndrome bits must be inverted whenever a correction is made. As shown, this is done as the syndrome is shifted to prepare for the correction of the next data block.

The decoder for the dual of this code, that is, the (42,14) 4-error-correcting code discussed in Example 12-5, is depicted in Fig. 12.8. Its operation is identical to that of the decoder of Fig. 12.7.

Error propagation is not a serious problem with self-orthogonal codes. It is shown[8] that if a decoding error is followed by a relatively short (a few constraint lengths) error-free interval, no additional errors beyond this interval will be made. This result is proved by noting that because a bit is changed only if more than half of a particular subset of the syndrome bits are 1s, feeding back prior corrections always reduces the weight of the syndrome. In the absence of additional errors it will be set to 0 within a few constraint lengths and no additional decoding errors will be made.

Self-orthogonal codes also possess the unusual property that their guaranteed error-correcting ability t is not diminished if the results of prior corrections are not used to modify the syndrome.[8]

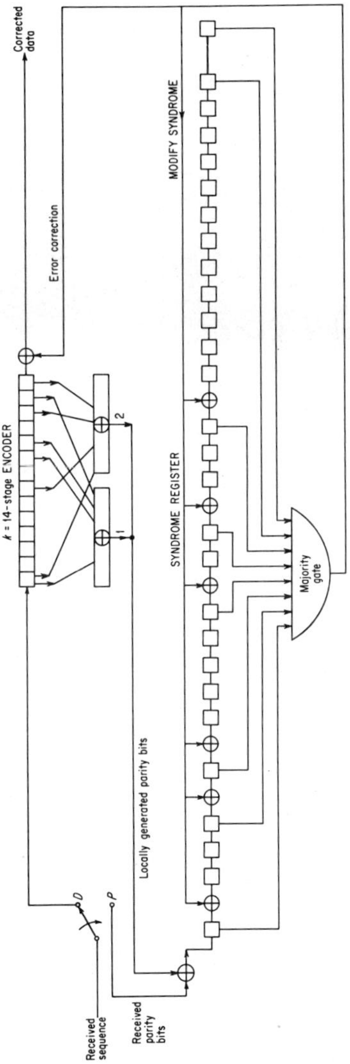

Fig. 12.8. Majority-logic decoder for the (42,14) self-orthogonal code of illustration 12.5.

In this case the code is capable of correcting all error patterns in which t or fewer errors occur in a span of $2n - n_0$, rather than n, bits. Thus complete elimination of error propagation can be purchased at the cost of a small decrease in performance. No modification of the decoding circuits of Figs. 12.4 and 12.6 beyond disconnecting the feedback connections from the majority gates to the SYNDROME REGISTER is required to implement this *definite decoding*, as it has been called.

The performance of self-orthogonal codes on the BSC is considered in Massey.[2] Although a meaningful comparison is difficult to make, it appears that short self-orthogonal codes perform just about as well as the corresponding block codes, but long codes perform considerably worse than BCH codes with the same minimum distance and efficiency.

12.2.3 Orthogonalizable Codes

The codes discussed in this section are taken from Massey,[2] where they are discussed in greater detail. They are closely related to the self-orthogonal codes of the preceding section but have several significant differences. On the one hand, for given values of code length and minimum distance the codes are generally more efficient than self-orthogonal codes and they can be implemented approximately as simply as the latter. On the other hand, no general procedure for constructing orthogonalizable codes is known; one must resort to poorly defined trial-and-error methods. These methods seem to break down completely unless $n_0 - k_0$ or k_0 is equal to 1 and are even unwieldy when $n_0 - k_0$ is equal to 1. As a consequence, useful classes of this type of code can be conveniently constructed with presently known techniques only for $R = 1/2, 1/3, 1/4, \ldots$. Another disadvantage inherent to orthogonalizable codes is that since feedback decoding must almost always be employed, decoding errors tend to propagate.

A self-orthogonal convolutional code with minimum distance d is one in which some set of $d - 1$ rows of its parity-check matrix is orthogonal on each of the k_0 information bits in the block being decoded. An *orthogonalizable* code with minimum distance d is one in which it is possible to form a set of $d - 1$ orthogonal check sums on each information bit in block $m - 1$ by taking sums of rows of the parity check matrix. At present no general rules are known for orthogonalizing a given parity-check matrix or for constructing matrices that can be orthogonalized. One is reduced to using the trial-and-error procedures mentioned earlier.

The following example serves to clarify the definition of an orthogonalizable code.

Example 12-7. A (12,6) orthogonalizable convolutional code with minimum distance 5 is specified by the basic parity-check matrix

$$\mathbf{h} = [\,101010000011\,] \tag{12.11}$$

The parity-check matrix of the code is, of course,

$$\mathbf{H} = \begin{bmatrix} 11 \\ 00 & 11 \\ 00 & 00 & 11 \\ 10 & 00 & 00 & 11 \\ 10 & 10 & 00 & 00 & 11 \\ 10 & 10 & 10 & 00 & 00 & 11 \end{bmatrix} \tag{12.12}$$

The four check sums specified by the first, fourth, and fifth row and the sum of the second and sixth rows are orthogonal. These are presented in Table 12.1.

Table 12.1. Orthogonal Check Sums in (12,6) Code.

Row No.	Check sum
1	11
4	10 00 00 11
5	10 10 00 00 11
2 + 6	10 01 10 00 00 11

The decoder for this code is shown in Fig. 12.9. Note that since the fifth-block data bit is checked by the parity bits in blocks 0, 1, 2, and 5, the syndrome bits associated with these blocks are inverted whenever a correction is performed.

Orthogonalizable codes with efficiency 1/2 and 1/3 are listed in Table 12.3 for various values of minimum distance. These are taken from Massey[2] where codes with $R = 1/5$ and $1/10$ are also tabulated. Massey also discusses two small classes of analytically constructed orthogonalizable codes, the *uniform codes* and the *Reed-Muller-like codes*. The efficiency of these codes is of the form 2^{-m} with very few codes defined for small m.

We mentioned earlier that an orthogonalizable code can be implemented as simply as a self-orthogonal code of the same efficiency and minimum distance. Since orthogonalizable codes are shorter in general, fewer stages of shift register are required. However, on the negative side, up to $d-1$ EXCLUSIVE-OR circuits may be required to sum various syndrome bits to form the orthogonal check sums.

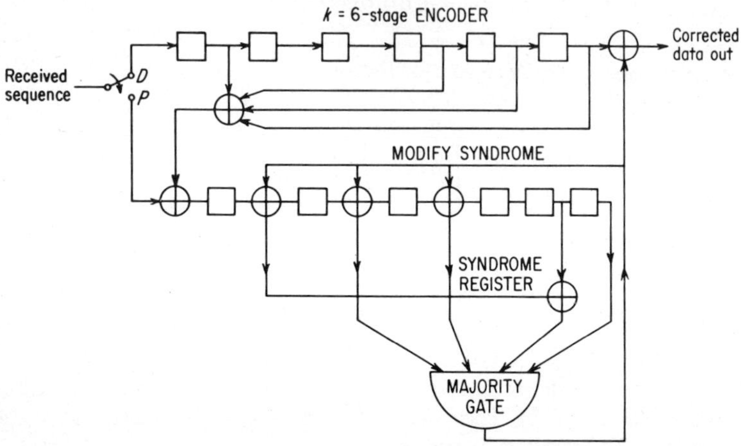

Fig. 12.9. Majority-logic decoder for the (12,6) orthogonalizable code of Example 12-7.

12.2.4 Probabilistic (Sequential) Decoding [11,20]

In Sec. 10.2.3 we saw that the standard array could be used as an optimum decoding table for a block code. In its modified form this array can also be used for the algebraic decoding of a convolutional code. In both cases, however, roughly 2^{n-k} (or 2^k, if $k < n/2$) decoder operations are necessary to decode a single block. In the case of a cyclic block code this says that 2^{n-k} decoder operations must be performed in n channel-bit times. As we have seen, because of the cyclic nature of codes of this type, this is a practical decoding procedure when n-k is not large, say, 15 or less. For a convolutional code, however, the 2^{n-k} decoder operations must be performed in n_0 channel-bit times. Furthermore, the implementation of the search is complicated somewhat because the codes does not possess the symmetry of a cyclic block code.

Probabilistic decoding has been proposed as a way of circumventing this exponential growth of complexity with block length while maintaining the simplicity and wide applicability of systematic search decoding. Although probabilistic decoding is applicable to more general codes, we shall continue to focus attention on the binary case. The cost of implementing this type of decoding seems to be considerably higher than that of algebraic decoding of convolutional codes and of the various procedures for decoding the random-error-correcting block codes of Sec. 11.2. However, probabilistic decoding applies to a considerably larger class of codes than these other decoding procedures and, in some situations, is the only workable technique.

As mentioned in Chap. 10, the probabilistic decoding of a single n_0-bit block is based on a section of the syndrome considerably longer than the $(n-k)$-bit section used in algebraic decoding. Let us denote by $N-K = M(n_0 - k_0)$ the length of this syndrome, where M is the number of blocks stored in the decoder. The symbol N denotes the decoding constraint length. A given n_0-bit block is decoded by making a succession of tentative decodings or estimates of the block. Since the decoder can store only M blocks, decoding must be completed exactly N-bit times after the block entered the decoder. (If the best estimate of the block available at the time it is leaving the decoder is judged to be a poor one, the decoder can assume that it has detected an error.[12,16] Then, as with block codes, the corrupted data can be deleted or retransmitted if system consideration permits).

The mechanism by which the estimates of a given block are formed can be illustrated by considering the behavior of a probabilistic decoder when no errors have occurred. The first block to enter the decoder is examined and tentatively decoded into the closest (in the Hamming sense) n_0-tuple for which the $n_0 - k_0$ parity bits in the block are satisfied. Since n_0 is ordinarily quite small, this decoding presents no problems. The distance between the received block and the estimate is recorded. Since no error occurred here, this distance is zero. The second block is decoded in the same manner into the closest n_0-bit block for which its $n_0 - k_0$ parity bits are satisfied. In general these check bits also check the information bits of the first block. When the $(M+1)$th block enters the decoder, the first block is ejected, and decoding has been completed. It is clear that in the absence of channel errors all blocks are decoded correctly in this manner.

Now assume that errors occur. Since the blocks are ordinarily quite short, the probability that a tentative decoding is erroneous is high. Since the check bits in many of the $M-1$ following blocks also check this erroneously estimate block, it is likely that several of these blocks will also be estimated in error. That is, the distance between the received sequence and the sequence of estimates is likely to increase as more blocks are examined. When this distance exceeds a prescribed threshold which increases linearly with the number of blocks examined, the decoder decides that some past block was estimated erroneously. A procedure for finding this error relatively quickly with high probability, known as the *Fano algorithm*, is explained in Wozencraft and Jabocs[17] and Fano.[19] A general discussion of the prominent characteristics of probabilistic decoding as exemplified by this algorithm follows.

How does one construct codes suitable for use with probabilistic decoding? Although some convolutional codes with good distance properties have been found by a computer search,[4,7] analysis shows and experiments confirm that for codes of at least moderate length this is not necessary since nearly all randomly selected codes perform adequately.

Because the effective number of bits involved in each decoding, N, is considerably larger than n, the code constraint length, the error rate produced by a convolutional code decoded probabilistically is considerably lower than that of the same code decoded algebraically or of an n-bit block code.[15] While this is perhaps interesting philosophically, its engineering significance is often negated because both the total decoding delay and amount of equipment needed to implement probabilistic decoding are, in most cases, considerably greater than required for other decoding procedures.

Probabilistic decoding is best suited to situations in which other, more simply implemented, techniques perform unsatisfactorily. Since practically all algebraic procedures deteriorate as n increases, probabilistic decoding is economically attractive when both high data rate and very low error rate must be attained simultaneously. In such situations the "buffer-overflow" problem becomes a major concern.

Unlike algebraic decoding, the time required to arrive at the last tentative decoding (effectively, to decode) of a block is a random variable. In low-noise periods little computation, i.e., few tentative decodings per block, must be performed. However, in high-noise periods the reverse is true and a considerable amount of time may be required to arrive at the final estimate of a block. Since data is being fed into and, presumably, out of the decoder at a constant rate, a queuing problem arises. Savage[14] has shown that a B-bit buffer for this queue will be exceeded with probability

$$cB^{-\alpha(R)}$$

where the coefficient is on the order of unity and the exponent is a linearly decreasing function of rate. Thus, the probability of buffer overflow is an algebraic function of the buffer size and consequently is rather insensitive to the size of the buffer.

These results show that for code rates less than some rate R_{comp} less than channel capacity C, the average number of computations per block increases algebraically with code length. However, for larger values of rate it increases exponentially. If and only if the rate is less than R_{comp} the second moment of this distribution of computation increases algebraically with code length; for higher rates the second moment, too, increases exponentially. Similar results hold for higher moments.

In situations where one would consider using probabilistic decoding, the probability of a buffer overflow is considerably higher than the probability of a decoding error. Since overflow can be detected by the decoder, it seems natural to augment forward-acting error correction with retransmission. Such systems are analyzed in Stiglitz.[16]

To summarize, the properties of probabilistic decoding are as follows:

1. Probabilistic decoding can be used to attain a low error rate and a high data rate on a wide variety of channels.
2. In numerous sophisticated analyses the properties of probabilistic decoding have been explored to the point where they are relatively well understood despite their inherent complexity.
3. The results of these analyses show, among other things, that probabilistic decoding is impractical for rates above $R_{\text{comp}} < C$ and that the probability of a decoding error is considerably smaller than the probability of failing to decode because of lack of time, i.e., of buffer overflow.
4. At present probabilistic decoding seems to be considerably more costly to implement than competitive algebraic techniques on binary channels.
5. On nonbinary channels if both a high data rate relative to capacity and a low error rate are desired, an expensive error control system is dictated and probabilistic decoding appears to be competitive with algebraic techniques.

12.3 Burst-error Correction[6, 21-26]

We shall now discuss convolutional codes that are able to correct a single burst of errors. Our concern will be predominantly with codes that correct all bursts of length not greater than some number b; codes able to correct most bursts of slightly greater length are discussed elsewhere.[21]

A convolutional code that is able to correct all bursts of length not greater than b is said to have *(type B1) burst-correcting ability* b.[6] A code capable of correcting all bursts which are confined to r consecutive n_0-bit blocks is said to have *type B2 burst-correcting* ability $b_2 = rn_0$. From an engineering point of view the parameter b is the more interesting of the two. However, type $B2$ codes have been analyzed to a considerably greater extent than type $B1$ codes. Also, as we shall see shortly, type $B2$ codes have some (type $B1$) burst-correcting ability. For this reason we shall consider type $B2$ as well as type $B1$ codes.

12.3.1 Bounds

A tight upper bound on the type $B2$ burst-correcting ability of an (n, k) convolutional code has been derived.[6] [See Eq. (12.17).]

Before discussing this bound consider the following relationships. A code with burst-correcting ability b clearly has type $B2$ burst-correcting ability

$$b_2 \geq n_0 \left\lfloor \frac{b}{n_0} \right\rfloor \tag{12.13}$$

Let $\overline{b_2}$ denote the previously mentioned upper bound on b_2. From Eq. (12.13) it follows that

$$b \leq \overline{b_2} + n_0 - 1 \tag{12.14}$$

This upper bound on b, though unattainable in practice (and thus probably loose) for b/n_0 on the order of unity, is quite tight for b_2 considerably greater than n_0. A code with type $B2$ burst-correcting ability b_2 has burst-correcting ability

$$b \geq b_2 - n_0 + 1 \tag{12.15}$$

Let us now derive the upper bound on b_2.

As a first step, we show that a type $B2$ code with $b_2 = n_0$ cannot have a nonzero code word that consists of a single burst confined to blocks $m-1$ and $m-2$. If we assume that such a code word exists, there are three cases to consider:

1. Block $m-2$ is all zero. However, provided no code word has weight one, two bursts are confined to the block $m-1$ and have the same syndrome; thus $b_2 \neq n_0$.
2. Neither block is all zero. However, then two bursts are each confined to a single block and have the same syndrome. Since one of the bursts is in the incorrect subset it follows that $b_2 \neq n_0$.
3. Block $m-1$ is all zero. A forward shift of this code word by one block produces another code word. This new word consists either of a single burst in block $m-1$ or two bursts in the 0th and $(m-1)$th blocks. By the arguments used in cases 1 and 2, respectively, $b_2 \neq n_0$.

Therefore, if $b_2 \neq n_0$ there cannot be a code word which consists of a single burst in the $(m-1)$th and $(m-2)$th blocks. But by an argument identical to that employed in the derivation of Eq. (11.70), unless the code has $2n_0$ or more parity checks, some code word does consist of a single burst confined to the block being decoded and the following block. Consequently

$$b_2 \leq \frac{n-k}{2} \tag{12.16}$$

where $b_2 = n_0$.

This argument can also be carried through for b_2/n_0 an integer greater than one although it is more involved.[6] This yields

$$n - k \geq 2b_2 - (n_0 - k_0)\left(\frac{b_2}{n_0} - 1\right) \tag{12.17}$$

These results show that although most of the codes discussed in the sequel are not truly optimal, they are so close to the bounds that little practical importance attaches to finding better codes. In fact it seems likely that it is the bounds on b, rather than the codes, which can be improved.

In the presentation of burst-correcting block codes of Sec. 11.3, the concept of interleaving was seen to be useful for generating a long burst-correcting code from a short one. Of course, convolutional codes can be interleaved with the same result. This reduces the problem of constructing good burst-correcting codes for all useful values of code length and efficiency to the simpler problem of constructing good codes for useful values of efficiency. This point and other aspects of interleaving are examined in Sec. 12.3.2.

12.3.2 Interleaving

When discussing the interleaving of block codes, we defined the interleaving degree i as one more than the number of bits between adjacent bits of a subcode word. In this definition we assumed tacitly that $i-1$ bits would be inserted between every pair of bits of every subcode word. This is called *bit interleaving*. If we had so desired, we could have sectioned subwords into an integral number of short blocks and inserted $i-1$ blocks between adjacent blocks in a subword. This technique, called *block interleaving*, is sometimes easier to implement. Provided the length of the interleaved block does not exceed the burst-correcting ability of the subcode, its performance may be superior to that attained with bit interleaving. With respect to type $B2$ burst-correcting convolutional codes, since b_2 is a multiple of the length of a block, it is convenient to interleave n_0-bit blocks. Throughout the remainder of this section the word *interleaving* will mean block interleaving and the *interleaving degree* will be interpreted as one more than the number of blocks between adjacent blocks of the basic code.

Interleaving an (n, k) block code to degree i produces an (ni, ki) block code regardless of how it is interleaved. Interleaving an (mn_0, mk_0) convolutional code produces not an $(mn_0 i, mk_0 i)$ code, but rather an $(mn_0 i - n_0 i + n_0,\ mk_0 i - k_0 i + k_0)$ code. This is because the number of blocks between the first and last blocks (inclusive) of the basic code is $im - i + 1$.

As mentioned earlier, interleaving reduces the general problem of constructing burst-correcting codes to that of constructing good short codes with useful values of efficiency. Two general approaches to this latter problem have been taken. The first is to find parity-check matrices that directly specify useful codes. With seemingly few useful mathematical techniques available, this has been largely a trial-and-error approach. Not surprisingly, few good codes have been found in this way; the ones that have are discussed in Sec. 12.3.3. The second approach is to construct good type $B2$ codes and then interleave them. This produces (type $B1$) burst-correcting codes for which b is bounded by Eq. (12.15). The problem of constructing type $B2$ codes has proved solvable and is discussed in Sec. 12.3.4.

12.3.3 Trial-and-error Codes

In order for an $(mn_0, m(n_0 - 1))$ convolutional code to have burst-correcting ability b, it is necessary and sufficient that no linear combination of columns selected from two sets of b adjacent columns of the parity-check matrix add to zero when at least one column is chosen from the first n_0 columns of the matrix. It has proved difficult to convert this complicated-sounding statement into a set of requirements on the structure of the parity-check matrix. However, two useful codes have been found by a trial-and-error search procedure.[23]

1. $R = 1/2$. The matrix

$$\mathbf{H} = \begin{bmatrix} 01 & & & \\ 10 & 01 & & \\ 10 & 10 & 01 & \\ 00 & 10 & 10 & 01 \end{bmatrix} \tag{12.18}$$

specifies an (8,4) convolutional code for which $b = 2$.

2. $R = 2/3$. The matrix

$$\mathbf{H} = \begin{bmatrix} 001 & & & & & \\ 110 & 001 & & & & \\ 100 & 110 & 001 & & & \\ 010 & 100 & 110 & 001 & & \\ 100 & 010 & 100 & 110 & 001 & \\ 100 & 100 & 010 & 100 & 110 & 001 \end{bmatrix} \tag{12.19}$$

specifies an (18,12) convolutional code for which $b = 3$.

No decoding technique specialized to these codes has been devised. However, these codes are so short that a simple syndrome-recognizing decoder is practical. For the (8,4) code the syndromes corresponding to correctable bursts are

Burst	Syndrome
10 0	0110
11 0	1110
01 0	1000
01 1	1011

Thus the single information bit in the $(m-1)$th block is inverted if and only if the last three bits of the syndrome are 110. The decoder for the (18,12) code is slightly more complex (i.e., there are now $2^{b-1}n_0 = 12$ different bursts to consider), but its principles are identical.

If this decoding procedure is employed, a correctable burst which overlaps into block $m-2$ can be completely corrected while block $m-2$ is being decoded. Feeding the results of this correction back to the syndrome simply means setting it to 0. As a result, errors do not propagate after an incorrect decoding because the feedback always reduces the weight of the syndrome. After the errors that caused the erroneous decoding have passed through the decoder, an n-bit guard space is all that is required to guarantee that the syndrome be set to 0.

The performance of codes of this type on a channel on which intererror intervals are independent random events distributed according to a Pareto distribution is considered in Richters.[3]

12.3.4 Type B2 Codes

The type $B2$ burst-correcting codes presented in this section are capable of correcting all bursts of errors confined to a single n_0-bit block, provided the burst is followed by an $(n-n_0)$-bit guard space. The codes are optimal in that they meet the upper bound on b_2 expressed in Eq. (12.16). A simply implemented decoding algorithm has been devised for them also.

Berlekamp[23]-Preparata[25]-Massey[24] (BPM) codes. For every $n_0 > 1$ there exists a $(2n_0^2, 2n_0(n_0-1))$ convolutional code with type $B2$ burst-correcting ability n_0. In such a code no code word can have one burst confined to the $(m-1)$th block while another burst is in some other n_0-bit block. That is,

$$\mathbf{v}\mathbf{H}^T \neq 0 \tag{12.20}$$

where $\mathbf{v}$ is of the form

$$\mathbf{c}_0 000 \ldots \mathbf{c}_j \ldots 0 \tag{12.21}$$

where c_0 represents a (nonzero) burst confined to the $(m-1)$th block and c_j represents a burst confined to the $(m-i-j)$th block.

The 0s represent all-zero n_0-tuples. If $\mathbf{H}$ has the form

$$\mathbf{H} = \left[\mathbf{B}_0\,\mathbf{B}_1\,\mathbf{B}_2 \ldots \mathbf{B}_{2n_0-1}\right] \tag{12.22}$$

where the $\mathbf{B}_i$ represent down-shifted, truncated versions of the matrix $\mathbf{B}_0$ [= $\mathbf{B}$ in Eq. (10.45)], Eq. (12.20) says that for the code to have type $B2$ burst-correcting ability n_0

$$\mathbf{c}_0\,\mathbf{c}_i\,[\mathbf{B}_0\,\mathbf{B}_i]^T \neq 0 \qquad 1 \leq i \leq 2n_0 - 1 \tag{12.23}$$

for all bursts $\mathbf{c}_0, \mathbf{c}_i$. There are two cases to consider.

1. When $n_0 \leq i \leq 2n_0 - 1$, the upper right quadrant of $[\mathbf{B}_0\,\mathbf{B}_i]$ is the all-zero matrix of order n_0. Choosing the upper half of $\mathbf{B}_0$ to be any nonsingular matrix of order n_0 thus guarantees that Eq. (12.23) holds. For ease of implementation we choose it to be the mirror image of the identity matrix, denoted $\bar{\mathbf{I}}$.
2. When $1 \leq i \leq n_0 - 1$, Eq. (12.23) holds provided the matrices $[\mathbf{B}_0\,\mathbf{B}_1]$, $[\mathbf{B}_0\,\mathbf{B}_2]$, ..., $[\mathbf{B}_0\,\mathbf{B}_{n_0-1}]$ are simultaneously nonsingular.

With the upper half of $\mathbf{B}_0$ specified as $\bar{\mathbf{I}}$, elementary row operations can transform $[\mathbf{B}_0\,\mathbf{B}_i]$ to the form

$$\begin{bmatrix} \bar{\mathbf{I}} & \mathbf{X}_i \\ 0 & \mathbf{Y}_i \end{bmatrix} \tag{12.24}$$

Thus $[\mathbf{B}_0\,\mathbf{B}_i]$ is nonsingular if and only if the n_0-by-n_0 matrix $\mathbf{Y}_i$ is nonsingular. In Berlekamp[23] and Preparata[25] it is shown that it is possible to choose $\mathbf{B}_0$ such that all $\mathbf{Y}_i$ are rendered simultaneously nonsingular. Furthermore, the proof is a constructive one so that the matrix $\mathbf{B}_0$ is specified for all values of n_0. This procedure yields

$$\mathbf{B}_0 = \begin{bmatrix} \bar{\mathbf{I}} \\ 00\ldots00 \\ 1 \\ 11 \\ 111 \\ 1011 \\ 10001 \\ 110001 \\ \vdots \quad \ddots \end{bmatrix} \tag{12.25}$$

That is, to construct a code with block length n_0, choose the rows of the upper half of the matrix $\mathbf{B}_0$ to be the mirror image of the identity matrix of order n_0; and choose the rows of the lower half to be the first n_0 rows of the lower half of $\mathbf{B}_0$ above. This construction is illustrated in

Example 12-8. It is desired to construct a type $B2$ burst-correcting code with efficiency 2/3 for which $b_2 = n_0 = 3$. By Eq. (12.16)

$$n \geq \frac{2b_2}{1-R} = 18 \tag{12.26}$$

The optimal, i.e., shortest-possible, code with these parameter values is the BPM code specified by the matrix

$$\mathbf{B}_0 = \begin{bmatrix} 001 \\ 010 \\ 100 \\ 000 \\ 100 \\ 110 \end{bmatrix} \tag{12.27}$$

The parity-check matrix of this code is

$$\mathbf{H} = \begin{bmatrix} 001 & & & & & \\ 010 & 001 & & & & \\ 100 & 010 & 001 & & & \\ 000 & 100 & 010 & 001 & & \\ 100 & 000 & 100 & 010 & 001 & \\ 110 & 100 & 000 & 100 & 010 & 001 \end{bmatrix} \tag{12.28}$$

We have seen that, in order for the code specified by $\mathbf{B}_0$ to be capable of correcting all type $B2$ bursts of length n_0 or less, it is necessary and sufficient that the two square matrices

$$[\mathbf{B}_0\,\mathbf{B}_1] = \begin{bmatrix} 001 & \\ 010 & 001 \\ 100 & 010 \\ 000 & 100 \\ 100 & 000 \\ 110 & 100 \end{bmatrix} \quad \text{and} \quad [\mathbf{B}_0\,\mathbf{B}_2] = \begin{bmatrix} 001 & \\ 010 & \\ 100 & 001 \\ 000 & 010 \\ 100 & 100 \\ 110 & 000 \end{bmatrix} \tag{12.29}$$

be nonsingular. Then

$$\mathrm{T}[\mathbf{B}_0\,\mathbf{B}_1] = \begin{bmatrix} 001 & \\ 010 & 001 \\ 100 & 010 \\ 000 & 100 \\ 000 & 010 \\ 000 & 111 \end{bmatrix} \quad \text{and} \quad \mathrm{T}[\mathbf{B}_0\,\mathbf{B}_1] = \begin{bmatrix} 001 & \\ 010 & \\ 100 & 001 \\ 000 & 010 \\ 000 & 101 \\ 000 & 001 \end{bmatrix} \tag{12.30}$$

where T is a nonsingular matrix. Thus the nonsingularity of $[B_0\, B_i]$, where $i = 1, 2, \ldots,$ is guaranteed by the nonsingularity of the lower-right quadrant matrices

$$X_1 = \begin{bmatrix} 100 \\ 010 \\ 111 \end{bmatrix} \quad \text{and} \quad X_2 = \begin{bmatrix} 010 \\ 101 \\ 001 \end{bmatrix} \tag{12.31}$$

Consequently the (18,12) code specified by H in Eq. (12.28) has $b_2 = 3$, and since $n - k = 6$, the code is optimal.

Decoding.[24] Assume that a single burst has occurred in the block to be decoded. Because the upper half of the matrix B_0 was chosen to be the mirror image of the identity matrix, the first n_0 bits of the syndrome in reverse order are identical to the error pattern which was added to the 0th block. The only additional information to decode a block, therefore, is whether the burst occurred in the $(m-1)$th block or elsewhere.

If the burst occurred in the $(m-1)$th block, the second half of the syndrome must be

$$s_2 = c_0[B_{02}]^T \tag{12.32}$$

where c_0 denotes the burst, B_{02} denotes the lower half of the matrix B_0, and s_2 is the second half of the syndrome. In other words, since the first half of the syndrome in reverse order $\bar{s}_1$ is identical to c_0, if

$$s_2 + \bar{s}_1[B_{02}]^T = 0 \tag{12.33}$$

the burst occurred in the first block; otherwise it did not.

Equation (12.33) represents n_0 linear equations involving the syndrome bits. A single multiple-input modulo-2 adder is all that is required to implement each equation. Then one n_0-input LOGICAL-OR gate is sufficient to determine whether or not all of the equations are satisfied.

This decoding technique is illustrated in Example 12-9.

Example 12-9. Consider the (18,12) code of Example 12-8. If a burst occurred in the $(m-1)$th block, by Eq. (12.33)

$$s_2 + \bar{s}_1 \begin{bmatrix} 011 \\ 001 \\ 000 \end{bmatrix} = 0 \tag{12.34}$$

The three binary equations thus specified are

$$s_{21} = 0 \quad s_{22} = s_{13}, \quad \text{and} \quad s_{23} = s_{12} + s_{13} \tag{12.35}$$

where s_{ij} represents the jth bit of the ith half of the syndrome.

The decoder for this code is shown in Fig. 12.10. It operates as follows. The syndrome is calculated as usual (see Sec. 12.1.2) and stored in the 6-stage SYNDROME REGISTER. If the relationships in Eq. (12.35) hold, the output of the inverter is a 1; otherwise it is a 0. If it is a 1 and if all the errors affecting the bits in the decoder are confined to a single 3-bit block, these errors occurred in the $(m-1)$th block. Furthermore, this burst is identical to the first 3 bits of the syndrome with the order reversed. Since we do not ordinarily correct parity bit errors, the two data bits affected by the burst can be corrected by adding to each the appropriate syndrome bit as shown.

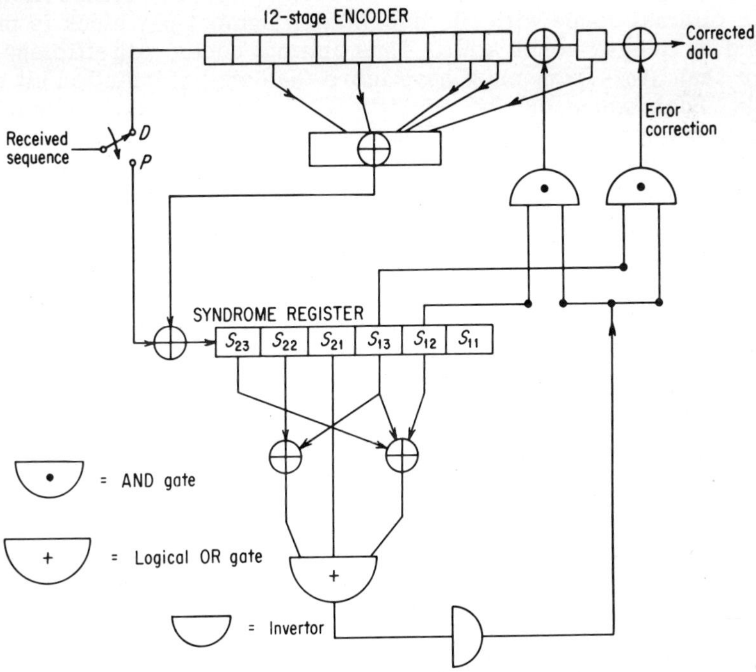

Fig. 12.10. Decoder for (18,12) Berlekamp-Preparata-Massey code of Example 12.9.

As with single-error-correcting codes, this correction information is fed back to the syndrome by setting it to zero; as a result, errors cannot propagate. (This is not shown in Fig. 12.10.)

Codes for which $n_0 - k_0 > 1$. The BPM codes of the preceding paragraphs have one parity bit per n_0-bit block. Thus they are defined for efficiencies of the form $(n_0-1)/n_0$. It would be desirable to construct optimal type $B2$ codes having one information symbol per block. For example, a code of this type having $n_0 = 3$ is specified by the parity-check matrix

$$H = \begin{bmatrix} 010 & & \\ 001 & & \\ 100 & 010 & \\ 000 & 001 & \\ 000 & 100 & 010 \\ 100 & 000 & 001 \end{bmatrix} \tag{12.36}$$

This can be verified directly, or the argument of the preceding section can be applied. The resulting (9,3) code has $b_2 = (n\text{-}k)/2 = 3$ and hence is optimal. The decoding technique for the BPM codes can also be used for this code. Unfortunately, the construction of other optimal codes with one information symbol per block is precluded by integer constraints. Thus optimal codes with efficiencies other than $(n_0 - 1)/n_0$ must have more than one information bit per block. Such codes can be constructed using the basic principles employed where $(n_0 - k_0) = 1$. That is, choosing the matrix B so that the appropriate set of matrices are simultaneously nonsingular guarantees that the code has type $B2$ burst-correcting ability. This technique is employed in Preparata.[25]

We can summarize the material on burst-correcting convolutional codes as follows:

1. For reasonably large values of interleaving degree, b is approximately equal to b_2. Thus very good, easily implemented codes are known for most practical values of efficiency unless b is small.
2. For small b the trial-and-error codes of Sec. 12.3.3 are best. Unfortunately, at present codes of this type are known only for efficiencies 1/2 and 1/3.
3. For small b and higher efficiency, the suboptimal codes of Hagelbarger[22] and Epley and Lin[26] are best. Because of their narrow range of applicability, a discussion of these codes is not presented here.

12.4 Codes for Correcting Bursts and Random Errors

The theory of this type of code is perhaps even less well developed than that of the corresponding block codes. There are basically two related techniques for constructing convolutional codes which can correct bursts and random errors. These are discussed in the following sections.

12.4.1 Basic Codes

Consider a self-orthogonal or orthogonalizable random-error-correcting (mn, m) code which has the properties that (1) no burst of length b or less which begins on the first information bit to be decoded affects more than $t-1$ of the syndrome bits used to decode that bit; and (2) no burst which begins anywhere else affects more than t of these bits. Such a code has, in addition to its random-error-correcting ability, burst-correcting ability b. Some authors have referred to codes of this type as b-*diffuse*.[9] The generalization of this concept to codes for which k_0 is greater than 1 is immediate. No theory for constructing these codes has been published, although it is not difficult to construct relatively short codes by hand, especially if $k_0 = 1$. An example of a b-diffuse code is provided in Example 12-10.

Example 12-10. The (20,10) code specified by the parity-check matrix

$$\mathbf{H} = \begin{bmatrix} 11 \\ 00 & 11 \\ 10 & 00 & 11 \\ 00 & 10 & 00 & 11 \\ 00 & 00 & 10 & 00 & 11 \\ 00 & 00 & 00 & 10 & 00 & 11 \\ 10 & 00 & 00 & 00 & 10 & 00 & 11 \\ 00 & 10 & 00 & 00 & 00 & 10 & 00 & 11 \\ 00 & 00 & 10 & 00 & 00 & 00 & 10 & 00 & 11 \\ 10 & 00 & 00 & 10 & 00 & 00 & 00 & 10 & 00 & 11 \end{bmatrix} \tag{12.37}$$

is capable of correcting all error patterns of weight two or less. For the first, third, and seventh rows and the sum of the sixth and tenth rows are orthogonal on the information bits in block 9. Also, the reader may check that no burst of length 4 or less results in a decoding error. Thus the code has simultaneous burst-correcting ability four in addition to its random-error-correcting ability.

In Table 12.3, as in Fig. 12.7, we see that there is a double-error-correcting, orthogonalizable (12,6) code. Since this code is very nearly optimal, one would expect that in order to add simultaneous burst-correcting ability while maintaining the code efficiency and random-error-correcting ability, a longer code would be required. This borne out by the last example.

12.4.2 Interleaved Codes

Interleaving a random-error-correcting convolutional code produces a code with both burst- and random-error-correcting ability. However, interleaving a basic code which itself corrects both types of errors may produce a better code. For example, interleaving the (20,10) code of Example 12-10 to degree 3 produces a (60,30) code for which b = 12. Interleaving the (12,6) orthogonalizable code of Examples 12-7 to degree 5 also produces a (60,30) code but one which has b = 10. Clearly the minimum distance of both codes is 5.

Interleaving a convolutional code to degree i produces another much longer, convolutional code. The implementation of the interleaved code is identical in principle to that of the basic code, except that the encoding and decoding shift registers are roughly i times as long.

12.5 Summary

In this chapter we considered the problem of correcting errors with a convolutional code. Although these codes are not as well developed as block codes, they share many of the basic properties of the latter: (1) they can be encoded very simply; (2) decoding is more difficult than encoding but several classes of codes have been found for which the amount of decoding equipment required is not excessive; (3) practical codes exist which are capable of correcting random errors, bursts, or both. As with block codes, random-error-correcting convolutional codes are less well understood and more costly to implement than burst-correcting codes.

Convolutional codes differ from block codes in that, because successive decodings are not independent, it is much more difficult to derive a meaningful expression for the probability of decoding error; also as a result, a past decoding error tends to cause subsequent decodings to be erroneous as well. That is, errors tend to propagate. Fortunately, for most of the practical codes known, this propagation is quite brief and consequently unimportant.

Probabilistic decoding of convolutional codes was mentioned, and it was pointed out that this very general and powerful technique is useful when both high code efficiency and low error rate must be obtained simultaneously.

In brief, convolutional codes, though not as well developed and more difficult to analyze than block codes, are competitive with the latter in many situations.

Appendix

Some Useful Convolutional Codes

This Appendix contains tables of the important parameters of self-orthogonal and orthogonizable convolutional codes. The following symbols are employed in the tables:

n_0 = code block length
m = 1 + largest integer in any of the sets of integers which specify the code
n = mn_0 = code length
R = code efficiency (rate)
d = minimum distance = $2t+1$, where
t = guaranteed random-error-correcting ability of the code

Table 12.2. Self-orthogonal Codes.[a]

$n_0 = 2$

$R = 1/2$ d	n	Sets of integers that specify code
3	4	(0, 1)
5	14	(0, 2, 5, 6)
7	36	(0, 2, 7, 13, 16, 17)
9	72	(0, 7, 10, 16, 18, 30, 31, 35)
11	112	(0, 2, 14, 21, 29, 32, 45, 49, 54, 55)
13	172	(0, 2, 6, 24, 29, 40, 43, 55, 68, 75, 76, 85)
15	256	(0, 5, 28, 38, 41, 49, 50, 68, 75, 92, 107, 121, 123, 127)
17	360	(0, 6, 19, 40, 58, 67, 78, 83, 109, 132, 133, 162, 165, 169, 177, 179)
19	434	(0, 2, 10, 22, 53, 56, 82, 83, 89, 98, 130, 148, 153, 167, 188, 192, 205, 216)
21	568	(0, 24, 30, 43, 55, 71, 75, 89, 104, 125, 127, 162, 167, 189, 206, 215, 272, 275, 282, 283)

$n_0 = 3$

$R = 1/3$ d	$R = 2/3$ d	n	Sets of integers that specify code
5	3	9	(0, 1)
			(0, 2)
9	5	42	(0, 8, 9, 12)
			(0, 6, 11, 13)
13	7	123	(0, 2, 6, 24, 29, 40)
			(0, 3, 15, 28, 35, 36)
17	9	261	(0, 1, 27, 30, 61, 73, 81, 83)
			(0, 18, 23, 37, 58, 62, 75, 86)
21	11	393	(0, 1, 6, 25, 32, 72, 100, 108, 120, 130)
			(0, 23, 39, 57, 60, 74, 101, 103, 112, 116)

$n_0 = 4$

$R = 1/4$ d	$R = 3/4$ d	n	Sets of integers that specify code
7	3	16	(0, 1)
			(0, 2)
			(0, 3)
13	5	80	(0, 3, 15, 19)
			(0, 8, 17, 18)
			(0, 6, 11, 13)
19	7	272	(0, 5, 15, 34, 35, 42)
			(0, 31, 33, 44, 47, 56)
			(0, 17, 21, 43, 49, 67)

Table 12.2 Self-orthogonal Codes.[a] (Continued)

n_0 = 4 (continued)			
$R = 1/4$ d	$R = 3/4$ d	n	Sets of integers that specify code
25	9	520	(0, 9, 33, 37, 38, 97, 122, 129)
			(0, 11, 13, 23, 62, 76, 79, 123)
			(0, 19, 35, 50, 71, 77, 117, 125)
n_0 = 5			
$R = 1/5$ d	$R = 4/5$ d	n	Sets of integers that specify code
9	3	25	(0, 1)
			(0, 2)
			(0, 3)
			(0, 4)
17	5	145	(0, 1, 26, 28)
			(0, 3, 13, 24)
			(0, 7, 19, 23)
			(0, 8, 17, 22)
25	7	490	(0, 22, 41, 57, 72, 93)
			(0, 14, 17, 61, 87, 95)
			(0, 39, 49, 51, 62, 69)
			(0, 9, 33, 37, 38, 97)

[a]This table is taken from the more complete version given in Robinson and Bernstein.[8] For a discussion of the codes tabulated here, see Sec. 12.2.2.

Table 12.3 lists orthogonalizable codes of efficiency 1/2 and 1/3 for various values of minimum distance. It is taken from Table II of Ref. [2]* where codes of efficiency 1/5 and 1/10 are also listed. For a discussion of these codes see Sec. 12.2.3.

Along with the usual parameters of d and n, the basic parity-check matrix h of each code is given. To save space it is given in the following form. Each row of the $(n_0 - 1)$-row matrix is given on a set of integers corresponding to the blocks in which a nonzero term appears in that row. The $n_0 - 1$ sets are distinguished by an exponent corresponding to the row number. Thus the matrix

$$\mathbf{h} = \begin{bmatrix} 000 & 000 & 000 & 100 & 110 \\ 100 & 100 & 100 & 000 & 101 \end{bmatrix}$$

* By permission of J. L. Massey.[9]

Table 12.3. Orthogonalizable Codes.

$R = 1/2$

d	n	Sets of integers which specify code	Rules for orthogonalization
3	4	(0, 1)	(0) (1)
5	12	(0, 3, 4, 5)	(0) (3) (4) (1, 5)
7	24	(0, 6, 7, 9, 10, 11)	(4, 8, 11)
9	44	(0, 11, 13, 16, 17, 19, 20, 21)	(0) (11) (13) (16) (17) (2, 3, 6, 19) (4, 14, 20) (1, 5, 8, 15, 21)
11	72	(0, 18, 19, 27, 28, 29, 30, 32, 33, 35)	(0) (18) (19) (27) (1, 9, 28) (10, 20, 29) (11, 30, 31) (13, 21, 23, 32) (14, 33, 34) (2, 3, 16, 24, 26, 35)
13	104	(0, 26, 27, 39, 40, 41, 42, 44, 45, 47, 48, 51)	(0) (26) (27) (39) (1, 13, 40) (14, 28, 41) (15, 42, 43) (17, 29, 31, 44) (18, 45, 46) (2, 3, 20, 32, 24, 47) (21, 35, 48, 49, 50) (24, 30, 33, 36, 38, 51)

$R = 1/3$

d	n	Sets of integers which specify code	Rules for orthogonalization
3	3	$(0)(0)^1(0)^2$	$(0^1)(0^2)$
5	9	$(0,1)^1(0,2)^2$	$(0^1)(0^2)(1^1)(1^2)$
7	15	$(0,1)^1(0,2,3,4)^2$	$(0^1)(0^2)(1^1)(2^2)(1^2 3^2)(2^1 4^2)$
9	24	$(0,1,7)^1(0,2,3,4,6)^2$	$(0^1)(0^2)(1^1)(2^2)(1^2,3^2)(2^1,4^2)(7^1)(3^1,5^1,6^1,6^2)$
11	33	$(0,1,9)^1$	$(0^1)(0^2)(1^1)(2^1,2^2)(9^1)(3^2,4^2)(3^1,5^1,5^2)$
		$(0,1,2,3,5,8,9)^2$	$(1^2,4^1,6^1,6^2)(8^1,8^2)(7^2,9^2,10^2)$
13	54	$(0,4,5,6,7,9,12,13,16)^1$	$(0^1)(0^2)(1^1,1^2)(4^1)(5^1)(2^2,6^1)(14^2)$
		$(0,1,14,15,16)^2$	$(7^1,10^1,11^1,11^2)(3^2,5^2,9^1)(6^2,8^2,12^1)(3^2,16^2,17^2)(4^2,10^2,12^2,16^1)$

is represented as $(0,1)^1\ (0,2,3,4)^2$. For $R = 1/2$ the exponent is always 1 and is omitted.

The information necessary to construct the decoders for these codes, that is, the rules for constructing the set of $d-1$ orthogonal check sums (for "orthogonalizing" the code) is also given in the table. Each set is composed of the numbers of the rows of the parity-check matrix (H) which must be added modulo-2 to give one of the orthogonal check sums. This number is given as a block number (base) and a row number (exponent). Thus the orthogonalization rules for the code discussed above are $(0^1)(0^2)(1^1)(2^2)(1^2,3^2)(2^1,4^2)$. Its parity-check matrix is

$$\mathbf{H} = \begin{bmatrix} 110 & & & & \\ 101 & & & & \\ 100 & 110 & & & \\ 000 & 101 & & & \\ 000 & 100 & 110 & & \\ 100 & 000 & 101 & & \\ 000 & 000 & 100 & 110 & \\ 100 & 100 & 000 & 101 & \\ 000 & 000 & 000 & 100 & 110 \\ 100 & 100 & 100 & 000 & 101 \end{bmatrix} \begin{matrix} 0^1 \\ 0^2 \\ 1^1 \\ 1^2 \\ 2^1 \\ 2^2 \\ 3^1 \\ 3^2 \\ 4^1 \\ 4^2 \end{matrix}$$

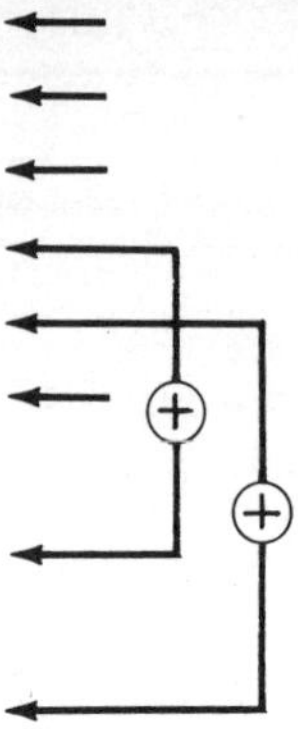

The row numbers are listed to the right of the matrix. The reader may wish to verify that the six check sums produced according to the given rules are othogonal.

Notes

Elias[28] invented convolutional codes. Almost from the first the investigation of these codes proceeded along two separate paths. Probabilistic decoding was first considered by Wozencraft[11] and later modified by Fano,[19] among others. Recent developments, as well as a complete bibliography, are given by Savage[13] and Ling.[18] The first practical, algebraically decoded, convolutional codes were Hagelbarger's recurrent burst-correcting codes.[22] Kilmer[30,31] worked on both burst and random-error-correcting codes and laid the foundations for the basic work of Wyner and Ash.[6] These authors provided a precise mathematical framework for algebraic convolutional codes; it is significant that most expositions of these codes, including that of the present authors, have employed their results. Massey[2] devised the first useful class of random-error-correcting convolutional codes, the self-orthogonal and orthogonalizable codes of Sec. 12.2. Robinson and Bernstein[8] systematized Massey's procedure for constructing self-orthogonal codes and in so doing improved on Massey's codes. The elegant burst-correcting codes of Sec. 12.3.4 were constructed independently by Berlekamp[23] and Preparata[25] using the mathematical framework set up by Wyner and Ash. Massey's[24] simply implemented decoding procedure for these codes makes them attractive for use in practical systems requiring burst-error correction.

References

Chapter 1

Noise and Signal Theory

1. Bendat, J. S.: "Principles and Applications of Random Noise Theory," John Wiley & Sons, Inc., New York, 1958.
2. Bennett, W. R.: "Electrical Noise," McGraw-Hill Book Company, New York, 1960.
3. Davenport, W. B., Jr., and W. L. Root: "Introduction to Random Signals and Noise," McGraw-Hill Book Company, New York, 1958.
4. Rowe, H. E.: "Signals and Noise in Communication Systems," D. Van Nostrand Company, New York, 1965.
5. Wainshtein, L. A., and V. D. Zubakov: "Extraction of Signals from Noise," Prentice-Hall, Inc., Englewood Cliffs, N.J., 1962.
6. Wax, N., ed.: "Selected Papers on Noise and Stochastic Processes," Dover Publications, Inc., New York, 1954.

Statistical-communication and Information Theory

7. Ash, R.: "Information Theory," Interscience Publishers, New York, 1965.
8. Abramson, N.: "Information Theory and Coding," McGraw-Hill Book Company, New York, 1963.
9. Fano, R. M.: "Transmission of Information," The M.I.T. Press, Cambridge, Mass., and John Wiley & Sons, Inc., New York, 1961.
10. Hancock, J. C.: "An Introduction to the Principles of Communication Theory," McGraw-Hill Book Company, New York, 1961.
11. Hancock, J. C., and P. A. Wintz: "Signal Detection Theory," McGraw-Hill Book Company, New York, 1966.
12. Harman, W. W.: "Principles of the Statistical Theory of Communication," McGraw-Hill Book Company, New York, 1963.

13. Helstrom, C. W.: "Statistical Theory of Signal Detection," Pergamon Press, New York, 1960.
14. Kotel'nikov, V. A.: "The Theory of Optimum Noise Immunity," McGraw-Hill Book Company, New York, 1960.
15. Lee, Y. W.: "Statistical Theory of Communication," John Wiley & Sons, Inc., New York, 1960.
16. Middleton, D.: "An Introduction to Statistical Communication Theory," McGraw-Hill Book Company, New York, 1960.
17. Reza, F. M.: "An Introduction to Information Theory," McGraw-Hill Book Company, New York, 1961.
18. Schwartz, M.: "Information Transmission, Modulation, and Noise," McGraw-Hill Book Company, New York, 1959.
19. Shannon, C. E., and W. Weaver: "The Mathematical Theory of Communication," The University of Illinois Press, 1949.

Communications Systems

20. Baghdady, E. J., Ed.: "Lectures on Communication System Theory," McGraw-Hill Book Company, New York, 1960.
21. Balakrishnan, A. V., Ed.: "Advances in Communication Systems," vols. 1–3, Academic Press, Inc., New York.
22. Bennett, W. R., and J. R. Davey: "Data Transmission," McGraw-Hill Book Company, New York, 1965.
23. Golomb, S. W., Ed.: "Digital Communications with Space Applications," Prentice Hall, Inc., Englewood Cliffs, N.J., 1961.
24. Schwartz, M., W. R. Bennett, and S. Stein: "Communication Systems and Techniques," McGraw-Hill Book Company, New York, 1966.
25. Wozencraft, J. M., and I. M. Jacobs: "Principles of Communication Engineering," John Wiley & Sons, Inc., New York, 1965.
26. Viterbi, A. J.: "Principles of Coherent Communication," McGraw-Hill Book Company, New York, 1966.

Coding Theory

27. Massey, J. L.: "Threshold Decoding," The M.I.T. Press, Cambridge, Mass., 1963.
28. Peterson, W. W.: "Error-Correcting Codes," The M.I.T. Press, Cambridge, Mass., and John Wiley & Sons, Inc., New York, 1961.

Chapter 2

1. Alexander, A. A., R. M. Gryb, and D. W. Nast: Capabilities of the Telephone Network for Data Transmission, *BSTJ*, vol. 39, p. 431, 1960.

2. Fano, R. M.: "Transmission of Information," The M.I.T. Press, Cambridge, Mass., 1961.
3. Townsend, R. L., and R. N. Watts: Effectiveness of Error Control in Data Communications over the Switched Telephone Network, *BSTJ*, vol. 43, p. 2611, 1964.
4. Gilbert, E. N.: Capacity of a Burst-Noise Channel, *BSTJ*, vol. 39, p. 1253, 1960.
5. Elliott, E. O.: Estimates of Error Rates for Codes on Burst-Noise Channels, *BSTJ*, vol. 42, p. 1977, 1963.
6. Elliott, E. O.: A Model of the Switched Telephone Network for Data Communications, *BSTJ*, vol. 44, pp. 89–110, 1965.
7. Berger, J. M., and B. Mandelbrot: A New Model for Error Clustering in Telephone Circuits, *IBM J. Res. & Dev.*, vol. 7, p. 224, 1963.
8. Sussman, S. M.: Analysis of the Pareto Model for Error Statistics on Telephone Circuits, *IEEE Trans. on Communication Systems*, vol. CS-11, p. 213, 1963.
9. Schwartz, M., W. R. Bennett, and S. Stein; "Communication Systems and Techniques," McGraw-Hill Book Company, New York, 1966.

Chapter 3

1. Holsinger, J. L.: Digital Communication over Fixed Time-Continuous Channels with Memory—with Special Application to Telephone Channels, *M.I.T. Lincoln Lab. Tech. Rept.* 366, Lexington, Mass., Oct., 1964.
2. Shannon, C. E.: Probability of Error for Optimal Codes in a Gaussian Channel, *BSTJ*, vol. 38, pp. 611–55, 1959.
3. Gallager, R. G.: A Simple Derivation of the Coding Theorem and Some Applications, *IEEE Trans. on Information Theory*, vol. IT-11, pp. 3–18, 1965.
4. Shannon, C. E.: The Mathematical Theory of Communication, *BSTJ*, July and Oct., 1948.
5. Shannon, C. E., R. G. Gallager, and E. R. Berlekamp: Lower Bounds to Error Probability for Coding on Discrete Memoryless Channels, *IEEE Trans. on Information and Control*, Jan., 1967.
6. Ebert, P. M.: Error Bounds for Parallel Communication Channels, *M.I.T. Res. Lab. of Electronics Tech. Rept.* 448, Cambridge, Mass., Aug. 1, 1966.
7. Slepian, D.: Bounds on Communication, *BSTJ*, May, 1963.
8. Ash, R.: "Information Theory," Interscience Publishers, New York, 1965.
9. Fano, R. M.: "Transmission of Information," The M.I.T. Press, Cambridge, Mass., and John Wiley & Sons, Inc., New York, 1961.

10. Wozencraft, J. M., and I. M. Jacobs: "Principles of Communication Engineering," John Wiley & Sons, Inc., New York, 1965.

Chapter 4

1. Middleton, D.: "An Introduction to Statistical Communication Theory," McGraw-Hill Book Company, New York, 1960.
2. Nyquist, H.: "Certain Topics in Telegraph Transmission Theory," *Trans. AIEE*, Apr., 1928.
3. Lender, A.: The Duobinary Technique for High-Speed Data Transmission, *1963 IEEE Winter General Meeting*.
4. Bennett, W. R., and J. R. Davey, "Data Transmission," McGraw-Hill Book Company, 1965.
5. Kretzmer, E. R.: Generalization of a Technique for Binary Data Communication, *IEEE Trans. on Communication Technology*, vol. COM-14, Feb., 1966.
6. Gibby, R. A., and J. W. Smith: Some Extensions of Nyquist's Telegraph Transmission Theory, *BSTJ*, Sept., 1965.
7. Sunde, E. D.: Theoretical Fundamentals of Pulse Transmission, I and II, *BSTJ*, May and July, 1954.
8. Wheeler, H. A.: The Interpretation of Amplitude and Phase Distortion in Terms of Paired Echoes, *Proc. IRE*, June, 1939.
9. Gerrish, A. M., and R. D. Howson, Multilevel Nyquist II Signaling, to be published.

Chapter 5

1. Helstrom, C. W.: "Statistical Theory of Signal Detection," Pergamon Press, New York, 1960.
2. Chang, R. W., and J. C. Hancock: On Receiver Structures for Channels Having Memory, *IEEE Trans. on Information Theory*, vol. IT-12, Oct., 1966.
3. Aaron, M. R., and D. W. Tufts: Intersymbol Interference and Error Probability, *IEEE Trans. on Information Theory*, vol. IT-12, Jan., 1966.
4. Gibby, R. A., and J. W. Smith: Some Extensions of Nyquist's Telegraph Theory, *BSTJ*, Sept., 1965.
5. Tufts, D. W.: Nyquist's Problem—The Joint Optimization of Transmitter and Receiver in Pulse Amplitude Modulation, *Proc. IEEE*, vol. 53, March, 1965.
6. Smith, J. W.: The Joint Optimization of Transmitted Signal and Receiving Filter for Data Transmission Systems, *BSTJ*, Dec., 1965.
7. Gerst, I., and J. Diamond: The Elimination of Intersymbol Interference by Input Pulse Shaping, *Proc. IRE*, vol. 53, July, 1961.

8. Doetsch, G.: "Theorie und Anwendung der Laplace Transformation," Birkhauser, Basel, Switzerland, 1958.
9. Schwarzlander, H., and J. C. Hancock: "Signal Optimization for Digital Communication over Channels with Memory," Symp. on Signal Transmission and Processing, Columbia Univ., May, 1965.
10. Slepian, D., and H. O. Pollak: Prolate Spheroidal Wave Functions, Fourier Analysis and Uncertainty, pt. I, *BSTJ*, Jan., 1961.
11. Landau, H. J., and H. O. Pollak: Prolate Spheroidal Wave Functions, Fourier Analysis and Uncertainty, pt. II, *BSTJ*, Jan., 1961.
12. Hancock, J. C., and P. A. Wintz: "Signal Detection Theory," McGraw-Hill Book Company, New York, 1966.
13. George, D. A.: Matched Filters for Interfering Signals, *IEEE Trans. on Information Theory*, vol. IT-11, Jan., 1965.
14. Berger, J., and D. W. Tufts: Optimum Pulse Amplitude Modulation, pts. I and II, *IEEE Trans. on Information Theory*, vol. IT-13, Apr., 1967.

Chapter 6

1. Kallman, H. E.: Transversal Filters, *Proc. IRE*, vol. 28, pp. 302–10, July, 1940.
2. DiToro, M. J.: A New Method of High-Speed Adaptive Serial Communication through any Time-Variable and Dispersive Transmission Medium, paper presented at *1st IEEE Ann. Comm. Conf.*, June 7–9; 1965 IEEE Internat'l. Conf., pp. 763–67.
3. Schreiner, K. E., H. L. Funk, and E. Hopner: Automatic Distortion Correction for Efficient Pulse Transmission, *IBM J.*, pp. 20–30, Jan., 1965.
4. Becker, F. K., L. N. Holzman, R. W. Lucky, and E. Port: Automatic Equalization for Digital Communication, *Proc. IEEE*, vol. 53, pp. 96–97, Jan., 1965.
5. Bogert, B. P.: Demonstration of Delay Distortion Correction by Time-Reversal Techniques, *IRE Trans. on Communication Systems*, vol. CS-5, pp. 2–7, Dec., 1957.
6. Lucky, R. W.: Automatic Equalization for Digital Communication, *BSTJ*, vol. 44, pp. 547–88, Apr., 1965.
7. Gersho, A.: Automatic Time-Domain Equalization with Transversal Filters, *1966 Nat. Electronics Conf.*, Oct 3–5, Chicago.
8. Feller, W.: "An Introduction to Probability Theory and Its Applications," John Wiley & Sons, Inc., New York, 1957.
9. Lucky, R. W.: Techniques for Adaptive Equalization of Digital Communication, *BSTJ*, vol. 45, pp. 255–86, Feb., 1966.
10. Coll, D. C., and D. A. George: A Receiver for Time-Dispersed Pulses, paper presented at the *1st IEEE Ann. Comm. Conf.*, June 7–9; *1965 Internat'l. Conf. Rec.*, pp. 753–57.

11. Kettel, E.: Ein Automatischer Optimisator fur Abgleich des Impulsentzerrers in einer Datenubertragung, *Arch. Ubertragung*, Germany, vol. 18, pp. 271–78, May, 1964.
12. O'Neill, J. R., and B. R. Saltzberg: An Automatic Equalizer for Coherent Quadrature Carrier Data Transmission Systems, *1966 IEEE Internat'l. Comm. Conf.*; June 15–17, Philadelphia.
13. Lucky, R. W., and H. R. Rudin: Generalized Automatic Equalization for Communication Channels, *Proc. IEEE* (Correspondence), vol. 54, pp. 439–40, Mar., 1966.

Chapter 7

1. Titchmarsh, E. C.: "Introduction to the Theory of Fourier Integrals," chap. 5, Oxford University Press, London, 1950.
2. Becker, F. K., E. R. Kretzmer, and J. R. Sheehan: A New Signal Format for Efficient Data Transmission, *BSTJ*, May and June, 1966.
3. Becker, F. K.: An Exploratory Multilevel Vestigial-Sideband Data Terminal for Use on High-Grade Voice Facilities, paper presented at the *1st IEEE Ann. Comm. Conf.*, June 7–9, Philadelphia; 1964 IEEE Internat'l. Conv. Rec., pp. 481–84.
4. Bennett, W. R., and J. R. Davey: "Data Transmission," McGraw-Hill Book Company, New York, 1965.
5. Voelcker, H. B.: Demodulation of Single-Sideband Signals via Envelope Detection, *IEEE Trans. on Communication Technology*, vol. COM-14, Feb., 1966.
6. Costas, J. P.: Synchronous Communications, *Proc. IRE*, vol. 44, pp. 1713–18, 1956.
7. Bennett, W. R.: Synthesis of Active Networks, *1955 Proc. Symp. on Modern Network Synthesis*, pp. 45–61.
8. Zador, P. L.: Error Probabilities in Data System Pulse Regenerator with DC Restoration, *BSTJ*, July and Aug., 1966.
9. Aaron, M. R., and M. K. Simon: Approximation of the Error Probability in a Regenerative Repeater with Quantized Feedback, *BSTJ*, Dec., 1966.
10. Fracassi, R. D., and F. E. Froehlich: "Wideband Data Station," *1966 IEEE Internat'l. Conv. Rec.*, part 1.
11. Becker, F. K., J. R. Davey, and B. R. Saltzberg: AM Vestigial Sideband Transmission Set Using Synchronous Detection, *AIEE Trans. on Comm. and Elec.*, May, 1962.
12. H. S. Black: "Modulation Theory," D. Van Nostrand Company, Inc., New York, 1953.
13. Voelker, H.: Toward a Unified Theory of Modulation, pts. 1 and 2, *Proc. IEEE*, vol. 54, Mar., 1966, and May, 1966.
14. Oswald, A. A.: Early History of Single-Sideband Transmission, *Proc. IRE*, vol. 44, Dec., 1956.
15. Sunde, E. D.: Ideal Pulse Transmission by AM and FM, *BSTJ*, vol. 38, Nov., 1959.

Chapter 8

1. Bedrosian, E., and S. O. Rice: Distortion and Crosstalk of Linearly Filtered Angle-Modulated Signals, to be published.
2. Rowe, H. E.: "Signals and Noise in Communication Systems," McGraw-Hill Book Company, New York, 1965.
3. Cohn, J.: A New Approach to the Analysis of FM Threshold Detection, *Proc. NEC*, vol. 12, pp. 221–36, 1956.
4. Rice, S. O.: Noise in FM Receivers, M. Rosenblatt, Ed., in "Time Series Analysis," John Wiley & Sons, Inc., New York, 1963.
5. Mazo, J. E., and J. Salz: Theory of Error Rates in Digital FM, *BSTJ*, Nov., 1966.
6. Bennett, W. R., and J. Salz: Binary Data Transmission by FM over a Real Channel, *BSTJ*, vol. 42, Sept., 1963.
7. Gabor, D.: Theory of Communication, *Jour. IEE*, vol. 93, p. 429.
8. Bennett, W. R., and J. R. Davey: "Data Transmission," McGraw-Hill Book Company, New York, 1965.
9. Anderson, R. R., W. R. Bennett, J. R. Davey, and J. Salz: Differential Detection of Binary FM, *BSTJ*, vol. 44, pp. 111–59, Jan., 1965.
10. Lender, A.: The Duobinary Technique for High-Speed Data Transmission, *AIEE Trans.* (Communication and Electronics), vol. 82, pp. 214–18, May, 1963.
11. Salz, J., and S. Stein: Distribution of Instantaneous Frequency for Signal plus Noise, *IEEE Trans. on Information Theory*, vol. IT-10, Oct., 1964.
12. Mazo, J. E., and J. Salz: Probability of Error for Quadratic Detectors, *BSTJ*, vol. 44, Nov., 1965.
13. Salz, J.: Performance of Multilevel Narrowband FM Digital Communication Systems, *IEEE Trans. on Communication Technology*, vol. COM-13, Dec., 1965.
14. Kotel'nikov, V. A.: "The Theory of Optimum Noise Immunity," sect. 4–9, pp. 39–42, McGraw-Hill Book Company, New York, 1960.
15. Meyerhoff, A. A., and W. M. Mazer: Optimum Binary FM Reception Using Discriminator Detection and IF Shaping, *RCA Rev.*, vol. 22, pp. 698–728, Dec., 1961.
16. Shaft, P. D.: Error Rate of PCM-FM Using Discriminator Detection, *IEEE Trans. on Space Electronics and Telemetry*, vol. SET-9, pp. 131–37, Dec., 1963.
17. Salz, J., and V. G. Koll: An Experimental Digital Multilevel FM Modem, *IEEE Trans. on Communication Technology*. vol. CS-4, June, 1956.
18. Akima, H.: The Error Rates in Multiple FSK Systems and the Signal-to-Noise Characteristics of FM and PCM-FS Systems, *NBS Tech. Rept.* 167.

19. Turin, G. L.: The Asymptotic Behavior of Ideal M-ary Systems, *Proc. IRE*, vol. 47, Jan., 1959.
20. Golay, M. J. E.: Note on the Theoretical Efficiency of Information Reception with PPM, *Proc. IRE* (Correspondence), vol. 37, p. 1031, Sept., 1949.
21. Chaffee, J. M.: The Application of Negative Feedback to Frequency Modulation Systems, *BSTJ*, vol. 18, pp. 403–37, 1939.
22. Jacobs, I.: Comparison of *M*-ary Modulation Systems, *BSTJ*, to be published.
23. Rice, S. O.: Mathematical Analysis of Random Noise, *BSTJ*, vol. 23.
24. Barnard, R. D.: A Note on a Special Class of One-Sided Distribution Sums, *BSTJ*, vol. 46, pp. 203–06, Jan., 1965.
25. Sunde, E. D., Ideal Pulse Transmission by AM and FM, *BSTJ*, vol. 38, Nov., 1959.
26. Carson, J. R., and T. C. Fry: Variable Frequency Electric Circuit Theory with Application to the Theory of Frequency-Modulation, *BSTJ*, vol. 16, pp. 513–40, Oct., 1937.
27. Crosby, M. G.: Frequency-Modulation Noise Characteristics, *Proc. IRE*, vol. 25, pp. 472–514, 1937.
28. Blachman, N. M.: The Demodulation of a Frequency-Modulated Carrier and Random Noise by a Limiter and Discriminator, *J. Appl. Phys.*, vol. 20, pp. 38 and 976, 1949.
29. Lawson, J. L., and G. E. Uhlenbeck: "Threshold Signals," M.I.T. Radiation Laboratory Series, vol. 24, chap. 13, McGraw-Hill Book Company, New York, 1950.

Chapter 9

1. Doelz, M. I., E. T. Heald, and D. L. Martin: Binary Data Transmission Techniques for Linear Systems, pt. 1, *Proc. IRE*, vol. 45, May, 1957.
2. Baker, P. A.: Phase Modulation Data Sets for Serial Transmission at 2000 and 2400 Bits per Second, part 1, *AIEE Trans.* (Communication and Electronics), July, 1962.
3. Hancock, J. C., and R. W. Lucky: Performance of Combined Amplitude- and Phase-Modulated Communication Systems, *IRE Trans. Communication Systems*, vol. CS-8, Dec., 1960.
4. Davenport, W. B., Jr., and W. L. Root: "Introduction to Random Signals and Noise," McGraw-Hill Book Company, New York, 1958.
5. Cahn, C. R.: Performance of Digital Phase-Modulation Communications Systems, *Ramo-Wooldridge Rept.* M1/0-905, Apr., 1959.
6. Lawton, J. G.: Theoretical Error Rates of Differentially Coherent Binary and Kineplex Data Transmissions, *Proc. IRE*, vol. 47, pp. 334–44, Feb., 1959.

7. Bennett, W. R., and J. Salz, Binary Data Transmission by FM over a Real Channel, *BSTJ*, vol. 42, p. 2395, Sept., 1963
8. Khvorostenko, N. P.: Noise Immunity of Multiplex Phase Telegraphy, *TRE J.*, vol. 19, Dec., 1964.
9. Bennett, W. R., and J. R. Davey: "Data Transmission," chap. 10, McGraw-Hill Book Company, New York, 1965.
10. Salz, J., and B. R. Saltzberg: Double Error Rates in Differentially Coherent Phase Systems, *IEEE Trans. on Communication Systems*, vol. CS-12, June, 1964.
11. Slepian, D., and H. O. Pollak: Prolate Spheroidal Wave Functions, pt. 1, *BSTJ*, vol. 40, pp. 43–63, Jan., 1961.
12. Slepian, D., and Mrs. E. Sonnenblick: Eigenvalues Associated with Prolate Spheroidal Wave Functions, *BSTJ*, vol. 44, pp. 1745–59, Oct., 1964.
13. Mazo, J. E., and J. Salz: Probability of Error for Quadratic Detectors, *BSTJ*, vol. 44, Nov., 1965.
14. Bussgang, J. J., and M. Leiter: Phase-Shift Keying with a Transmitted Reference, *IEEE Trans. on Communication Technology*, vol. COM-14, Feb., 1966.
15. Kac, M., and A. J. F. Siegert: On the Theory of Noise in Radio Receivers with Square Law Detectors, *J. Appl. Phys.*, vol. 18, pp. 383–97, 1947.
16. Emerson, R. C.: First Probability Densities for Receivers with Square Law Detectors, *J. Appl. Phys.*, vol. 24, pp. 1168–80, 1953.

Chapter 10

1. Peterson, W. W.: "Error-Correcting Codes," The M.I.T. Press, Cambridge, Mass., 1961.
2. Shannon, C. E., Mathematical Theory of Communication, *BSTJ*, vol. 27, 1948.
3. Elias, P.: Coding for Noisy Channels, 1955 *IRE Conv. Rec.*, pt. 4, pp. 37–44.
4. Slepian, D.: A Class of Binary Signalling Alphabets, *BSTJ*, vol. 35, pp. 203-34, 1956.
5. Hamming, R. W.: Error Detecting and Error Correcting Codes, *BSTJ*, vol. 29, pp. 147–60, 1950.
6. Prange, E.: Cyclic Error Correcting Codes in Two Symbols, *AF Cambridge Res. Ct.*, Tech. Rept. AFCRC-TN-58-156, Bedford, Mass., 1957.
7. Massey, J. L.: "Threshold Decoding," The M.I.T. Press, Cambridge, Mass., 1963.
8. Wyner, A. D., and R. B. Ash: Analysis of Recurrent Codes, *IEEE Trans. on Information Theory*, vol. IT-9, pp. 143–56, 1963.

9. Wozencraft, J. M.: "Sequential Decoding for Reliable Communication," Sc. D. dissertation, Mass. Inst. Tech., Cambridge, 1957.
10. Abramson, N. M., and B. Elspas: Double-Error-Correcting Coders and Decoders for Non-Independent Binary Errors, paper presented at the *UNESCO Information Processing Conf.*, Paris, 1959.
11. Golay, M. J. E.: Notes on Digital Coding, *Proc. IRE*, (Correspondence), vol. 37, p. 657, 1949.

Chapter 11

1. Benice, R. J., and A. H. Frey, Jr.: An Analysis of Retransmission Systems, *IEEE Trans. on Communication Technology*, vol. COM-12, pp. 135–46, 1964.
2. Benice, R. J., and A. H. Frey, Jr.: Comparisons of Error Control Techniques, *IEEE Trans. on Communication Technology*, vol. COM-12, pp. 146–55, 1964.
3. Peterson, W. W.: "Error-Correcting Codes," The M.I.T. Press, Cambridge, Mass., 1961.
4. Hamming, R. W.: Error Detecting and Error Correcting Codes, *BSTJ*, vol. 29, pp. 147–60, 1950.
5. Plotkin, M.: Binary Codes with Specified Minimum Distance, *IRE Trans. on Information Theory*, vol. IT-6, pp. 445–50, 1960.
6. Gramenopolous, N.: "An Upper Bound for Error-Correcting," M.S. thesis, Elec. Engrg. Dept., Mass. Inst. Tech., Cambridge, 1963.
7. Johnson, S. M.: Improved Asymptotic Bounds for Error-Correcting Codes, *IEEE Trans. on Information Theory*, vol. IT-9, pp. 198–205, 1963.
8. Varsharmov, R. R.: Estimate of the Number of Signals in Error Correcting Codes, *Dokl. Akad. Nauk (USSR)*, vol. 117, pp. 739–41, 1957.
9. Gilbert, E. N.: A Comparison of Signaling Alphabets, *BSTJ*, vol. 31, pp. 504–22, 1952.
10. Sacks, G. E.: Multiple Error Correction by Means of Parity Checks, *IRE Trans. on Information Theory*, vol. IT-4, pp. 145–47, 1958.
11. Bose, R. C., and D. K. Ray-Chaudhuri: On a Class of Error-Correcting Binary Group Codes, *IEEE Trans. on Information and Control*, vol. IC-3, pp. 68–79, 1960.
12. Bose, R. C., and D. K. Ray-Chaudhuri: Further Results on Error-Correcting Binary Group Codes, *IEEE Trans. on Information and Control*, vol. IC-3, pp. 279–90, 1960.
13. Hocquenghem, A.: Codes Correcteurs d'erreurs, *Chiffres*, vol. 2, pp. 147–56, 1959.

14. Peterson, W. W.: Encoding and Error-Correction Procedures for the Bose-Chaudhuri Codes, *IRE Trans. on Information and Theory*, vol. IT-6, pp. 459–70, 1960.
15. Chien, R. T.: Cyclic Decoding Procedures for Bose-Chaudhuri-Hocquenghem Codes, *IEEE Trans. on Information Theory*, vol. IT-10, pp. 357–63, 1964.
16. Berlekamp, E. R.: On Decoding Binary Bose-Chaudhuri-Hocquenghem Codes, *IEEE Trans. on Information Theory*, vol. IT-11, pp. 577–80, 1965.
17. Berlekamp, E. R.: Practical BCH Decoders, *IEEE Trans. on Information Theory*, vol. IT-13, 1967.
18. Berlekamp, E. R.: "Monograph on BCH Codes," McGraw-Hill Book Company, New York, to be published.
19. Massey, J. L.: Step-by-Step Decoding of the Bose-Chaudhuri-Hocquenghem Codes, *IEEE Trans. on Information Theory*, vol. IT-11, pp. 580–85, 1965.
20. Reed, I. S., and G. Solomon: Poylnomial Codes over Certain Finite Fields, *JSIAM*, vol. 8, pp. 300–04, 1960.
21. Zierler, N.: A Class of Cyclic Linear Error-Correcting Codes in p^m Symbols, *M.I.T. Lincoln Lab. Group Rept.* 55-19, Lexington, Mass., 1960.
22. Forney, G. D., Jr.: On Decoding BCH Codes, *IEEE Trans. on Information Theory*, vol. IT-11, pp. 549–57, 1965.
23. Gilbert, E. N.: Gray Codes and Paths on the n-Cube, *BSTJ*, vol. 37, 1958.
24. Bennett, W. R., and J. R. Davey: "Data Transmission," McGraw-Hill Book Company, New York, 1965.
25. Reed, I. S.: A Class of Multiple-Error-Correcting Codes and the Decoding Scheme, *IRE Trans. on Information Theory*, vol. 38–49, 1954.
26. Massey, J. L.: "Threshold Decoding," The M.I.T. Press, Cambridge, Mass., 1963.
27. Mitchell, M. E., et al.: Coding and Decoding Operations Research, *GE Advanced Electronics Ct. Final Rept. on Contr.* AF 19(604)-6183, AFCRL8, 1961.
28. Townsend, R. L., and E. J. Weldon, Jr.: Self-Orthogonal Quasi-Cyclic Codes, *IEEE Trans. on Information Theory*, vol. IT-13, 1967.
29. Muller, D. E.: Application of Boolean Algebra to Switching Circuit Design and to Error Detection, *IRE Trans. on Electronic Computers*, vol. EC-3, pp. 6–12, 1954.
30. Weldon, E. J., Jr.: Difference-Set Cyclic Codes, *BSTJ*, vol. 45, 1966.
31. Kasami, T., S. Lin, and W. W. Peterson: Some Results on Cyclic Codes Which Are Invariant Under the Affine Group, *AF Cambridge Res. Ctr. Rept.* 9, Proj. AF19(628)-4371, Bedford, Mass., 1966.

32. Weldon, E. J., Jr.: Non-Primitive Reed-Muller Codes, *IEEE Trans. on Information Theory*, vol. IT-13, 1967.
33. Weldon, E. J., Jr.: Euclidean Geometry Cyclic Codes, *Proc. Symp. on Combinatorial Mathematics*, University of North Carolina, Chapel Hill, 1967.
34. Graham, R. L., and J. MacWilliams: On the Number of Information Symbols in Difference-Set Cyclic Codes, *BSTJ*, vol. 45, 1966.
35. Veblen, A., and W. Bussey: Finite Projective Geometries, *Amer. Math. Soc. Trans.*, vol. 7, pp. 241–59, 1906.
36. Veblen, O., and J. W. Young: "Projective Geometry," vol. 1, Blaisdell Publishing Co., New York, 1910.
37. Carmichael, R. D.: "Groups of Finite Orders," Dover Publications, Inc., New York, 1937.
38. Prange, E.: Some Cyclic Error-Correcting Codes with Simple Decoding Algorithms, *AF Cambridge Res. Ctr. Tech. Rept.* AFCRC-TN-58-156, Bedford, Mass., 1958.
39. Prange, E.: The Use of Coset Equivalence in the Analysis and Decoding of Group Codes, *AF Cambridge Res. Ctr. Tech. Rept.* AVCRC-TR-59-164, Bedford, Mass., 1959.
40. Rudolph, L. D., and M. E. Mitchell: Implementation of Decoders for Cyclic Codes, *IEEE Trans. on Information Theory*, vol. IT-10, pp. 259–60, 1964.
41. MacWilliams, J.: Permutation Decoding of Systematic Codes, *BSTJ*, vol. 43, pp. 485–505, 1964.
42. Kasami, T.: A Decoding Procedure for Multiple-Error-Correcting Cyclic Codes, *IEEE Trans. on Information Theory*, vol. IT-10, pp. 134–39, 1964.
43. Nesenbergs, M.: A Combinatorial Problem and a Simple Decoding Method for Cyclic Codes, *IEEE Trans. on Information Theory*, vol. IT-10, pp. 250–51, 1964.
44. Reiger, S. H.: Codes for the Correction of "Clustered" Errors, *IRE Trans. on Information Theory*, vol. IT-6, pp. 16–21, 1960.
45. Fire, P.: A Class of Multiple-Error-Correcting Binary Codes for Non-Independent Errors, *Sylvania Report* RSL-E-2, Sylvania Electronic Defense Lab., Reconnaissance Systems Div., Mountain View, Calif., 1959.
46. Gallager, R. G.: Private communication.
47. Kasami, T.: Cyclic Codes for Burst-Error-Correction, *J. Inst. Elec. Commun. Engrs. (Japan)*, vol. 45, pp. 9–15, 1962.
48. Corr, F., and E. Gorog: Les Codes Capables d'Assurer une Securite Contre les Erreurs dans la Transmission de Donnees, *Onde Elec. (France)*, vol. 43, pp. 117–27, 1963.
49. Elspas, B., and R. A. Short: A Note on Optimum Burst-Error-Correcting Codes, *IRE Trans. on Information Theory*, (Correspondence), vol. IT-8, pp. 39–42, 1962.

50. Elspas, B.: Design and Instrumentation of Error-Correcting Codes, *Rome Air Dev. Ctr. Tech. Rept.* RADC-TDR-62-511, Rome, N.Y., ASTA DOC. 294 957, 1962.
51. Kasami, T.: Optimum Shortened Cyclic Codes for Burst-Error-Correction, *IEEE Trans. on Information Theory*, vol. IT-9, pp. 105–09, 1963.
52. Kasami, T., and S. Matoba: Some Efficient Shortened Cyclic Codes for Burst-Error-Correction, *IEEE Trans. on Information Theory*, vol. IT-10, pp. 252–53, 1964.
53. Zetterberg, L. H.: Cyclic Codes from Irreducible Polynomials for Correction of Multiple Errors, *IEEE Trans. on Information Theory*, vol. IT-8, pp. 13–21, 1962.
54. Melas, C. M., and E. Gorog: A Note on Extending Certain Codes to Correct Error Bursts in Longer Messages, *IBM J. Res. & Dev.*, vol. 7, pp. 151–52, 1963.
55. Stone, J. J.: Multiple Burst Error Correction, *Information and Control*, vol. 4, pp. 324–31, 1961.
56. Gross, A. J.: Augmented Bose-Chaudhuri Codes Which Correct Single Bursts of Errors, *IEEE Trans. on Information Theory*, vol. IT-9, p. 122, 1963.
57. Abramson, N.: Error Correcting Codes from Linear Sequential Networks, presented at the *4th London Symp. on Information Theory*, 1960.
58. Burton, H. O., and E. J. Weldon, Jr.: Cyclic Product Codes, *IEEE Trans. on Information Theory*, vol. IT-11, pp. 433–40, 1965.
59. Peterson, W. W.: On the Weight Structure and Symmetry of BCH Codes, *AF Cambridge Res. Ctr. Tech. Rept.* 1, Proj. AF19 (628)-4371, Bedford, Mass., 1965.
60. Gilbert, E. N.: Synchronization of Binary Messages, *IEEE Trans. on Information Theory*, vol. IT-6, pp. 470–77, 1960.
61. Stiffler, J. J.: Synchronization of Binary Messages, *IEEE Trans. on Space Electronics and Telemetry*, vol. SET-8, pp. 112–17, 1962.
62. Ullman, J. D.: Near Optimal, Single Synchronization Error-Correcting Code, *IEEE Trans. on Information Theory*, vol. IT-12, pp. 418–25, 1966.
63. Barker, R. H.: Group Synchronizing of Binary Digital Systems, in W. Jackson, Ed., "Communication Theory," Academic Press, Inc., New York, 1953.
64. Bose, R. C., and J. G. Caldwell: Synchronizable Error-Correcting Codes, *Information and Control*, to be published.
65. Weldon, E. J., Jr.: A Note on Synchronization Recovery with Extended Cyclic Codes, *IEEE Trans. on Information Theory*, vol. IT-13, to be published.
66. Tong, S. Y.: Synchronization Recovery Techniques for Binary Cyclic Codes, *BSTJ*, vol. 45, pp. 561–96, 1966.

67. Gorenstein, D., W. W. Peterson, and N. Zierler: Two Error-Correcting Bose-Chadhuri Codes are Quasi-Perfect, *Information and Control*, vol. 3, pp. 291–94, 1960.
68. Abramson, N.: A Class of Systematic Codes for Non-Independent Errors, *IRE Trans. on Information Theory*, vol. IT-5, pp. 150–57, 1959.
69. Brown, D. T., and W. W. Peterson: Cyclic Codes for Error Detection, *Proc. IRE*, vol. 49, pp. 228–35, 1961.
70. Campopiano, C. N.: Bounds on Burst Error Correcting Codes, *IRE Trans. on Information Theory*, vol. IT-8, pp. 257–59, 1962.
71. Corr, F.: Multiple Burst Detection, *Proc. IRE* (Correspondence), vol. 49, p. 1337, 1961.
72. Elliott, E. O.: Estimates of Error Rates for Codes on Burst-Noise Channels, *BSTJ*, vol. 42, pp. 1977–98, 1963.
73. Elspas, B.: A Note on P-nary Adjacent-Error-Correcting Codes, *IRE Trans. on Information Theory*, vol. IT-6, pp. 13–15, 1960.
74. Cheilik, P.: A Note on Utilization of Cyclic Codes, *IEEE Trans. on Information Theory*, vol. IT-12, pp. 286–91, 1966.
75. Corr, F. P.: A Shortened Cyclic Code for Multiple Errors, *IEEE Trans. on Information Theory*, vol. IT-12, p. 404, 1966.
76. Forney, G. D., Jr.: Generalized Minimum Distance Decoding, *IEEE Trans. on Information Theory*, vol. IT-12, pp. 125–31, 1960.
77. Golay, M. J. E.: Notes on Digital Coding, *Proc. IRE* (Correspondence), vol. 37, p. 657, 1949.
78. Gorog, E.: Some New Classes of Cyclic Codes Used for Burst Error Correction, *IBM J. Res. & Dev.*, vol. 7, pp. 102–11, 1963.
79. Meggitt, J. E.: Error Correcting Codes for Correcting Bursts of Error, *IBM J. Res. & Dev.*, vol. 4, pp. 329–34, 1960.
80. Melas, C. M.: A New Group of Codes for Correction of Dependent Errors in Data Transmission, *IBM J. Res. & Dev.*, vol. 4, pp. 58–64, 1960.
81. Melas, C. M.: A Cyclic Code for Double Error Correction, *IBM J. Res. & Dev.*, vol. 4, pp. 364–66, 1960.
82. Mitchell, M. E.: Simple Decoders and Correlators for Cyclic Error-Correcting Codes, *IEEE Trans. on Communication Systems*, vol. CS-10, pp. 284–91, 1962.
83. Peterson, W. W.: Binary Coding for Error Control, *Proc. NEC*, vol. 16, pp. 15–21, 1960.
84. Prange, E.: Cyclic Error-Correcting Codes in Two Symbols, *AF Cambridge Res. Ctr. Tech. Rept.* AFCRC-TN-57-103, Bedford, Mass., 1957.
85. Prange, E.: The Use of Information Sets in Decoding Cyclic Codes, *IRE Trans. on Information Theory*, vol. IT-8, pp. 85–89, 1962.

86. Myravaagnes, E. R.: The Weight Distribution of Two Cyclic Codes of Length 73, *IEEE Trans. on Information Theory*, vol. IT-11, p. 316, 1965.
87. Levy, J. E.: Self-Synchronizing Codes Derived from Binary Cyclic Codes, *IEEE Trans. on Information Theory*, vol. IT-12, pp. 286–91, 1966.
88. Riordan, J.: "Combinatorial Analysis," John Wiley & Sons, Inc., New York, 1958.
89. Singer, J.: A Theorem in Finite Projective Geometry and Some Applications to Number Theory, *Amer. Math Soc. Trans.*, vol. 39, pp. 377–85, 1938.
90. Slepian, D.: A Class of Binary Signaling Alphabets, *BSTJ*, vol. 35, pp. 203–34, 1956.
91. Slepian, D.: Some Further Theory of Group Codes, *BSTJ*, vol. 39, pp. 1219–52, 1960.
92. Stiffler, J. J.: Comma-Free Error-Correcting Codes, *IEEE Trans. on Information Theory*, vol. IT-11, pp. 107–12, 1965.
93. Yale, R. B.: Error Correcting Codes and Linear Recurring Sequences, *M.I.T. Lincoln Lab. Tech. Rept.* 34-77, Lexington, Mass., 1958.
94. Zierler, N.: On a Variation of the First Order Reed-Miller Codes, *M.I.T. Lincoln Lab. Tech. Rept.* 34-80, Lexington, Mass., 1958.
95. Zierler, N.: Linear Recurring Sequences, *J. SIAM*, vol. 7, pp. 31–48, 1959.
96. Rudolph, L. D.: Threshold Decodable Codes, *IEEE Trans. on Information Theory*, vol. IT-13, to be published.
97. Kasami, T., Hsu and Chien, R. T.: Codes for a Compound Channel, *IEEE Trans. on Information Theory*, vol. IT-13, to be published.
98. Gallager, R. G.: Low Density Parity Check Codes, *IRE Trans. on Information Theory*, vol. IT-8, pp. 21–28, 1962.

Chapter 12

1. Freiman, C. V., and J. P. Robinson: A Comparison of Block and Recurrent Codes for the Correction of Independent Errors, *IEEE Trans. on Information Theory*, vol. IT-11, pp. 445–49, 1965.
2. Massey, J. L.: "Threshold Decoding," the M.I.T. Press, Cambridge, Mass., 1963.
3. Richters, J. S.: Application of Pareto Error Statistics to Hagelbarger Codes, *IEEE Trans. on Information Theory*, vol. IT-11, pp. 571–77, 1965.
4. Bussgang, J. J.: Some Properties of Binary Convolutional Code Generators, *IEEE Trans. on Information Theory*, vol. IT-11, pp. 90–100, 1965.

5. Robinson, J. P.: An Upper Bound on the Minimum Distance of a Convolutional Code, *IEEE Trans. on Information Theory*, vol. IT-11, pp. 567–71, 1965.
6. Wyner, A. D., and R. B. Ash: Analysis of Recurrent Codes, *IEEE Trans. on Information Theory*, vol. IT-9, pp. 143–56, 1963.
7. Lin, S., and H. Lyne: On the Generators of Binary Tree Codes, *IEEE Trans. on Information Theory*, vol. IT-13, to be published.
8. Robinson, J. P., and A. J. Bernstein: A Class of Recurrent Codes with Limited Error Propagation, *IEEE Trans. on Information Theory*, vol. IT-12, 1966.
9. Massey, J. L.: Advances in Threshold Decoding, in "Advances in Communications Systems," vol. 2, A. V. Balakrishnan, Ed., Academic Press, Inc., New York, to be published.
10. Hagelbarger, D. W.: "Recurrent Codes for the Binary Symmetric Channel," lecture notes from *Univ. of Michigan Summer Conf. on Coding Theory*, 1962.
11. Wozencraft, J. M.: Sequential Decoding for Reliable Communication, *M.I.T. Res. Lab. of Electronics Tech. Rept.* 325, Cambridge, Mass., 1957.
12. Wozencraft, J. M., and B. Reiffen: "Sequential Decoding," The M.I.T. Press, Cambridge, Mass., 1961.
13. Savage, J. E.: Progress in Sequential Decoding, in "Advances in Communications Systems," vol. 2, A. V. Balakrishnan, Ed., Academic Press, Inc., New York, to be published.
14. Savage, J. E.: The Computation Problem with Sequential Decoding, Elec. Engrg. Dept., Mass. Inst. Tech., Cambridge, 1965.
15. Viterbi, A. J.: Error Bounds for Convolutional Codes and an Asymptotically Optimum Decoding Algorithm, *IEEE Trans. on Information Theory*, vol. IT-13, to be published.
16. Stiglitz, I. G.: "Sequential Decoding with Feedback," Sc. D. dissertation, Elec. Engrg. Dept., Mass. Inst. Tech., Cambridge, 1963.
17. Wozencraft, J. M., and I. M. Jacobs: "Principles of Communication Engineering," John Wiley & Sons, Inc., New York, 1965.
18. Lin, S.: Some Results on Sequential Decoding, *Information and Control*, to be published.
19. Fano, R. M.: A Heuristic Discussion of Probabilistic Decoding, *IEEE Trans. on Information Theory*, vol. IT-9, pp. 64–74, 1963.
20. Jordan, K. L., Jr.: The Performance of Sequential Decoding in Conjunction with Efficient Modulation, *IEEE Trans. on Communication Technology*, vol. COM-14, pp. 283–96, 1966.
21. Gallager, R. G.: Binary Codes for Burst Error Correction, *IEEE Internat'l. Symp. on Information Theory*, 1966.
22. Hagelbarger, D. W.: Recurrent Codes: Easily Mechanized, Burst-Correcting, Binary Codes, *BSTJ*, vol. 38, pp. 969–84, 1959.

23. Berlekamp, E. R.: Note on Recurrent Codes, *IEEE Trans. on Information Theory*, vol. IT-10, pp. 257–59, 1964.
24. Massey, J. L.: Implementation of Burst-Correcting Convolutional Codes, *IEEE Trans. on Information Theory*, vol. IT-11, pp. 416–22, 1965.
25. Preparata, F. P.: Systematic Construction of Optimal Linear Recurrent Codes for Burst Error Correction, *Calcolo*, vol. 2, pp. 1–7, 1964.
26. Epley, A. W., and L. Lin: On the Burst-Error Correcting Recurrent Codes, *AF Cambridge Res. Ctr. Rept.* 10, AFCRL Cont. AF19(628)-4379, Bedford, Mass., 1966.
27. Berlekamp, E. R.: A Class of Convolutional Codes, *Information and Control*, vol. 6, pp. 1–13, 1963.
28. Elias, P.: Coding for Noisy Channels, *IRE Conv. Rec.*, pt. 4, pp. 37–47, 1955.
29. Freiman, C. V., J. P. Robinson, and A. D. Wyner: Error Control through Coding, vol. 2: Recurrent Burst-Error-Correcting Codes, *Rome Air Dev. Ct. Tech. Rept.* RADC-TDR-64-149, DOC. AD609467, Rome, N.Y., 1964.
30. Kilmer, W. L.: Some Results on Best Recurrent-Type Binary-Error-Correcting Codes, *IRE Conv. Rec.*, pt. 4, pp. 135–47, 1960.
31. Kilmer, W. L.: Linear-Recurrent Binary Error-Correcting Codes for Memoryless Channels, *IRE Trans. on Information Theory*, vol. IT-7, pp. 7–12, 1961.
32. Massey, J. L., and R. W. Liu: Application of Lyaponuv's Direct Method to the Error-Propagation Effect in Convolutional Codes, *IEEE Trans. on Information Theory*, vol. IT-10, pp. 248–50, 1964.
33. Peterson, W. W.: "Error-Correcting Codes," The M.I.T. Press, Cambridge, Mass., 1961.
34. Wyner, A. D., On the Equivalence of Two Convolutional Code Definitions, *IEEE Trans. on Information Theory*, vol. IT-11, pp. 600–01, 1965.

Index

560,-